International Society of Fire Service Instructors

TRAINING OFFICER'S DESK REFERENCE

Edited by Brian J. Ward

JONES & BARTLETT LEARNING

World Headquarters
Jones & Bartlett Learning
5 Wall Street
Burlington, MA 01803
978-443-5000
info@jblearning.com
www.jblearning.com

International Society of Fire Service Instructors
14001C Saint Germain Drive, Suite 128
Centreville, VA 20121
800-435-0005
www.isfsi.org

Jones & Bartlett Learning books and products are available through most bookstores and online booksellers. To contact Jones & Bartlett Learning directly, call 800-832-0034, fax 978-443-8000, or visit our website, www.jblearning.com.

Substantial discounts on bulk quantities of Jones & Bartlett Learning publications are available to corporations, professional associations, and other qualified organizations. For details and specific discount information, contact the special sales department at Jones & Bartlett Learning via the above contact information or send an email to specialsales@jblearning.com.

Production Credits
Chief Executive Officer: Ty Field
President: James Homer
Chief Product Officer: Eduardo Moura
Vice President, Publisher: Kimberly Brophy
Director of Sales, Public Safety Group: Patricia Einstein
Executive Editor: William Larkin
Senior Development Editor: Alison Lozeau
Production Editor: Cindie Bryan
Senior Marketing Manager: Brian Rooney
Art Development Editor: Joanna Lundeen
Art Development Assistant: Shannon Sheehan
VP, Manufacturing and Inventory Control: Therese Connell
Composition: Cenveo® Publisher Services
Cover Design: Kristin E. Parker
Rights and Photo Research Coordinator: Ashley Dos Santos
Printing and Binding: Edwards Brothers Malloy
Cover Printing: Edwards Brothers Malloy

Library of Congress Cataloging-in-Publication Data
Training officer's desk reference / edited by Brian Ward.
pages cm
"International Society of Fire Service Instructors."
Includes bibliographical references and index.
ISBN 978-1-4496-3249-6 (casebound)—ISBN 1-4496-3249-1 (casebound)
1. Fire fighters—Training of—Handbooks, manuals, etc. 2. Fire prevention--Study and teaching—Handbooks, manuals, etc. 3. Fire extinction—Study and teaching—Handbooks, manuals, etc. I. Ward, Brian (Firefighter) II. International Society of Fire Service Instructors.
TH9120.T73 2016
363.37'80683—dc23

2014033491

6048

Printed in the United States of America
18 17 16 15 14 10 9 8 7 6 5 4 3 2 1

BRIEF CONTENTS

TABLE OF CONTENTS

ACKNOWLEDGMENTS

Jones & Bartlett Learning would like to thank the contributors and reviewers of *Training Officer's Desk Reference*.

Captain Kevin B. Czarnecki
Wilton Fire Department
Wilton, Connecticut

Earl "Rob" Freese III, MS
Executive Director, Public Safety Training & Certification
Bucks County Community College
Doylestown, Pennsylvania

Captain Gene McDaniel
Phoenix Fire Department
Phoenix, Arizona

Matt Mead
Shepherd Tri-Township Fire Department
Shepherd, Michigan

William J. Pfeifer
State Fire Marshal Training Division
Grand Island, Nebraska

Chief Howard Rice, Jr.
Falmouth Fire-EMS Department
Falmouth, Maine

Derek Rosenquist
Estes Valley Fire Protection District
Estes Park, Colorado

Robert Simmonds
Vaughan Fire and Rescue Service
Vaughan, Ontario, Canada

Charles A. Williams, NREMT-P, Instructor III
Midwest Training Solutions
Blue Earth Fire Department
Blue Earth, Minnesota

ABOUT THE AUTHORS

Managing Editor

■ Brian Ward

Brian holds various roles within the fire service, including career and volunteer status. He serves as the Chief of Emergency Operations /Training Manager for Georgia Pacific—Madison (GA) industrial facility, and Training Officer for the City of Gray (GA) Volunteer Fire Department. Previously, Brian served in several roles with Gwinnett County Fire and Emergency Services as the Training Officer for the Career Development Division, which oversaw the development of 850 personnel. Brian received his bachelor's degree in Fire Safety and Technology Engineering from the University of Cincinnati and is Certified Georgia Smoke Diver #741. Brian has received several honors including Fire Department Management from NACO, Distinguished Service Award, Fire Fighter of the Year, Top 20 CTBS Instructor and National Seal of Excellence for Leadership and Safety from the National Fallen Firefighters Foundation and Everyone Goes Home. Brian has served on the Honeywell Advisory Council and the FEMA AFG Grant Criteria Committee. He was the past Chairman of the Metro Atlanta Training Officers Association and Everyone Goes Home State of Georgia Advocate. Currently, he serves as an International Society of Fire Service Instructors—Director At Large and Lead Live Fire Credentialing Instructor. He has authored over 30 nationally publicized articles and research papers, including speaking engagements with FDIC and Firehouse. Lastly, Brian thanks his wonderful wife for allowing him to do all of this!

Contributing Authors

■ Chad Abel

Chad Abel served for over 12 years as Division Chief of Education and Safety with the Fishers (IN) Fire Department. He has been in the fire service for 26 years. Chad is a past President of the Indiana Fire Instructors Association and later served as the IFIA liaison to the International Society of Fire Service Instructors. He currently is a Captain on a Ladder Company.

■ Bryan Altman

Bryan is a Lieutenant/Assistant Training Officer with Worth County (GA) Fire/Rescue. He is a 10-year veteran of the fire service. He spent 2011 serving as a full-time instructor with the Georgia Fire Academy teaching the Basic Fire Fighter Program and assisted with live fire classes; he is still an adjunct instructor with GFA. He taught at the Metro Atlanta Fire Fighters Conference in 2012, he is a Contributing Author for FireTrainingToolbox.com, and he has been published in *Fire Engineering* magazine. He is certified as a Fire Fighter II, Georgia F.L.A.M.E.S. #159, Acting Officer, Live Fire Instructor, Basic Fire Fighter Instructor, Fire Instructor II, P.O.S.T. Instructor, and has a diploma in Fire Science.

■ Alan Brody

Alan Brody began his fire service career in 1972 as an Explorer Scout and has worked his way through several fire departments to his current position as the Training Officer for the Mnjikaning Fire Rescue Service, Rama, Ontario, Canada, and a volunteer fire fighter with the Orillia Fire Department in Ontario. He is a certified Ontario Instructor, Public Fire and Life Safety Educator, Company Officer, Fire Fighter, and Incident Safety Officer. He is a member of the Ontario Association of Fire Training Officers and teaches at the Ontario Fire College as an Associate Instructor.

■ David Bullard

David is a 16-year member of the fire service and currently serves as a Lieutenant and co-assigned to the Training Division with the Columbia County Fire Rescue (GA) and a part-time fire fighter with Grovetown (GA) DPS. In addition, he serves on the Board of Directors for the Georgia State Fire Fighter's Association and was Secretary of IFSTA's *Pumping Apparatus Driver Operator, Third Edition*, Curriculum Committee, Chairs the SEAFC Communications Committee, and is an ER Supervisor with HEPACO LLC. As an active instructor for the Georgia Fire Academy and in private industry, David is constantly engaged with teaching live fire, leadership, HAZMAT, and Fire Fighter Survival/RIT classes.

Chad Cobb

Michael "Chad" Cobb is a Field Safety Professional for Amerisafe Workers' Compensation. Previously, Chad worked as the accreditation manager for ProBoard Certifications and division manager at Georgia Fire Fighter Standards and Training Council. Chad has an associate's degree in Fire Science, a bachelor's degree in Occupational Health and Safety, and a master's degree in Management with a concentration in Change Leadership. Chad served on the Georgia Search and Rescue Team and responded to the aftermath of Hurricane Katrina for search and rescue operations. Chad served as a fire fighter with Macon-Bibb County Fire Department (GA), City of Griffin Fire Rescue (GA), and currently serves as a volunteer fire fighter with Monroe County Fire and Emergency Services (GA).

Cameron Gasaway

Cameron is a second-generation fire fighter and serves as an aircraft crash rescue specialist for the United States Marine Corps reserves. His full-time employment is working as a Project Manager for Gasaway Consulting Group, where he coordinates the company's fire service consulting projects and administers the website and social media presence.

Richard Gasaway

Rich Gasaway served 33 years on the front lines as a fire fighter, EMT-Paramedic, and fire chief, and he earned a Doctor of Philosophy degree while studying the neuroscience of how individuals, teams, and organizations develop and maintain situational awareness and make decisions in high-stress, high-consequence, time-compressed environments.

Dr. Gasaway is widely considered to be one of the nation's leading authorities on first responder situational awareness and decision making. His contributions have been featured and referenced in more than 350 publications. His leadership and safety programs have been presented to more than 43,000 first responders worldwide.

Nita Ham

Nita began her EMS career in 1982, after graduating from the Paramedic Technology Program at Gadsden State Community College in Gadsden, Alabama. She relocated to Atlanta and has been heavily involved in both EMS operations and education in Georgia and on a National level for the past 26 years. Nita has worked as a field medic for Grady EMS, Mid-Georgia Ambulance Service, Medical Center of Central Georgia EMS, and Monroe County EMS. Nita has also been a licensed EMS instructor since 1986 and worked for the Department of Technical and Adult Education as an instructor and an EMS Program Field Training Officer between 1986 and 1993. She held the position of EMS Program Manager at the Georgia Public Safety Training Center (GPSTC) in Forsyth, GA, between 1993 and 2003. Nita has also served as Training and Education Manager at Grady EMS, Program Director for the Region 3 Office of EMS, and a General Manager for Rural Metro Ambulance Service. Nita is now employed with Emanuel Medical Center in Swainsboro, GA, through a grant with the Office of Rural Health and is responsible for developing continuing education programs for rural providers.

Skip Heflin

Skip Heflin is the Training and Professional Development Chief for Hall County Fire Services. In his role, Skip plans, provides, and oversees all training activities for the department. This includes new recruit training, annual training, special teams training, and EMS certification and recertification training. Skip has been in the fire service for 25 years and is an NREMT-P, Task Force Leader with the Georgia Search and Rescue Task Force 1, and a Type 3 Incident Commander with the state incident management team. He considers himself a lifelong student of the profession.

Jason Hoevelmann

Jason Hoevelmann is a deputy chief/fire marshal with the Sullivan Fire Protection District, a combination department, and a career captain and training officer with the Florissant Valley Fire Protection District in North St. Louis County. His experience spans more than 20 years, and he has been an instructor for more than 15 years. Hoevelmann is an adjunct instructor for the St. Louis County Fire Academy and a State Certified Fire Officer II. He holds an associate's degree in paramedic science and a bachelor's degree in fire service administration. Hoevelmann is a state advocate for the Everyone Goes Home initiative, a Board Member for the International Society of Fire Service Instructors, and is on the technical committee for Professional Fire Officer Qualifications for NFPA.

Leigh Hubbard

Leigh Hubbard is the President/CEO of TradeUp, an association consulting and management firm in Virginia. Leigh uses her leadership skills to facilitate strategic planning, management of board functions, operating procedures, administration, educational programming, development of policies and procedures, fundraising, public relations, and financial analysis and budgeting. Leigh bases her success in organizational leadership on being a big-picture, goal-oriented strategist. She is passionate about understanding organizational challenges and customizing her responses to member relations and board development. She brings a business planning perspective to association product and program development. Currently Leigh serves as the Executive Director of the International Society of Fire Service Instructors in Centreville, VA. Her executive leadership of the Society has included the successful launch of multiple international programs including the Live Fire Instructor credential and the development and oversight of the Principles of Modern Fire Attack initiative.

■ Brian P. Kazmierzak

Brian P. Kazmierzak, EFO, CTO, is the Chief of Training for the Penn Twp. Fire Dept. in Mishawaka, IN. He has a bachelor's degree in fire service administration from Southern Illinois University and serves as the Director of Operations for www.FireFighterCloseCalls.com and the Webmaster for www.ModernFireBehavior.com. Brian was the recipient of the 2006 F.O.O.L.S. International Dana Hannon Instructor of the Year Award, the 2008 Indiana Fire Chiefs Training Officer of the Year Award Recipient, and the 2011 ISFSI/FDIC George D. Post Fire Instructor of the Year. In addition, Brian completed the National Fire Academy's Executive Fire Officer Program in 2006, is a CPSE credentialed Chief Training Officer, serves as a Director at Large for the ISFSI, and is on the UL FSRI PPV Research Study Panel. Brian has been a student of the fire service since 1991.

■ Scott Maddison

Deputy Fire Chief Scott (Scotty) Maddison is a 34-year veteran of the Fire/EMS service in both the military and civilian sectors. Deputy Chief Maddison started his career in 1980 on a volunteer department in Rawlins, WY. He graduated from the U.S. Air Force Fire Protection School at Chanute AFB in November 1984 while serving as a member of the Wyoming Air National Guard. Later, he graduated from Combat Medic/EMT while serving on active duty with the U.S. Army. Scott has held the following leadership and training positions during his career, including military positions as Squad Leader and NCOIC/Trainer, EMS Assistant Supervisor/Training Officer for Burns EMS, and Chief of Training for Laramie County Fire District #4 (WY). Deputy Chief Maddison has spent 15 years of his career on rural agencies, developing training programs that meet the needs of the farmers and ranchers serving the local agencies and the certifying authorities. He holds numerous DOD and State of Wyoming Certifications including Wyoming State Fire Instructor III.

■ Kevin Milan

Kevin Milan is the Assistant Chief of Preparedness for South Metro Fire Rescue (CO) overseeing training, special operations, and emergency management. He is a graduate of the National Fire Academy (NFA) Executive Fire Officer Program and possesses Chief Fire Officer Designation through the Center for Public Safety Excellence. Kevin holds a Master of Science in Executive Fire Service Leadership and is pursuing a PhD in Fire and Emergency Management through Oklahoma State University. Milan is Past President of the Colorado Training Officers & Fire Marshal's Associations, former director for the International Society of Fire Service Instructors, State Advocate for the National Fallen Firefighters Foundation, a Governors appointee to the Colorado Fire Safety Advisory Board, and recipient of two NFA outstanding research awards.

■ Philip Oakes

Prior to taking his current position with the National Association of State Fire Marshals in 2011, Philip served as the State Training Director for the Wyoming State Fire Marshal's Office for 5 years. In this capacity, he was responsible for coordinating the activities of the State Training Division, as it provided training, fire investigation, and incident reporting assistance throughout the State of Wyoming. During this time, the Training Division received accreditation and two national awards for excellence in certification. Philip has been in the fire service since 1993, serving as a fire fighter, Public Information Officer, Engineer, Training Officer, and Operations Chief. He currently serves as the Chief of Laramie County Fire District #4, a post he has held for the past 5 years. In the past, he has worked as an EMT-B and as an Emergency Services Dispatcher. He currently serves as the Assistant Volunteer Ambulance Director for Burns EMS. He has also served as an elected councilman and mayor for the community of Burns, WY, where he currently lives with his wife and two children. Philip is a certified Fire Officer III, Fire Service Instructor III, Hazardous Materials Incident Commander, Fire Inspector, and Fire Investigator, and he holds degrees in both Business Administration and Fire Science. Philip has served as a book reviewer and contributor for several book publishers and magazines, as well as serving on the Board of Directors for the International Society of Fire Service Instructors.

■ Steve Pegram

Steve Pegram was appointed as Chief of Goshen Township Fire and EMS in 2009. In addition to his duties as Chief in Goshen, Chef Pegram is active in both local and national fire service organizations. Chief Pegram currently serves as the President of the International Society of Fire Service Instructors (ISFSI). Chief Pegram is also an active member of the International Association of Fire Chiefs (IAFC) Safety and Health section, and currently represents this organization on a National Fire Protection Association IAFC task force reviewing the design and implementation of a safety and training program for fire fighters on hybrid and electric cars. Chief Pegram was born and raised in Brooklyn, NY. However, he started his fire service career in the suburbs of Philadelphia, PA, while still in high school. Soon after high school, Chief Pegram moved to New Jersey where he attended college and served as fire inspector for Princeton, NJ, as well as Fire Chief in Pennington, NJ. In 1997, Chef Pegram moved to Mason, OH, where he was the Deputy Fire Chief until October 2000. From 2000 to 2007, Chief Pegram served as the Training Officer and Shift Commander for the Loveland Symmes FD in Loveland, OH, and from 2007 to 2009 he served as the Deputy Fire Chief for the City of Xenia, OH.

Chief Pegram has an associate's degree in Fire Science from Mercer College in Trenton, NJ, and a bachelor's degree in Organizational Management from Wilberforce University. Chief Pegram is a 1998 graduate of the National Fire Service Staff and Command School, and he has completed multiple courses through the National Fire Academy. Chief Pegram has also served on the editorial board for Fire Rescue Magazine for 10 years, has published more than 25 articles pertaining to fire service operations, and teaches locally and nationally on fire and EMS related topics. Chief Pegram and his wife reside in Goshen with their two children and a dog.

■ Rich Rodewald

Rich Rodewald began his fire service career in 1996 with the Council Bluffs Fire Department (IA) and currently serves as department Drill Master, coordinating and delivering training to the 100-member career department. He has worked as a Field Staff Instructor for the Iowa Fire Service Training Bureau for more than a decade and is a regular speaker at fire schools and conferences throughout the State of Iowa. Rodewald is a member of the International Society of Fire Service Instructors (ISFSI), as well as the Iowa Chapter of the ISFSI, and serves as a cadre instructor for the ISFSI's Live Fire Instructor credentialing program. He has an associate's degree in Fire Protection Science from Iowa Western Community College and a bachelor's degree from Cornell College (IA). In 2014, Rich was honored with the Floyd Wm. "Bill" Nelson Instructor of the Year award by the Iowa Society of Fire Service Instructors.

■ David M. Wall

David M. Wall has been the Director of the Georgia Fire Academy, a Division of the Georgia Public Safety Training Center, since November 2004, and is responsible for statewide training programs in basic and advanced firefighting skills, chief fire officer programs, structural fire control instructor programs, hazardous materials, technical rescue, supervisor and manager training, and incident command. The fire academy staff and adjuncts provide training at a main campus and statewide to over 12,000 students annually. He came to this position after serving as Chief of the Morrow Fire and Emergency Medical Services in Clayton County, GA, a position he held for 11 years. Previously, he had served as a fire academy staff member and prior to that was with Henry County GA Fire Rescue (GA), holding positions as a fire fighter, paramedic, instructor, and eventually lieutenant. His career began in 1976 as an EMT and volunteer fire fighter. For over 15 years, he served as a court-appointed special advocate (CASA) for children in the Clayton County Juvenile Court. Chief Wall earned a master's degree in Public Administration from Georgia College and State University and a bachelor of science degree from Brenau College. He has completed the Executive Fire Officer (EFO) Program at the National Fire Academy and is a Chief Fire Officer Designee (CFOD). He is currently in a doctoral program, seeking the degree of Doctor of Education in organizational leadership. He is studying naturalistic decision making, situation awareness, and is doing research on Company Officer–level decision making and expertise. However, the most important position Davis has held is as the father of three, Melissa, 20 (college student), Adam, 26 (police officer), and Rachel, 28 (dance teacher);

grandfather of Norah Ruth (6 months), and husband of 30 years to his best friend, Robin.

Brian Wolfe

Brian currently serves as a Captain with the Gwinnett County Department of Fire and Emergency Services at one of the busiest stations in Gwinnett County. Brian has served the emergency services since 1994 and has been with Gwinnett County since 1998. Prior to his service with Gwinnett County, Brian began his fire service career with the Roswell Fire Department in Roswell, GA, where he continues to serve as a station captain. Brian received his undergraduate degree in Political Science from the University of Richmond. He continued his education by obtaining his master's degree from Eastern Kentucky University in Safety, Security, and Emergency Management with a dual concentration in Fire and Emergency Services and Emergency Management. Brian has maintained his dedication as a student of the fire service by teaching and assisting with the development of current and future fire fighters through the Gwinnett Fire Academy, Roswell Fire Department, Metro Atlanta Fire Fighters Conference, and other training opportunities. Most recently, Brian has been accepted to the Executive Fire Officer Program at the National Fire Academy to further his knowledge, skills, and abilities as a fire officer as an executive level officer.

PREFACE

Congratulations on the opportunity to read this manual! It is our intention to provide you with a comprehensive guide that will assist you in your day-to-day training operations. It does not matter if you are a training officer for a major metropolitan fire academy or the lone instructor in a rural environment; the business of training is essential to the success of every emergency services organization. You are to be commended for reaching out for new information and best practices to make your fire fighters and emergency responders safer and more efficient.

This book is the result of the hundreds of years of experience accumulated by our authors. We have assembled a fantastic team of training professionals who have proven their expertise in training and adult education in the field and in the classroom. They have worked to organize and assemble the key training concepts essential to your success. I would like to personally thank them for their hours in front of a computer screen creating this work. Their contributions will benefit instructors and fire fighters for many years to come.

It is important to recognize that you are not alone in your training mission. The members and staff of the International Society of Fire Service Instructors (ISFSI) stand ready to assist you. Our organization was founded in the 1960s on the premise of instructors working together to share information to improve fire service training and fire fighter safety. We specialize in providing our members valuable resources they can use to improve their programs and enhance learning for their students. If you are not already a member, we encourage you to learn more about our organization at www.isfsi.org.

Finally, we would like to dedicate this text to fire fighters and instructors who have died in the line of duty. We hope that, through their tragic losses, we can learn how to do our jobs more safely and to better protect our communities. Fire departments in the new millennium are faced with ever-growing challenges, such as reductions in funding, light-weight building construction practices, more volatile fuel loading in buildings, and a host of man-made and natural threats. The instructor and training officer will serve at the forefront of the battle to ensure our fire fighters receive the necessary training so that at the end of the day, everyone goes home.

Stay safe and God bless,
Eddie Buchanan, Past President
International Society of Fire Service Instructors

FOREWORD

Thank you for taking the time to acquire and look through *Training Officer's Desk Reference*. This book is going to be a bit different from the textbooks or manuals you have read or reviewed throughout your career. This manual is a collection of articles, writings, and various works from members of the International Society of Fire Service Instructors and placed into categories, in order to help the training officer/instructor. Many times new training officers can find themselves overwhelmed by the seemingly endless number of tasks that they have been assigned to do, or just not have enough time to get it all done, and properly research their material.

This book is designed with these problems in mind. Rather than read the book in the traditional cover to cover method you can simply go to the applicable section that applies to you, read it, and move on. Then, when something else occurs, simply consult the appropriate section of the manual. Also keep in mind, there may be multiple articles on one topic. Some people may like one presentation style, while others prefer another. In order to help as many people as possible, key elements are presented in several different styles, and you should find at least one that suits you.

Additionally, the point of keeping the sections short and to the point is so that you, the training officer, can read through the material quickly and efficiently, and get right back out there doing what you do best with all the new information.

The International Society of Fire Service Instructors would like to thank all of you for the job that you do and for the time and effort you have put into making not only your department but the fire service as a whole safe and successful.

Sincerely,
Board of Directors
International Society of Fire Service Instructors

CHAPTER

1 Introduction

Brian Ward, Brian Wolfe, Phillip Oaks, and David Bullard

Responsibility

Congratulations! You have just been promoted to the position of training officer of your fire department. For good or bad, you are it, and a lot of people are counting upon you to be very successful at your job. From the newest recruit to the Chief of the Department to the public you serve, all of them need you to make your fire department the best it can be (**FIGURE 1-1**). This book goes a long way towards helping you take care of all of their needs.

Now, if that was not stressful enough, you have to consider the needs of management, your board of directors, the politicians who oversee your agency, the Chief, the other officers, the fire fighters, the state/provincial agencies, and your recruits. It is very likely you cannot do this all yourself. Learn the skills of management and delegation early (if you have not already), and get organized and stay organized. There are a million ways to get organized and several styles of delegation. Find what works for you and your organization and use it. The one greatest piece of advice that could be offered is to learn where to find and use resources. It is not always about knowing everything, but knowing where to go to obtain the correct information. These are just some of the items that you will be responsible for to some degree; other areas that a training officer needs to be competent in include budgets, documentation, records, supervisory responsibilities (**FIGURE 1-2**), and a whole host of other situations you will be facing.

Training Officer Obligations

A training officer must meet multiple obligations in order to maintain integrity within a training program. The instructor has an obligation first and foremost to the fire service, and also to the administration of an organization, to the trainee/student, and to himself/herself. For many recruit trainees, this is the first opportunity to instill professionalism and pride in the recruit. In fireground terms, "As the first line goes, so goes the fire—as the initial training goes, so goes the new recruit." This provides a reputation based on tradition and new technology that can be embraced by the recruit trainee. For an experienced trainee, the training officer can share knowledge of new technology and new, safer approaches to time-honored traditions that need to be amended. For instance, a new recruit has no concept of riding tailboard on a fire apparatus, but the veteran has a mindset against the use of seat belts. This simple concept alone enables the recruit to understand the importance of wearing a seat belt, while the veteran is instructed on the benefits of wearing a seat belt through career longevity, career advancement, and the potential for medical disability or retirement due to potential injury from seat belt noncompliance. Understanding these vastly different mentalities and backgrounds will help you succeed as a training officer. Bringing a sense of pride and professionalism to the department will assist you in endeavors similar to these. In addition, positively

FIGURE 1-1 Day 6 of the Georgia Smoke Divers Program.
Courtesy of Brian Ward/Georgia Smoke Divers Program.

influencing fire fighters and sharing information is the mainstay of a training officer's obligation to the fire service.

Administrative obligations can overlap with other essential areas, but they are limited to the organization. A training officer can increase departmental pride and professionalism, departmental preparedness, efficiency on incident scenes, the quality of service delivery, and safety. Training officers have the unique opportunity to foster and guide appropriate change within the specified guidelines of an organization. However, the most important aspect of a training officer in relation to an organization is the opportunity to provide a positive role model for behavior. Training officers have an inherent responsibility to know the policies and procedures of an organization and adhere to them. This concept is otherwise known as *leading by example*. This responsibility affords them the chance to show trainees and veterans the benefits of following established policies and procedures.

The main audience of any training officer is the student/trainee/recruit. The primary obligation of any training officer is to the student who needs the guidance to start, refresh, or advance a career. Part of a training officer's duties is helping the student to develop personal pride and professionalism for the job. As members of the fire service, we only have one chance to make a first impression. Our first assignment can make or break an employee if the training officer does not instill the pride and professionalism of the fire service into the student. Students should be taught to perform the job properly, efficiently, and to the highest standard possible. Training officers are obligated to prevent failure and prepare students for advancement. Training officers should illustrate the concept of safety, instill an attitude of continuous improvement and quality, and provide a positive role model for the students.

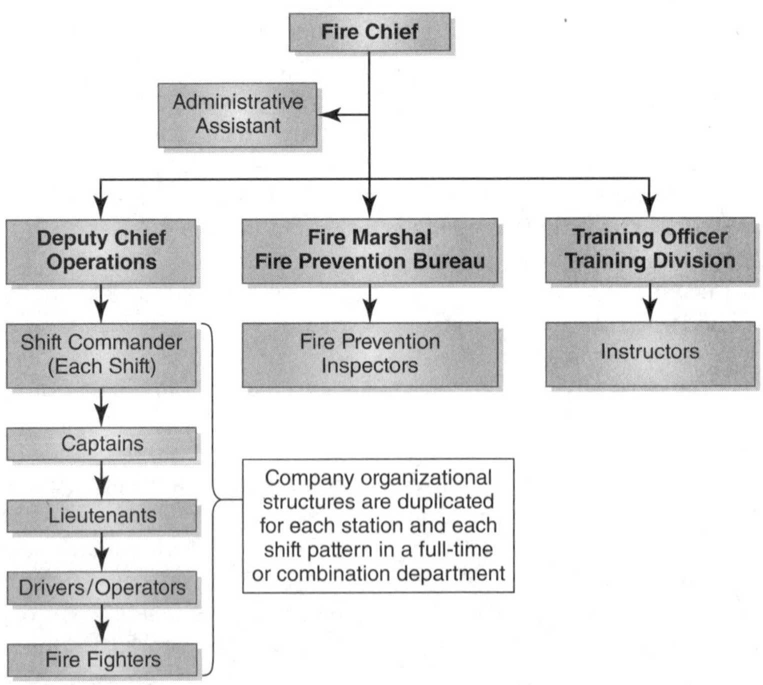

FIGURE 1-2 Sample organizational chart.

Q DOES OPS DRIVE TRAINING OR DOES TRAINING DRIVE OPS

The training officer must direct the individual training needs of each student.

Training officers are often looked to as experts in certain areas of instruction. Instructors should maintain an increased level of education about certain subjects, but they also should accept the fact that they cannot know everything for every subject being addressed in a training program as diverse as a fire recruit program, paramedic program, or any type of comprehensive curriculum. While a training officer may be proficient in specific subjects such as ventilation, fire streams, pump theory, or fire behavior, he or she cannot possibly be expected to be the expert in all aspects of the emergency services discipline. However, whatever the training officer teaches, he or she must be enthusiastic and knowledgeable about that subject, which is carried over to the students. In order to maintain that level of expert status, the training officers are obligated to maintain continual improvement and personal development, demonstrate safety at all times, keep up-to-date on current trends in research, and maintain the highest standard for their students and their teaching methods. Training officers should routinely set a positive example and perform by the standards of their department, which means that they must know departmental policies and procedures. Teaching the opposite of a standard or proven research does not promote a positive example of setting a standard and executing against that standard.

The training officer has multiple obligations that compete against each other on occasion, but more often than not, they are aligned to achieve the same purpose; namely, competent instruction. While the stakeholders vary within each organization, the same principles can be applied in order to address all of the training officer's obligations (**FIGURE 1-3**).

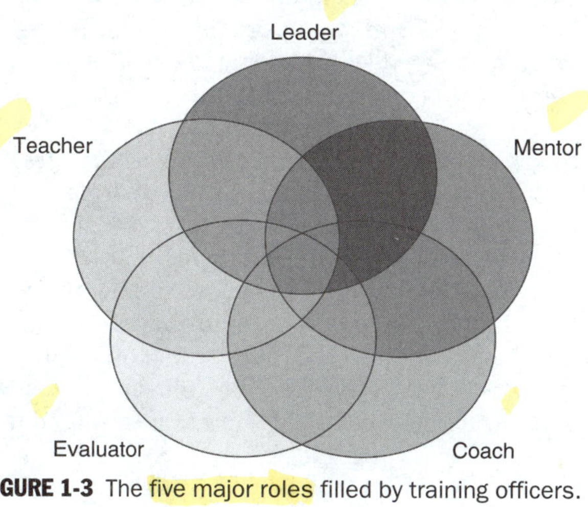

FIGURE 1-3 The five major roles filled by training officers.

Influence

Keep in mind you now are the primary change agent for your department. Generally, the assistant, deputy, and fire chiefs can affect formal change throughout a department. While training officers have some of the same capabilities, their impact comes from the informal changes of a good training officer or mentor developing the fire fighters. When fire fighters get into trouble, they do not think of the chief to save themselves—they remember what was taught to them by their training officers and instructors, and that is what keeps them alive. When an officer thinks of how to deal with the public, they fall back on the training they received during the officer school. This is not only a personal feeling; there is research that proves the philosophy that how we train is directly related to how we will react in similar circumstances. In the chapter *The Application Step*, Dr. Richard Gasaway discusses the importance of training to build a fire fighter's decision-making and situational awareness skills. The end result is that your department is better prepared for whatever they will face.

You now have the most important position in the fire service and the one position where the most change can happen. You will be developing the future and guiding the present. Make the best of it.

Legal Concerns

Accepting the responsibility of the training officer position carries a tremendous amount of accountability to ensure that drills are conducted safely and correctly. As you are probably aware, each year approximately 10 percent of the line-of-duty deaths (LODDs) occur during training events because of lack of knowledge and skills, complacency, attitude, or unforeseen circumstances. It is your job as the training officer to ensure that learning occurs, and in order for that to take place, the safety of the students has to be the highest priority. If a student is scared for their life, they will not gain any knowledge or skills. On the other hand, most of the time they are unaware of the hazards of the situation that they are in. This is where the training officer comes in. Understanding the standards, regulations, and laws that govern or are considered the national consensus is imperative of this position. Ignorance of the law will not save you, even with the greatest of intentions. There are numerous resources where information can be obtained that can guide you through these situations and assist you in your decisions. Several of these resources will be discussed throughout this book and templates for training topics are provided in the appendices.

The first piece of advice that should be given to any training officer is to research your drill before you conduct it. Quick drills are great and absolutely have their place in the fire station; however, we have to ensure that they are conducted according to policies and standards. Most the time, common sense and good judgment will guide you in the right direction, but understanding the policies and standards should help fill in the gaps in you plans.

This is not an Instructor or Live Fire manual; however, we feel that it is vital to mention National Fire Protection Association (NFPA) 1403, *Standard of Live Fire Training*, and live fire drills specifically, as they are one of the greatest hazards in the training environment. Of those 10 percent of LODDs, a good majority of them occurred during "smoke drills" or live fire training. Several incidents across the country have captured our attention over the last few years, from the volunteer department to the largest of career departments, concerning NFPA 1403 and live fire training. Several of these departments had the policies in place, but failed to adhere to them, while others were not aware of national consensus standards. Regardless of the reason, lives were lost and others were impacted for the rest of their lives. This is what every training officer fears, and it is the reason that we must research, study, and follow the best practices established in our occupation. Resources for additional information include:

- NFPA 1403, *Standard of Live Fire Training*
- *Live Fire Training: Principles and Practice* by the International Association of Fire Chiefs, International Society of Fire Service Instructors, and National Fire Protection Association
- *Live Fire Training Credential/Hands On Training*; International Society of Fire Service Instructors

Although not as publicized as live fire concerns, there are numerous other areas that training officers should be aware of and understand the ramifications of their actions and the actions of their subordinates. Hazing, harassment, and the punishment of students have tarnished the careers of many instructors and training officers in the fire service. This is a paramilitary structure, and we want to graduate tough fire fighters, but we have to be smart about it as well. This job is absolutely not for everyone, but we should not set out to fail students. It is our job to mentor them and set them up for success. If they choose to not complete the drill, that is their decision, as long as we as instructors did everything that we could for them. An important concept to remember is to document everything within a training program; as in the medical field: if it is not documented, it did not happen.

Additional laws and standards that the training officer should be aware of include: legislative statues, administrative laws, and regulations.

Case Study

One situation that demonstrates the importance of documentation occurred during a gear acclimation drill at a recruit school, which involved walking in gear for 15 minutes. While conducting the drill, the training officer noticed that one of the recruits was pushing on their face piece while breathing air. The training officer asked the student why he was holding his mask in this fashion, and the student replied that he felt the mask was pulling on his face. The instructor continued to monitor the situation throughout the drill. After the drill, the training officer went so far as to have the student fit-tested a second time and had the mask inspected by the self-contained breathing apparatus (SCBA) maintenance personnel. It was determined that the mask was correctly fitted and operational. Roughly 1 week later, the students were conducting "cold smoke drills" in a confidence-building maze, and the student was witnessed ripping his mask off in the maze. The student was quickly escorted outside, assessed for any medical conditions, and questioned by the instructors; the student replied that the mask was uncomfortable. Although this drill ended without injuries because safety precautions were in place, it could have been a different story. If the student would had been hurt or had smoke inhalation injuries, further investigations would have been conducted. In this case, the training officer had created a pattern of documentation of this particular student surrounding the gear acclimation drill, fit test, and operation of the face piece and SCBA. This documentation would be extremely important throughout the investigation to show the due regard of the training staff.

SAFETY

■ Legislative Statutes

Legislative statutes are created through federal, state/province, county, and city legislative bodies and are in place to ensure that the health and general welfare of the public is protected. These come into effect during training, state/provincial certification processes, hiring, and promotions. Within the states/provincial governments, states/provinces create statutes that become statutory laws. Statutory laws regulate matters such as taxes, industry, commerce, and professions. Examples of federal laws that impact the hiring and training of fire personnel include:

- **Title VII of the Civil Rights Act of 1964:** Covers equal opportunity employment and education practices to prevent discrimination based on race, color, religion, sex, or national origin; also prohibits sexual harassment. Applies to all employers.
- **Section 1983 of Civil Rights Act of 1964:** Allows a complainant to not only sue a department or employer, but also to sue the alleged aggressor personally. Applies only to public officials.
- **Americans With Disabilities Act:** Provides for reasonable accommodation for learners with disabilities.

A job analysis should be conducted to determine and document the jobs and tasks that every fire fighter must be capable of performing. In order for a fire department to be liable, the plaintiff must prove that they have a disability, that they are qualified for the job otherwise, and the employer either refused to make reasonable accommodations or that their disability was the decision for not employing them. As a training officer, you are not allowed to ask about disabilities, but you can establish a physical agility test (PAT) that is proven and documented to replicate fireground activities. Such tasks as raising ladders, stretching hoses, and performing searches with appropriate fire fighter gear can be included; however, it needs to be documented as to how these tasks are related to the fireground. Also, these tasks should be exact replicas and consistent between tests. For example, if the department carries 24-foot extension ladders, the PAT should include a 24-foot extension ladder, not a 35-foot ladder. A fire department's PATs should also reference NFPA 1001, *Standard for Fire Fighter Professional Qualifications* in order to be valid.

- **Privacy Act of 1979 or Buckley Amendment:** Prevents disclosure of personal or learner information to unauthorized parties, such as the posting of grades by name.
- **Age Discrimination in Employment Act (ADEA):** Prohibits discrimination against persons older than the age of 40 years unless there exists a recognized qualification requirement.
- **Equal Pay Act:** Requires that all persons, regardless of gender, be paid at the same rate when performing the same job.
- **Copyright Act of 1976:** Protects of the use of written words, photographs, and some art work.

The Copyright Act of 1976 is divided into two general categories: works protected by the law and those considered to be part of the public domain. The Copyright Act sets out specific time frames pertaining to copyrighted material. Any work published before 1923 is considered to be in the public domain; works published between 1923 and 1978 have a 95-year copyright protection from the date of actual publication; and for works published after 1977, the author, artist, or photographer has a copyright for life plus 70 years. Violation of the copyright law can result in liability,

Case Study

Gary Knapp et al. v. City of Columbus, F.3d. (6th circuit; July 6, 2006, unpublished opinion) addressed the issue of promotional testing of fire fighters with attention deficit/hyperactivity disorder (ADHD). Three Columbus, Ohio, fire fighters with ADHD requested accommodations from the City of Columbus for the Lieutenant's and Captain's promotional assessments. They asked for an alternate site that was quieter, additional time to conduct the test, and verbal explanations of test directions. Each of the fire fighters was prescribed Ritalin for his or her disability, and the city refused to accommodate them. Due to the fire fighters admitting that they were able to conduct their jobs and personal life with the assistance of Ritalin, two separate courts dismissed the case in favor of the city.

with the creator of the work being eligible to recover any damages suffered or lost profits.

The fact that some material is protected by copyright does not necessarily prohibit you from using it in your training, because a "fair use" exception exists relative to the protection afforded by copyrights. Examples of fair use include copying materials for the purpose of criticizing, commenting, news reporting, teaching, and research—copies for the class in a training session generally fall into this category. Ultimately, fair use depends on the nature of the copyrighted work and how much of the work is copied; for example, a training officer would not be protected if they purchased one textbook and made enough full copies for each student. Regardless of which method is used, it should include a notice indicating who holds the copyright to the work.

Administrative Law

Administrative laws and regulations are rules or directives issued by an administrative agency, such as Occupational Safety and Health Administration (OSHA), which is part of the executive branch of the government. Not all states have OSHA state plans, and each state is allowed to develop their own set of regulations; however, these laws must meet or exceed federal regulations. In addition, whether you are employed by the government versus a private industry fire department greatly affects your involvement with OSHA, even if you are a non-OSHA state. If you research OSHA fines and fire departments, you will find several instances where fire departments have been found in violation of numerous policies, typically surrounding fireground and/or training activities in which an individual was injured or killed. Not only can a fire department be held liable, but individual employees can as well. In the next section we will discuss the difference between public and private law and how an individual employee can be affected. It is imperative that we follow the laws, regulations, and standards that are set forth for the fire service. As a training officer, your standard operating procedures (SOPs) should address these areas for incident operations and training activities and ensure that your instructors are adhering to them. The typical responses of "This will never happen here," or "I'll be careful," can set the example; do not allow yourself to fall into this trap. Examples of administrative law are the Georgia Fire Fighters' Standards and Training Council: State Certification and the Federal Communications Commission.

Public and Private Law

The difference between public and private law can be confusing; however, it is important to understand the difference. An individual can be charged under one or both of these. Public law refers to administrative, federal, state/provincial, and local law. It requires proof beyond reasonable doubt and is generally referred to as criminal law; sometimes, jail time or fines are assessed.

Private law refers to torts, civil wrongs, and the intent to compensate the victim. Private law requires proof with a preponderance of evidence. An individual can be found not guilty of public law, but found guilty of a tort or private law and forced to pay some type of restitution.

Within private law, there are intentional and unintentional torts. Intentional torts include assault, battery, and defamation, while unintentional torts are when an individual may not intend to perform a wrong against another, but the individual may act improperly or fail to act appropriately because of ignorance or forgetfulness. There is a legal duty to exercise a degree of care in any act, because a certain regard for and certain actions toward others are expected. This is where negligence comes into play for the training officer if he or she fails to exercise the same degree of care that a person of ordinary prudence would in a similar circumstance. A reasonably prudent person is one who exercises those qualities of attention, knowledge, intelligence, and judgment required by society.

If the instructor's or training officer's conduct falls below the standard and a student is hurt, the instructor may be liable for the damages. This can be avoided by teaching the students the correct method prior to any live drills, paying close attention to each student's actions, and having the correct instructor-to-student ratio. When we discuss sequencing instruction when creating lessons, it is not just about creating a path: It is about teaching students to crawl before they walk. It is teaching students to wear personal protective equipment (PPE)/SCBA and explaining fire behavior, tool usage, and fire attack prior to actually conducting a live fire drill. This ensures that they have been introduced to all of the tools needed in order to be successful. This also sets the stage for the students to notice events that should not be occurring or are abnormal. As a member of the fireground, we state that all fire fighters are safety officers, but first we must give them the tools to be a safety-minded individual.

Legal Definitions

gross negligence An act (or a failure to act) that is so reckless that is shows a conscious, voluntary disregard for the safety of others. This act falls somewhere between a mere accident that causes harm and acting with an intent to harm.

liability Legal responsibility.

malfeasance Consciously partaking in unlawful conduct.

misfeasance/negligence Unintentional or accidental conduct

willful and wanton conduct An act that, if not intentional, shows an utter indifference or conscious disregard for the safety of others; for example, failure to follow a written protocol.

Legal Protection

The greatest protection an instructor can have is to follow laws, regulations, and standards. In addition, there are three items that every training officer should consider before conducting a drill: (1) foreseeability, (2) having a plan, and (3) following the plan.

1. **Foreseeability.** Review the situation or training drill and consider what could go wrong. What is the frequency and/or severity of something going wrong and what can you do to mitigate it? Items to consider include:
   - Potential for and degree of harm that may result from the condition.
   - Likelihood that harm will occur because of the condition.
   - Availability of alternative methods and/or appropriate equipment to correct the condition.
   - Burden of removing or changing the condition.
   - Adequate resources (in place) to perform safely.
   - Standards, laws, and regulations reviewed, discussed with all participants, and adhered to.
2. **Having a plan.** Based off of the laws, regulations, standards, and what you found through practicing foreseeability, develop a plan and discuss it with everyone involved (students and instructors).
3. **Following the plan.** The greatest plans are of no use if they are not followed and documented. Ensure that all instructors and students are aware of their roles and ensure that they follow them. For individuals who choose to step outside of the plan, it should be documented and they should be asked to leave the training drill. This is similar to an incident scene, where any individual who freelances is a danger to themselves and everyone involved. More than once has a fire department had perfect SOPs/standard operating guidelines (SOGs), but several members failed to execute against them and someone was injured, or worse. As the training officer, you are the foundation to preventing this from happening.

Another avenue to explore is to continuously learn and gather new information. Certifications and qualifications require you to be held to a higher standard, and in conjunction with these certifications and qualifications, you should have the knowledge, skill set, and experience to ensure that you are conducting drills appropriately. Other methods of minimizing liability include:

- Ensuring that all assisting instructors meet the organization's (authority having jurisdiction [AHJ]) qualifications and requirements.
- Checking equipment regularly for safe operating conditions.
- Maintaining written objectives and documenting each training session.

Case Study

A fire department had conducted several live fire drills throughout its history with no newsworthy situations. The fire department had developed SOPs and SOGs for live fire training that covered all aspects of the drill. The one area of concern, however, was the follow-through of the plan. During the actual live fire, several violations of their own SOPs and NFPA 1403 were documented. There were a total of 33 serious violations, meaning there was a substantial probability that death or serious physical harm could result. In addition, 17 violations were labeled as willful, meaning that the employer committed an intentional and knowing violation or acted with plain indifference to or in careless disregard of employer responsibilities. As a result, an LODD occurred, and the entire departmental process of live fire training was investigated, including the disciplinary action of the training chief and two training officers.

- Providing learners with a written course description and briefing so they understand all requirements.
- Making sure trainees are physically fit and prepared for the tasks.
- Instructing and testing all learners with the safe operation of equipment and in a controlled environment.
- Not leaving learners unattended while they are practicing potentially dangerous skills.
- Respecting learner privacy.
- Documenting everything.

Attributes of a Training Officer

As the training officer, everything that you do will be magnified, for better or worse. In order to make the most of this opportunity and to assist with building your credibility, there are a few attributes every training officer should learn. These attributes will help define who you are and increase your level of productivity, and as a training officer juggling multiple priorities, these attributes are vital. This is not an all-inclusive list; however, we will discuss some of the vital ones, including:

- Teamwork
- Communication and interpersonal skills
- Time management and organizational skills
- Holding effective meetings
- Computer skills and developing metrics
- Servant-based leadership (SBL)

Teamwork

Establishing teamwork on the incident is vital to the success of any operation. As a training officer, it is incumbent upon you to incorporate the teamwork aspect into your training program. It does not matter if the training program is an 8-hour course or a 40-plus-hour program. During the training course, the training officer incorporates didactic and practical exercises in order to introduce, develop, and reinforce the lessons learned.

While carrying out this task, it is important for the training officer to show patience for the students/trainees. Specifically, keep in mind that one trainee may have some experience and be more advanced than another trainee, who has never seen or been exposed to a concept before. Patience can help the training officer avoid time wasters and a lack of credibility by allowing that first-time trainee the opportunity to make a mistake so the behavior can be corrected. According to *Tips for Teams: A Ready Reference for Solving Common Team Problems*, it is often "easier to jump in and correct a mistake, rather than letting the person you are training learn." Patience will also enable the training officer to redirect experienced trainees to a more in-depth exercise covering the same concept being learned. A proven tactic to use in this situation is to give a more challenging task to an experienced trainee, such as assisting the inexperienced trainees; this allows for a win-win situation and keeping both students involved.

Another method of incorporating teamwork into the training program is to build on that difference in experience levels. While the training officer should be well-versed in the topic of instruction, this is not always the case. The training officer can either be the evaluator or take the role of a student and allow a trainee who is more experienced to lead a particular exercise. According to *The Inspirational Trainer: Making Your Training Flexible, Spontaneous and Creative*, "as a general rule, trainers should admit what they don't know, and if you are able to assess weak areas of a program in advance, then these admissions can then be built into the lesson." This honesty regarding lack of knowledge allows the training officer to be a part of the team while motivating learning and garnering respect and credibility. When a training officer does not know the answer to a question, the training officer should admit their lack of knowledge, research the topic, and provide an answer as soon as possible, such as after the next break or by the end of the day.

Communication and Interpersonal Skills

The way that we communicate can help reduce confusion and the duplication of efforts. Understanding the flow of information from the sender to the receiver—the message and feedback—and some delivery tips will positively impact communication. Remember as well that communication is a two-way street, and both participants should be allowed to engage in the conversation. This ensures that there is an agreement between what was stated and what was heard.

Interpersonal skills are acquired skills. Learning to listen and communicate can be difficult at times, and each person is different. Learn the traits of the people you work with and try to pick up on how individuals respond in certain situations. These observations will help you determine what to say and how to say it according to the situation and that individual's behavior pattern. Fire fighters are a certain breed and are passionate about what they do. Generally, it is not what you say, but how you say it that makes the difference in most communications. When communicating, try to:

- Be short and concise, but clear.
- Consider the background, perspective, and culture of the listener.
- Check for understanding and allow for an open channel of communication.
- Consider your body movements and eye contact.
- Use tact.
- Remember that good communication is two-way.
- Remember that humor can break the ice, but it should not be at the expense of a student.

Relate topics to a common interest or field of play. For example, explain blood pressure effects on the body in relation to plumbing in a house; eventually, continual elevated pressures in piping can lead to early failure, just like our blood vessels in similar situations.

Time Management and Organizational Skills

Training officers juggle numerous priorities: personal life, fire department administration, fire fighters, and many more. In short, your time is stretched. Good time management skills will ensure that you stay on top of what has to be completed and that initiatives do not get left behind. In addition, if you use good time management and prioritize tasks according your organizational skills, your time will be positively affected. A good tip for time management is to learn how to prioritize tasks and to write them down. The use of the Urgent and Important Matrix is one method of doing this. This method was created by President Dwight Eisenhower and coined by Steven Covey in his *Seven Habits of Highly Effective People*. The four areas of the matrixane:

- **Urgent and Important:** Generally these items are time-sensitive or considered an "emergency." Manage these items as they come up; do not delay.
- **Not Urgent but Important:** These items create value and generally apply to long-term goals. This is the area where you want to spend most of your time.
- **Urgent but Not Important:** Routine task. This is the area where items need to be completed, but generally these are tasks that are scheduled on a continuous basis. These tasks are generally best to delay or delegate.
- **Not Urgent and Not Important:** These tasks take away from where you create the most value. Eliminate these time wasters.

Holding Effective Meetings

Have you ever felt that you were in a meeting about having a meeting? How do you stop this from occurring? As a training officer, you will be establishing and sitting in on numerous meetings. Here are a few tips for consolidating meetings and ensuring that value is created, thus assisting you toward a long and prosperous career:

- Produce an agenda and email it to attendees 3 days prior to the meeting. Ask for input and agreement on topics.
- Revise the agenda and email the revision to attendees 24 hours prior to the meeting.
- Start the meeting 5 minutes past the hour and finish 5 minutes before the hour, for a total of 50 minutes, if possible. Consider the attention span of most adults.
- The most effective time of day is approximately 10 A.M; avoid meetings after lunch, if possible.
- Differentiate between required individuals and optional individuals.
- Review the agenda within the first 5 minutes of the meeting.
- Have a time keeper and stay on time and topic.
- If a topic goes long, decide if more time is needed or if a select few individuals need to meet afterwards.
- Summarize the meeting in the last 5 minutes.
- Send meeting minutes to all involved within 72 hours.

Computer Skills

It goes without saying that in order to succeed in today's technology-driven world, computer skills are mandatory. Regardless of which programs are used, you need to be competent in their use. You can gain this competency either through trial and error, or through educational or training programs.

As a training officer, you will be asked to run reports, perform correlations, create spreadsheets and databases, develop presentations, and develop metrics. Understanding how to group and filter information in spreadsheet programs can save you numerous hours trying to manually gather information. Most vendor programs that are used to run reports and to track training can be "data dumped" or converted for use in common applications. If you are purchasing a program, ensure that these programs are compatible (i.e., ensure that they will they talk to each other) with those currently in use. This may require some assistance from your information technology department, but it will pay large dividends later on.

Developing Metrics

The programs discussed so far are only a small part of the big picture. As you become more involved in administrative duties, managing budgets, and tracking training results, you should become more competent in illustrating what you are trying to put into words. Developing metrics is nothing more than producing a chart or a graph that displays the correlation among several items. One common metric would be to track the number of fires that occur in November through March and compare it to the number of fires that occur in April through October. You could also use metrics to show inadequate station coverage, caused by long travel distances or excessive out-of-the-station times.

As a training officer, it is important to learn how to justify your training budget; consider using metrics to explain it. Metrics are simply one practice of measuring the performance of a particular item. Another example could be to show the cost of running a recruit school. Say that a 10-student recruit school cost $20,000 over a 4-month period; if the number of recruits is increased, the overall cost will rise, but the per-student cost might decrease based on instructor staffing. If you are planning to hire 25 recruits this year, does it make more sense to split them or do one large class? Of course, several factors, such as timing, resources, and available instructors, play a role in this decision; however, with all other variables even, it would make sense to do one large class. As training officers you have to be able to understand this and prove it to others. This is part of knowing your audience and what proof is your audience looking for.

Process

The first step in developing performance metrics is to involve the people who are responsible for the work to be measured, because they are the most knowledgeable about the work. Once these people are identified and involved, it is necessary to:

1. Identify critical work processes and customer requirements.
   - What jobs or task are you required to perform? What are the community, NFPA, OSHA, and SOP expectations that you perform against a certain standard or benchmark?
2. Identify critical results desired and align them to customer requirements.
   - Identify the criteria in the standard or the benchmark that you should be meeting. The next step is to identify who your customers are; they could be citizens, fire fighters, staff, or students. Then, align your customers with the criteria or benchmark that has been established.
3. Develop measurements for the critical work processes or critical results.
   - Determine which form of measurement you need to use in order to show the appropriate data for a particular process. Measurements can be in dollars, time, test scores, or any other system of measuring in use (gallons of water or fuel, etc.).
4. Establish performance goals, standards, or benchmarks.
   - These may be predetermined for you, such as in an NFPA document, response times, or in local policies. However, if you determine that there is a wide gap between the desired standard and where you currently are, it may be wise to use incremental steps to improve. If the standard is felt to be unrealistic, the chances for success are limited. However, if over time you increase the standard in steps, people have the chance to adjust, and they will feel a sense of accomplishment moving forward.

Another important point to consider when establishing goals or benchmarks is the use of SMART goals:

- S – Specific
- M – Measurable
- A – Accurate
- R – Realistic
- T – Timed

Several examples are listed below on how metrics could be used; these have all been created using standard spreadsheet software.

1. The overtime percentage chart shows month to month, with the average of the last 3 years on the left (**FIGURE 1-4**). If you are using adjunct instructors on their off days and overtime costs are exceeding your budgets, this correlation may help justify the addition of a new staff instructor to reduce overtime. If the training officer was attempting to prove this, the next step would be to show a cost–benefit analysis of reducing the overtime in exchange for one additional staff instructor.
2. Another example for using metrics could be to justify the need for training in specific disciplines

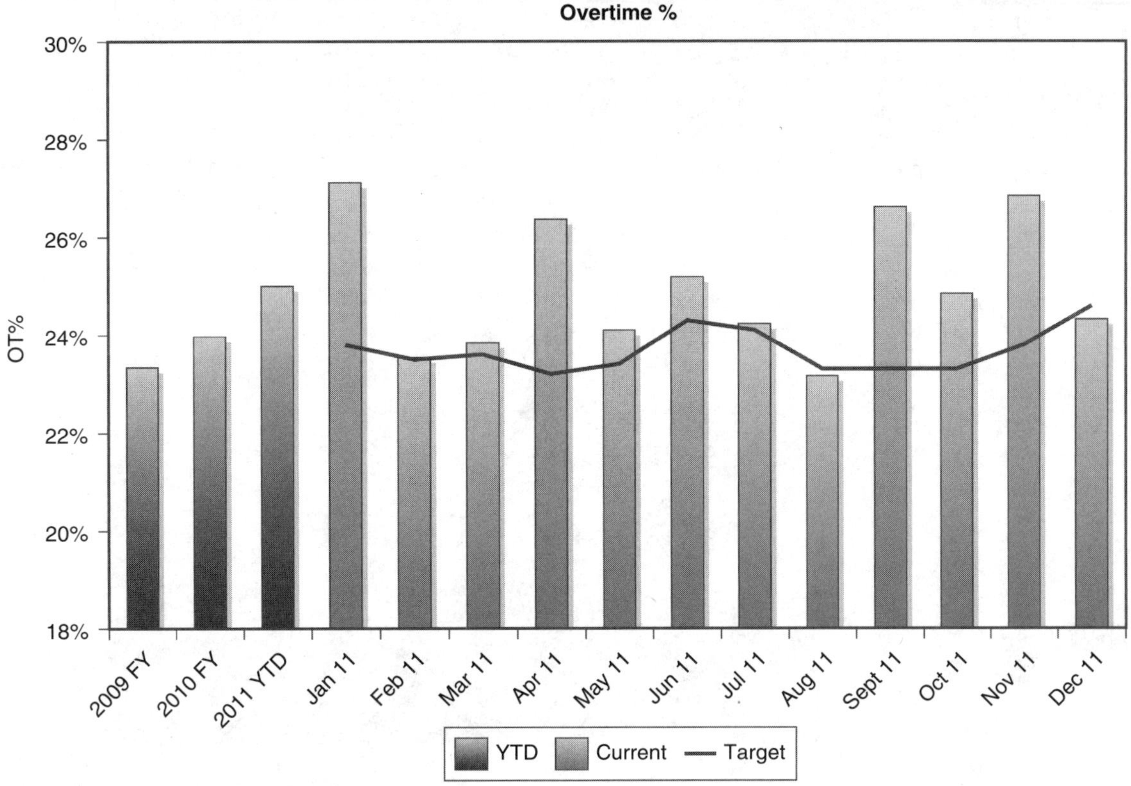

FIGURE 1-4 An overtime percentage chart.
Courtesy of Brian Ward.

based on the growing community need (**FIGURE 1-5**). If you already have the information in some form of spreadsheet or database, create whichever type of graph or chart that you wish from the data. Through analyzing this chart, a few items can be determined in a quick glance:

- Station 5 seems to have remained fairly consistent in their chest pains call volume. This could be because it is a rural area or the demographics of this community simply have not changed over time.
- Station 4 has since a drastic increase in chest pains. This could be because of the construction of nursing homes, assisted living homes, or retirement communities. This may also be cause for additional training on large commercial structures inhabited with elderly citizens.
- The remaining stations have seen slight increases in chest pains calls, which is typical for areas with a growing or aging population.
- Although it is not specifically a training item, if the chest pains calls of station 4 have risen, the likelihood of the overall call volume has risen as well. This would be important information to gather when considering the replacement of apparatus based on hours and mileage.

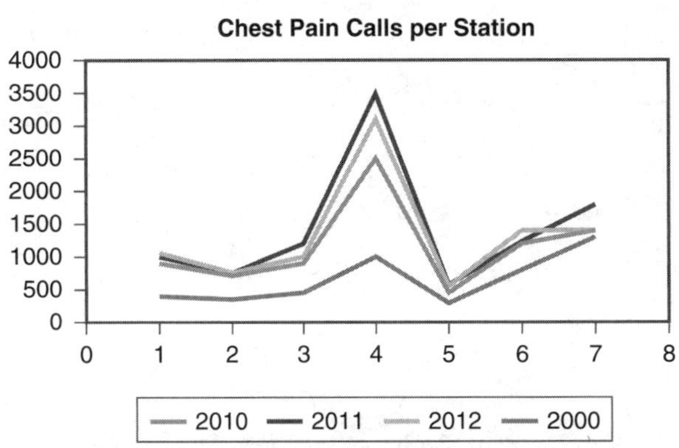

FIGURE 1-5 A chart reflecting the number of chest pain calls per station.
Courtesy of Brian Ward.

3. The data in **FIGURE 1-6** and **FIGURE 1-7** show additional spreadsheets. In Figure 1-7, the data are selected on the left side, then copied and pasted into the Y values box for the Series name "2010."

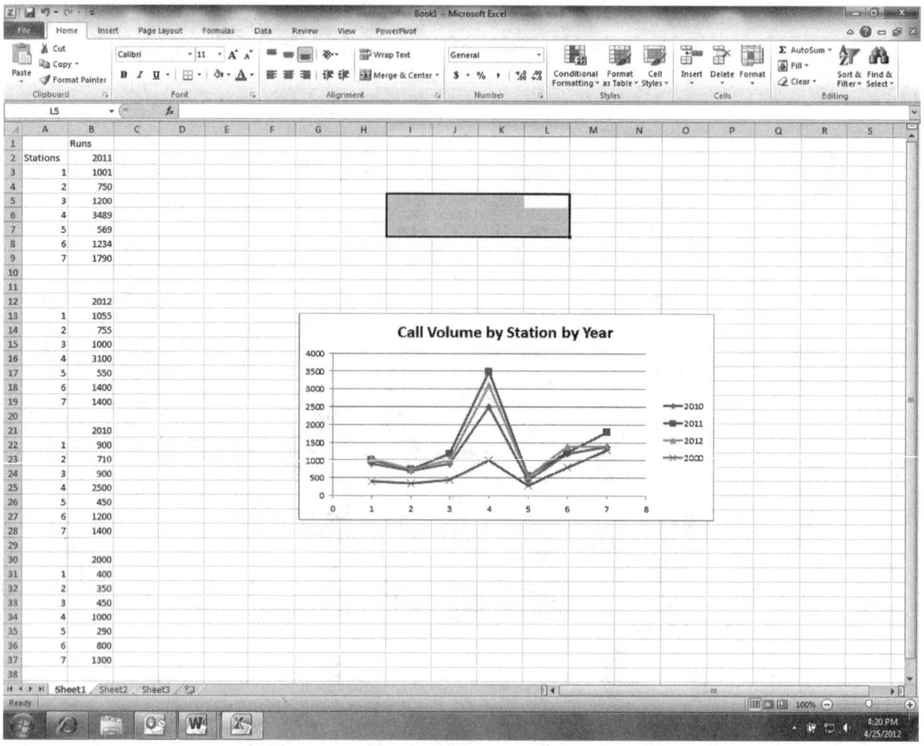

FIGURE 1-6 A chart reflecting the call volume per station per year.

Courtesy of Brian Ward.

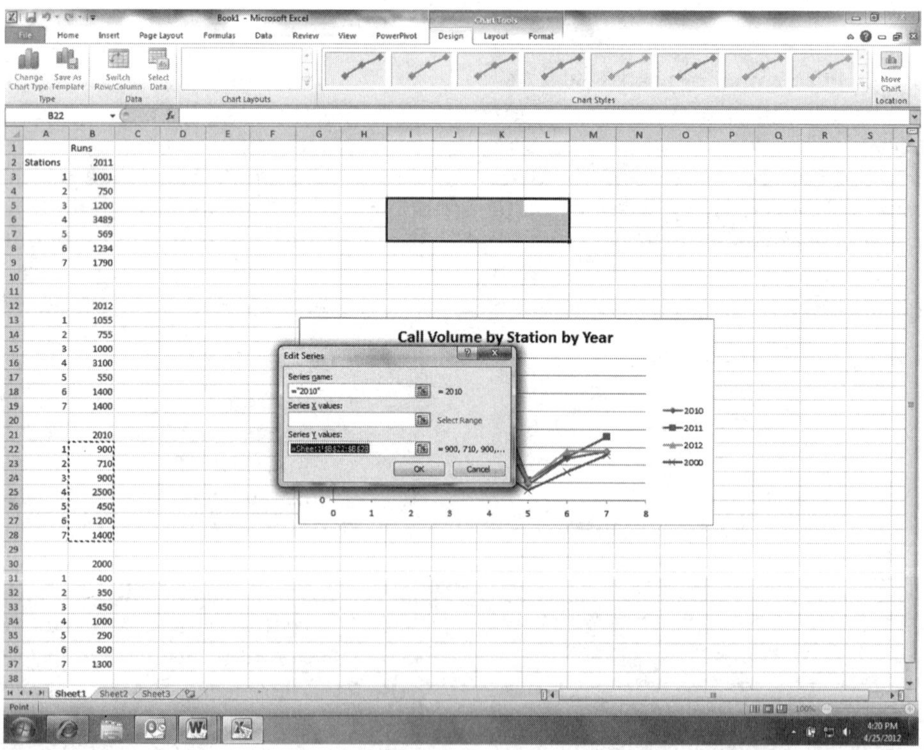

FIGURE 1-7 A chart reflecting the call volume per station per year, tracking by series a specific item.

Courtesy of Brian Ward.

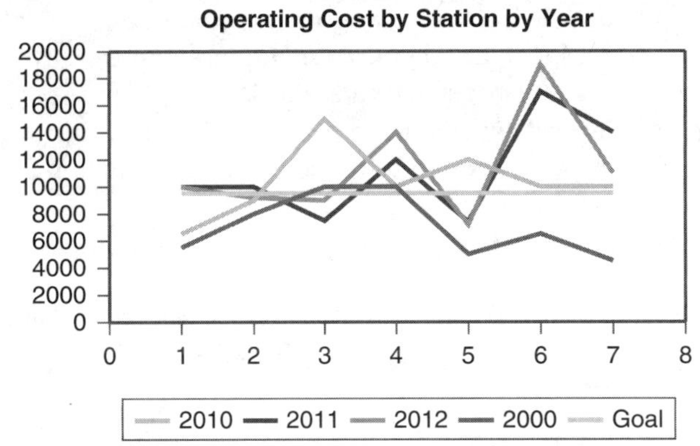

FIGURE 1-8 A chart reflecting the operating cost per station by year, in which "Goal" represents the targeted operating costs for the year 2012.

Courtesy of Brian Ward.

The Series name should be the name of the item that you are tracking. This will automatically plot your numbers. Repeat the process for each year represented. Additional information on how to develop spreadsheets, graphs, and charts can be found searching through the Microsoft Office training tutorial website.

4. The data in **FIGURE 1-8** show the operating cost per station by year and includes the current goal for 2012.

■ Servant-Based Leadership

As the training officer, you have many different styles of leadership to choose from. One of the most effective is that of servant-based leadership (SBL). This style allows a training officer to work towards the needs of the department and its individuals. With attention to a variety of needs, greater success may be achieved. When choosing to utilize SBL, the training officer must understand the principles of this leadership style: everyone has the potential to be a strong contributor in an organization, and individuals work best within a nurturing, valuing environment. This type of environment allows students to trust, risk, create, and measure up to the expectations of others, thus becoming more valuable assets in the organization. The primary driving forces behind continuous improvement are positive self-esteem and trust. Effective, long-lasting, personal growth and learning are accomplished interactively on the job. Although individuals may feel helpless at times and unable to exert enough influence to actually change the organization, they can, in fact, make a difference.

While the basic ideas of SBL are not new, the conceptualization of the style into mainstream recognition occurred in the early 1970s. It is now a fairly common leadership style, with many benefits for both the leader and the follower. The advantages come from the attention given to the needs of developing the student/member of the organization. Above all, the underlying principle is that one must have a desire to lead others and place their interests above one's own self-serving goals and principles.

A training officer operating on the principles of SBL will have to strive to maintain attributes of a leader whose focus is the needs, betterment, and empowerment of their students. The training officer must be a receptive listener as well. He or she must recognize that everyone will not share the same beliefs and that all people deserve empathy. Foresight and intuition are key to anticipating the needs and helping to shape the future direction of the organization. The training direction should not be at the whim of the training officer, but rather geared toward the needs of the students or individuals in the organization. In order to meet the needs of the organization, the training officer would also need to possess some strong persuasive talents and receive a "buy in" of the members. When problems arise or when difficult situations are faced, the ability to provide a healing influence can help to steer the recovery process. As a visionary, the training officer must be able to develop new concepts and deliver the reasoning to provide influence based on those concepts. In conjunction with being a visionary, the ability to contemplate and understand the trends, paradigms, and desires of those to be led or trained is essential. Servant-based leaders also recognize that they work for the needs of others and will often function in a pivotal role of binding the organization together and obtaining a sense of community among the members.

Often times when students in the fire service are taught skills, especially basic skills, they are only given information on how to perform the skill. While this enforces muscle memory, it often does not address the student's ability to think and understand the critical reasoning as to why that particular skill exists, when it is warranted, and why it is effective. In essence, SBL for a training officer makes sure that not only the "how" of a skill is presented, but also the "why." By understanding the goals of training and education, and supporting the mission statement and beliefs of the organization, the leader may then adapt the teaching process to the individuals. Every student learns differently, and every person has different strengths and weaknesses, so a blanket method of instruction may be ineffective. The SBL principles give guidance to tailoring the learning toward the needs of each student/member and improving upon their weaknesses. With attentive diligence, the training officer can greatly improve the skills, decision-making abilities, and desire to grow and learn of an organization. In order for progress and growth in the organization to occur, the students and members must be empowered to continuously seek improvement for themselves and the organization. A training officer should rate empowerment as an important goal to foster the development and progression of the students/members. With empowerment, the sense of community and a commitment to organizational goals can be greatly enhanced.

The benefits to the training officer can be very positive. By addressing the individual's needs, training officers will in return have more productive students who achieve greater success in learning. In the role of mentor, the guidance and development of students and members will ensure strong and visionary leadership in the future, maximizing individual growth and output. The training officer is often tasked with the delivery of new concepts and programs for the organization. By establishing an ability to influence, nurture, and understand the needs of each member, the training officer can not only provide guidance to the organization's leadership on the need for new concepts, but may already be in a position to deliver the message and ingrain it throughout the organization.

Key Points

- The obligations of a training officer are endless; however, you can create the greatest impact when mentoring and developing fire fighters and officers.
- Remember: as a training officer, you are one of the most influential change factors in your department.
- Review the laws that govern your state/province and local jurisdiction, and understand the NFPA national consensus standards.
- Three rules that every training officer should follow are practicing foreseeability, having a plan, and following the plan.
- Attributes of a good training officer include but are not limited to: teamwork, communication and interpersonal skills, time management and organizational skills, holding effective meetings, and practicing SBL.

References

Bennett, L. *ADA, Fire Service Law*. New York, NY: Brady Books, Pearson; 2008.

Covey, S. *The Seven Habits of Highly Effective People*. New York, NY: Simon & Schuster; 1989.

Fisher, K., S.R. Rayner, and W. Belgard. *Tips for Teams: A Ready Reference for Solving Common Team Problems*. New York, NY: McGraw-Hill; 1995.

International Association of Fire Chiefs, International Society of Fire Service Instructors, and National Fire Protection Association. *Live Fire Training: Principles and Practice*. Burlington, MA: Jones & Bartlett Learning; 2012.

International Society of Fire Service Instructors. *Live Fire Training: Credential/Hands On Training*. Centreville, VA.

Jackson, P.Z. *The Inspirational Trainer: Making Your Training Flexible, Spontaneous and Creative*. London: Kogan Page; 2001.

National Fire Protection Association (NFPA) 1001, *Standard for Fire Fighter Professional Qualifications*. 2013. http://www.nfpa.org/codes-and-standards/document-information-pages?mode=code&code=1001. Accessed December 9, 2013.

National Fire Protection Association (NFPA) 1403, *Standard of Live Fire Training*. 2012. http://www.nfpa.org/codes-and-standards/document-information-pages?mode=code&code=1403. Accessed December 9, 2013.

CHAPTER

2 You as the Training Officer

Brian Ward and Phillip Oaks

Understanding and Looking the Part

As a training officer, you assume the face of the organization or department for the field or operations personnel. It is your responsibility to interpret and train on new standard operating procedures (SOPs), best practices, and equipment. Typically, the command staff or fire chief develops and/or delegates the job of developing the above criteria for the department. Once the development phase is complete, the criteria must be implemented throughout the department. In essence, training must take place to ensure skill competence and understanding before the personnel or students can be expected to carry out the new directive.

To embody the role as a training officer, a few things must take place. The keys to being an effective training officer are preparation and to look the part. First, the training officer must have accurate and detailed information. Consider the questions that may be asked by the students and develop appropriate answers. The second step is to convince the operations personnel or students to "buy" the idea being communicated. In essence, the training officer is attempting to sell a particular idea, whether it is why fire fighters need to use a new piece of equipment or why they must follow a new safety standard. This task is typically accomplished either by the training officer or a small staff that reports to the training officer. In order to gain the students' acceptance, the training officer must sell the information in a convincing and truthful manner. This is where the matter of appearance comes in, as your body language, tone of voice, and choice of words must show that you fully support the concept you are selling. This task can be difficult at times, especially if the training officer does not agree with the direction of the department. However, training officers must remember that they are the face of the organization and that it is their responsibility to follow orders—as long as the directive is not unsafe, immoral, or unethical. The third step is to reinforce the idea through additional materials, remedial training, and accountability.

Accountability

Accountability is a delicate word in the fire service, but it is of great importance to an individual's reputation and appearance. Every officer and every fire fighter has a different interpretation of accountability. However, every training officer must remember the importance of holding himself or herself accountable. Failure to be accountable for your own actions will create distrust, discontent, and a lack of respect from colleagues and the operations personnel. With the honor of becoming a training officer comes a great responsibility. While it is important for all individuals to do what they say they will do, it is absolutely imperative that training officers follow this guideline.

Assertiveness Versus Aggressiveness

In addition to accountability, understanding the difference between assertiveness and aggressiveness could essentially be the difference between being successful or unsuccessful in persuading the students to buy into the training. Certain individuals rule with an iron fist.

Although these types may get their way, they ultimately will lose respect and create a barrier between themselves and their students. While the fire department occasionally thrives from having personnel who are aggressive on the fireground, this approach does not work in the classroom, nor does it work on a personal level.

Unlike aggressiveness, assertiveness allows an agreement to be made in a tactful and respect manner between the training officer and students. Aggressiveness, in contrast, is asserting your rights, beliefs, and thoughts without any regard for other individuals. As there is no single solution to every problem, there is no one individual with all of the answers. Attempting to force individuals, without explaining why, to believe as you do will not earn you any respect, nor will it lead to a consensus among the students. While being aggressive on the fireground can be an admirable trait, if it is uncoordinated or not part of the overall objective, freelancing can be deadly. The goal is to take the boldness of being aggressive, combine it with the respect for others, and turn it into assertiveness.

An assertive individual is one who asserts their rights, beliefs, and thoughts while respecting the same of others. All leaders and training officers should consider the differences of each individual and how their ideas could bring value. You must embody the theories, values, and concepts that you teach. Be the assertive individual who behaves consistently with their attitude toward safety-related items. Be the assertive individual who believes that training for both "ordinary" and "extraordinary" incidents will mentally and physically prepare us for whatever call we may run. Finally, be the assertive individual who performs the job as a public servant because you believe in servant-based leadership.

Pride and Passion in Your Profession

Having a sense of pride and passion in our profession is paramount for an individual to be successful. However, how you share this pride and passion will determine whether you succeed or fail. When you do decide to fight a battle, remember to lean toward being more assertive than aggressive. While you want to present your beliefs with confidence and boldness, much more success will come if you have an understanding of and respect for the opinions of others. Furthermore, it is not enough to lead with passion alone. If you are preparing to make a stand on a belief or idea, make sure you know the material by researching every aspect of it, even those that oppose your opinion. Lastly, follow what your heart tells and guides you to do. Nothing that you do will matter if your heart is not in it.

Understanding the pride and passion required of a leader will help the training officer gain the trust needed to make the program successful. This mentality goes hand in hand with the demeanor with which you carry yourself. In essence, the pride and passion you show toward your profession and what you do is part of your appearance and how others will perceive you.

Physical Appearance and Professionalism

Just as your reputation will always precede you, the training officer's appearance establishes his or her first line of credibility. The attitude and demeanor with which a training officer carries himself or herself will precede the classroom portion. In the moment that the training officer walks into a classroom or meeting, the only thing that matters is the presence exhibited by the training officer. The physical appearance of an individual is often underestimated, and the proper amount of attention is not given. You may think that it does not matter if your shirt is clean, tucked, and pressed, but these details make a difference. A training officer's physical appearance sets the tone for the rest of the class. It also influences whether he or she loses or gains the students' respect. For example, consider the training officer who walks into a classroom well-groomed, with shoes shined, pants and shirt pressed, a perfect gig line (buttons and belt bucket inline), and no pendants (strings clipped). What is your first impression? Now compare this to the training officer with scuffed shoes, gig line out of place, and disheveled hair. The students' first impression of the second training officer will likely leave much to be desired in the appearance department.

The old cliché, a picture is worth a thousand words, should come to mind when deciding how to present yourself to a class. As a training officer, you may be extremely knowledgeable and considered *the* subject matter expert on a particular topic; however, all of this can be lost before you even introduce yourself if you do not dress and present yourself in a professional manner. It is your choice on how you wish to be perceived. Remember that first impressions last a lifetime. The students may not remember everything that was said or taught during your time together, but they will remember how you presented and carried yourself.

Training officers who take their time to ensure their attire and appearance are professional reinforces several assumptions that demonstrate their level of professionalism. The first assumption is that the training officer wishes to be there and intends to teach or present some information in a professional manner. If the training officer wishes the class to act with a professional attitude, the training officer must take the first step and lead by example. Establishing this level of professionalism demonstrates his or her commitment to the class and students. The students will appreciate his or her commitment, and most will respond accordingly with the same level of professionalism.

The second assumption is that the training officer takes pride in this role and its responsibilities, and he or she recognizes the amount of time and effort it takes to adequately prepare for a class. It is imperative that training officers prepare for each class, no matter how many times they have previously presented the material. If training officers take the time to prepare their image with precision, it is a good bet that they prepared their materials as well.

The third assumption is that the training officer understands the importance of limiting distractions and creating a learning environment. A bad learning environment and endless distractions can ruin a potentially wonderful class before it even starts. Creating the perfect learning environment will allow the greatest amount of information and knowledge to be transferred from the training officer to the students. To maximize this transfer, several things must take place, with the training officer's appearance the first item of concern.

■ Fitting the Part

In addition, it is essential to match the right training officer to the class that will be taught. If physical fitness is part of the curriculum, then the individual presenting this portion of the class must meet some minimum standards of fitness. The training officer teaching the students how to develop and maintain their physical fitness should look the part. As you can imagine, the appearance of the training officer teaching this portion of the class will affect the seriousness of his or her message. The class attendees are more likely to take the information being presented seriously if the person teaching it appears to be competent in this area. If the training officer does not take this section of the class seriously, then the students will dismiss it. The best training officers will attempt to persuade students to accept their beliefs by reaching into the students' affective domain and presenting the information in a heartfelt and dedicated manner. It is important for training officers to understand that their ability to affect change as trainers not only requires them to demonstrate their knowledge or expertise in a given area, but that they must also demonstrate a level of dedication to the subject being taught. To maximize the students' understanding of the material, they need to consider how the material is being presented, and looking the part has a direct impact on the presentation.

If the training officer wishes to leave a lasting impression, taking these first few steps toward class preparation will go a long way in the minds of their colleagues and students. The simple steps of dressing appropriately and taking pride in their uniform will set the tone before the class starts. In return for the amount of dedication that the training officer gives to the class, the students will respond in an appropriate and positive manner. For a training officer, the end goal of every class should be that the students leave with the desire to take another class from him or her in the future. This desire should always stem from what and who the training officer represents, never from what the training officer required them to do.

Physical Appearance of the Classroom

The physical appearance of the training officer is only a small part of the overall picture when dealing with appearance. Physical appearance is important and sets the initial tone of the class; however, other items, such as the appearance of the classroom, must also be considered. The training officer and the classroom make up one whole. The students will not only remember the training officer, but also the environment in which they were trained. General Robert E. Lee was notorious for basing field promotions on the orderliness and organization of each individual's camp. The philosophy behind his decision was that if a solider could keep his camp in a well-structured manner during war time, then the solider was more likely to function in an organized fashion as an officer. Experience has shown that this philosophy holds true in most cases for the fire department, especially when concerning training officers. Imagine your students walking into a classroom with everything set up—the firefighting gear tested and prepared for them, the seats arranged in an orderly fashion, and the audiovisual equipment set up and ready to go. In this scenario, a few thoughts should come to mind.

First, the training officer had enough forethought to prepare for the class in order to start on time. Simple tasks such as setting up the classroom ahead of time lead to excellent time management skills and reduce the amount of downtime for the students. Given the number of tasks and duties expected of the training officer on a day-to-day basis, the last thing he or she wants or needs is to waste time. The students expect the same from the

Case Study

As a training officer, I was responsible for coordinating and teaching the instructor development series. I felt that it was my responsibility as an instructor to lead by example and exemplify the effort that an instructor must have in order to present in a class. Commonly, this required me to put in extra time after hours or before class, testing the operating systems for all of the computers, putting out books and student manuals, and making names plates if I did not know all of the students. There have even been times when I spent my own money on items such as extension cords and surge protectors in order to limit tripping hazards in the classroom.

training officer and every class that they attend. A few minutes of prep time before the class goes a long way in ensuring a successful class in the end.

Second, maintaining a well-organized and structured classroom demonstrates a concern for the students and their success. The students will greatly appreciate the thoughtfulness that the training officer shows during the class. In return, the students will respond in a positive manner.

Third, the training officer has set the tone for class expectations, in that he or she has invested the time to prepare the class for the students and therefore expects the same level of preparation from the students. If the training officer has made a point to establish order and a certain level of expectations, the students will know what to expect for the remainder of the class. There is no excuse for subpar work from the class, because you have set the bar high.

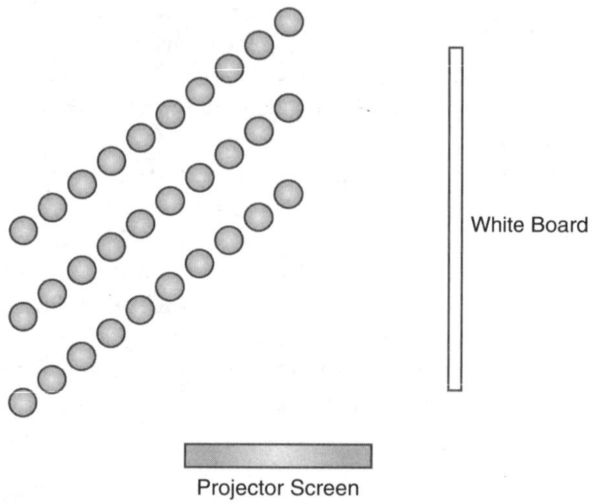

FIGURE 2-1 The training classroom set up in a modified chevron style.

Courtesy of Brian Ward.

Seating Styles and Arrangement

There are several other items that should be considered as well. What is the seating arrangement of your classroom? Are students positioned so they can see around the classroom? Is the classroom arranged according to your teaching style? A personal preference used in many training classrooms is the modified chevron style (**FIGURE 2-1**), which works well if teaching boards are on different walls of the classroom. This arrangement limits the students' needs to move as the instructor switches to different displays, such as from a PowerPoint presentation to a white board. By reducing the students' movement, the arrangement helps limit distractions in the classroom.

Another item you will need to consider is how to balance the personal needs of the students (such as bathroom breaks) with the need to maintain order and minimize distractions in the classroom. You will also need to consider the arrangement of the classroom and how it lends itself to students exiting and entering the classroom when class is in session. What are wstudents to do when they have to leave the classroom? Is there a way for them to exit without disrupting the classroom, or do they have to walk in front of the presentation to exit? It is best to establish exit routes from the classroom in order to limit distractions. The seating style you use should provide at least one exit route that leads straight to the outside without distracting the class. Although these are basic items, they must be planned for.

In addition to the seating arrangement, does your classroom provide the option of using different chairs? If a course is known to be laden with lectures, the need for seating options drastically changes: "The student will only understand what the rear-end can withstand." How comfortable the students are plays a large role in the number of breaks they need to take and how much information is transferred. If possible, opt for using the most comfortable chairs available while still making the classroom look uniform.

As you can see, there are several factors that play a part in creating a positive appearance. The physical appearance of the training officer is an important part of creating a successful learning experience, but it is only one piece of the puzzle. The appearance and set up of the classroom is also important. The training officer must not underestimate these guidelines. The legitimacy and authority of a training officer is affected not only by his or her physical appearance, but also by the classroom environment in which students are expected to learn. To create an ideal environment, place yourself in the students' shoes and note the pros and cons of the situation, then change the cons if possible. As a training officer, being cognizant of the manner in which you present yourself and the course will help you gain the respect and credibility you need to be successful. In some situations, your appearance is all that you have to work with, so be good at it.

Not an Expert at Everything

Training officers are considered, in many cases, to be experts in their field. They have spent many years training, both in the field and in the classroom. However, today's fire service is charged with many responsibilities. It has become responsible for hazardous materials response, emergency medical response, technical rescue, prevention, public safety, and emergency management. It is increasingly difficult to become an expert in every aspect of our profession.

One of the most important things that the training officer must know is the limitation of his or her own abilities and expertise. The safety of our fire fighters and of the people that we serve must always come first. In basic instructor classes it is emphasized that you should not teach beyond your capabilities and certification levels. As training officers, we must be ever diligent to abide by those rules. We just cannot do it all.

As we will discuss in the section on "Networking," the contacts that we make can be very valuable. Situations will arise where you will need to call upon those contacts to provide resources that you, the fire service, or your department lack. It is perfectly appropriate and expected to call in experts to teach technical subject matter or courses that we are not as familiar with.

However, do not fall into the trap in thinking that if the subject matter you are teaching is not for a certificate course, you can do a little research and teach it. While you may have a solid base of knowledge in the subject, you will not have the "real world" experience needed to supplement your knowledge. When presenting the subject matter, it is important, in any discipline, to be able to draw from those "real world" experiences.

In addition, bringing in another expert provides a great opportunity for your students to learn from someone with a different perspective and set of experiences. We have to be creative when designing and presenting training programs. The organization that you work for hears and trains with you almost every day. To get the class's attention, it is good to have a fresh face and a new voice. Not only can this be accomplished for technical classes, but it can also be done for the everyday training that is required by most fire departments.

Take time to research, using your networking contacts, to find out which people in the field are good at the particular discipline(s) you are looking for. Other agencies and training groups and academies can offer suggestions and advice about which instructors are good and which are great. Use the relationships you have built through networking to give your agency the best training available and to meet the ever-changing needs of the fire service. It is okay if this means that the best training is not from you for a particular subject.

Networking

As a fire service professional, it is very likely that you speak and interact with other fire service professionals on a regular basis. Stories are shared, experiences are related and compared, and solutions are given for problems. This is called networking. It is interacting with like-minded people that have knowledge and expertise in the same field or one very similar.

When considering the value of a conference or symposium, do not discount the networking that will take place at these types of events. In addition to the value of actual content being presented, there is often much to be gained from the networking opportunities that the average conference, class, or symposium provides. However, just because the potential for networking is there, this does not mean that it will occur. You have to be willing to put yourself out there and approach people you want to work with.

When looking for opportunities to network, look at who is likely to attend the event and what the subject matter is. The presenters also need to be considered: having lunch, a drink, or dinner with a fire service leader can open doors to new and valuable networking contacts. Try to sit and mingle with people you do not already know. We are here to learn, so the opportunity to meet new and interesting people should not be wasted. Simply walk up to these individuals and introduce yourself; it is that easy.

Networking can link you with experts in the industry who are more than willing to share their expertise, suggestions, and advice. Staying in touch with new contacts is even easier today with the advent of social networking websites, such as Facebook, and LinkedIn; email; and texting. Still useful is the old-fashioned phone, which adds a personal touch the social networking websites tend to lack. Without attending functions and events, it is very difficult to make these contacts.

Being a fire service leader means being involved. Most fire fighters will contribute to their community and organizations, but we need to consider being involved in the fire service as a whole. Look into and investigate the multiple organizations and associations that are available to fire service personnel. Many of these organizations are discipline-specific, such as the International Association of Fire Chiefs (IAFC) for Chief Officers, the Fire Department Safety Officers Association (FDSOA) for Safety Officers, and the International Society of Fire Service Instructors (ISFSI) for Instructors/Training Officers, and they provide their members with valuable resources, such as the ability to network with others in the organization. Becoming a member of such organizations provides the opportunity to attend larger conferences and events. In addition, these organizations link their members through the use of a common mission.

As mentioned previously, technology has created a new way to network and connect with other fire service personnel. Social networking sites and blogs have opened new portals and avenues for easily networking with others, regardless of geography. Many of these are free sites and platforms, and they provide an easy way to find people that have the same interests, mainly the fire service, as you. The ISFSI has introduced a new online platform that allows its members to post questions to a blog and receive real-time feedback on the blog or via email. In addition, members have the ability to search for members by region, state/province, and organization. This feature allows you to search for members that may have an interest in the same topics or a similar background as you.

As with the online training programs and resources, you should use caution when using these social networking sites and refrain from putting private information in a public format. Care also needs to be taken to use credible sites and programs. You also may need to get permission from your employer before utilizing these sources in a professional setting.

Many blogs have specific themes that are fire service-related. Many sites will post training clips, opinions, news, and other fire-related items. Most will also have a comments section, where people can respond to postings and share their thoughts and opinions. Some of these people could be a potential resource for information. You may use such sites to help you expand your contact list. This is just one example of how networking is not limited to the classroom or conference room anymore—it has expanded to the Internet.

Networking is a valuable tool for the training officer. Subjects that are specialized or more technical can be researched and presented with the assistance of those who are a part of your network. Do not limit your colleagues to those that are near or within driving distance. In today's fire service, it is easy to have a network of professional associates that are spread out across the country, such as the ISFSI.

Case Study

In the Metro Atlanta area, local fire departments conduct monthly meetings at different departments through the Metro Atlanta Training Officers Association (MATO). Approximately 30 departments make up the Metro Atlanta territory. There is a vast amount of knowledge and experience that convenes at this monthly meeting. As a young training officer, this was a great chance for me to learn from others and learn who to call when I had an issue or a question that I needed answered.

This association was in the beginning of its rebirth after a few years of being quiet. For this reason, very few people ran for any of the office positions. I did and was elected as the vice chairman. I served that year under a very experienced deputy chief from the city of Covington, Georgia. Once our terms expired, he elected to step down and gave me his full support for chairman of the association. A year earlier, no one in this association knew my name, and by 2009, I was the uncontested chairman of MATO. In a period of 2 years, this experience allowed me to meet several individuals who have since mentored me and provided numerous opportunities for me. This was a two-fold situation: it allowed me to gain knowledge about being a training officer and it allowed me to create a bank of contacts that I still use today, several years later.

If there are several departments in your area, you are encouraged to reach out to them and attempt to create a similar scenario to the case study, or you can simply meet once a month. The participating departments can rotate hosting the meeting and providing topics for each month's agenda. In the MATO setup, we elected officers, and there was also a Metro Atlanta Fire Chiefs Association (MAFCA) that we reported to once a month and that kept the records of our $25 annual dues and funds up-to-date. Your area does not have to extend the association this far; however, meeting the instructors and training officers in your area can be mutually beneficial. Difficult economic times can provide your department with extreme hardships concerning staffing when conducting certain drills. This association would be a great way to build relationships and find avenues to assist each other when training.

Never Stop Learning

As an instructor or training officer, we spend a great deal of our time teaching, creating programs, writing curricula, and looking for training opportunities for our fire fighters. What can easily get lost and forgotten are opportunities for the trainers themselves to train on new skills and disciplines. So much of the training officer's job is focused on providing realistic, meaningful, and relevant training for others that he or she forgets about the opportunities for personal development.

This is a problem if it is not recognized and addressed. The person who is most responsible for remedying this problem is the training officer. It typically is not an issue of a lack of desire to continue learning and training, but rather a matter of time and responsibility. It is easy to get so wrapped up in providing training and educational opportunities for others that we forget our own needs.

A simple way to get back into the saddle is to participate in some of the training that is provided in your department. There is always that officer or assistant that is highly regarded and can take your place as the evaluator or facilitator while you participate in the class. Not only does this allow you to be involved in training, but it will also gain you some credibility and respect from those who are normally trained by you.

Attend as many conference and symposia as you reasonably can. This is important in that it will provide new perspectives and teach new techniques to take back to your jurisdiction. At the same time, it will help you get some valuable training and experience. It also gives you an opportunity to witness other instructors and their methods of delivery. While at these trainings, pay attention to not only the content, but also to how the instructors deliver information. Use this opportunity to increase your knowledge in a specific discipline of the fire service and improve your instructional skills at the same time.

Meet with neighboring agencies and the training officer of each agency. They are facing the same issues that you are, and it is a great way to share ideas and opportunities. Provide free access to each other's training evolutions so that each training officer can attend training outside of his or her own department. This helps you build a great rapport with your neighboring departments, and it provides another opportunity to see how others are doing things differently from you.

In today's technically connected world, there are an unlimited number of resources available at the click of a mouse. There are websites from training academies, trade journals, blogs, and associations that offer training materials and ideas for free. Some will provide more structured training opportunities at a cost, but they deliver the training right to your computer screen. Although these media are abundant and easy to find, use caution in choosing which ones you use. Anybody can write something or offer advice without any qualifications. Use credible sources and be diligent in your research.

As a training officer there should be a desire to continue your education. It is recommended that some of the education pursued be of college-level classes; a degree is optimal. The job of the training officer not only requires you to be proficient in fire and emergency services, but it also requires critical thinking for supervision, budgeting, evaluation, and other administrative duties. Get out of your comfort zone and challenge yourself with higher education programs.

These higher education programs and degree programs are also easy to find. There are many reputable programs that offer accredited degrees. Degree options are numerous, including fire science, fire service administration, political science, and many others. Find a discipline that is appropriate for your career ambitions and current position. It is always a good idea to research trends in the fire service and what the future needs of our profession will require when considering an advanced degree. Two specific examples of college programs that work well with a typical fire fighter schedule are the University of Cincinnati in Ohio and Columbia Southern University in Alabama.

One of the simplest ways to continue to learn is to read fire service periodicals, journals, magazines, and other literature, such as *Fire Engineering Magazine*, *Journal of Emergency Medical Services*, and leadership publications, to name a few. Find subjects directly related to the fire service, such as those written by fire service leaders. Also, read books on subjects that are indirectly related to the fire service, such as teaching methods, business, administration and leadership, and accounting, to name a few.

Books and magazines are easy to get your hands on, and they can be read over and over again. Another great thing about reading material is that it can be shared with others. Find a good read and hand if off to someone else for their own educational purposes. A technique that works well is to make your own notes in the side spaces of the book or magazine, underline specific comments, or highlight points that you find important. Sharing a book you read with someone else can also become part of the teaching and learning process.

Learning never ends. It is, and should be, a dynamic process for the rest of your career and life. For training officers, extra effort may be needed to get out from behind the podium and back out into the chairs of a classroom because they are accustomed to performing in the role of an educator. Be creative when looking for training and educational opportunities; they are not always in the obvious places. Be aggressive with participation in training evolutions and drills, as these will build on your skill sets and help you establish credibility among those you teach. Most importantly, make the effort to keep learning. Nobody else but you can dictate how much or how little you decide to learn.

Key Points

- As a training officer, you assume the face of the organization or department.
- The first step that a training officer must remember is to hold himself or herself accountable.
- Aggressiveness is protecting your rights, beliefs, and territory without any regard for other individuals.
- An assertive individual is one who protects his or her rights, beliefs, and territory while respecting others.
- The training officer's appearance establishes their first line of credibility.
- Creating the perfect learning environment will allow the largest amount of information and knowledge to be transferred from the training officer to the students.
- The physical appearance of the training officer should be consistent with part of the curriculum that he or she is teaching. This affects the legitimacy of the information being presented and the reception and interpretation of the students.

References

Abrashoff, M. *It's Your Ship: Management Techniques from the Best Damn Ship in the Navy*. New York, NY: Hachette Book Group, Business Plus; 2012.

Covey, S. *The Seven Habits of Highly Effective People*. New York, NY: Simon & Schuster; 1989.

Crocker III, H.W. *Robert E. Lee on Leadership: Executive Lessons in Character, Courage, and Vision*. New York, NY: The Crown Publishing Group, Three Rivers Press; 2000.

International Society of Fire Service Instructors (ISFSI). http://www.isfsi.org/default.aspx. Accessed September 7, 2014.

Klein, G. *Sources of Power: How People Make Decisions*. Cambridge, MA: The MIT Press; 1999.

Klein, G. *The Power of Intuition: How to Use Your Gut to Make Better Decisions at Work*. New York, NY: The Crown Publishing Group, Crown Business; 2004.

Lasky, R. *Pride & Ownership: A Firefighter's Love of the Job*. Wilsonville, OR: PennWell, Fire Engineering Books & Videos; 2006.

Richards, C. *Certain to Win: The Strategy of John Boyd, Applied to Business*. Bloomington IN: Xlibris, Corp; 2004.

Salka, J. *First In, Last Out: Leadership Lessons from the New York Fire Department*. New York, NY: Penguin Group, Portfolio Trade; 2005.

U.S. Marine Corps Staff. *Warfighting*. North Charleston, NC: CreateSpace Independent Publishing Platform; 2014.

CHAPTER

3 Creating and Evaluating Your Own Program

David Bullard and Skip Heflin

Creating Your Own Training Program

Just as a fire department creates a mission statement to provide guidance in the delivery of its core services, the training division should create a plan that includes long- and short-term goals. Such plans are typically called training programs and allow the organization to document the methods and procedures by which they will meet their training goals. The training program should be in alignment with the fire department's overall mission statement and strategic plans. These training programs should be accessible by all members of the department and should be periodically revisited to ensure they are still meeting the continuously changing demands of the organization and align with the overall goals of the fire chief. Training programs are a great way to meet required mandates.

The training officer is typically charged with the administration of the training programs within the department. Depending on the organizational structure of the training division, he or she may also deliver the programs of study or classes/courses offered in the training programs.

Benefits of Establishing a Formal Training Program

There are many benefits to establishing a formal training program for your department. A formal training program can:

- Provide consistency and continuity across courses and among instructors
  - It can be confusing for the instructors who are delivering these training programs to switch from one program of study to the other. For example, completing practical skills for emergency medical services (EMS) may require one set of rules, while completing fire-based practical skills may require a completely different testing instrument and/or procedure. Evaluation methods and passing requirements can be easily confused. A formal training program provides a system that ensures the instructor will deliver the course of study according to the previously established guidelines for a particular program.
- Provide uniformity
  - The use of instructors outside of the formal training division is common among departments, particularly with the budget reductions currently faced by many departments. Having a formal training program helps to ensure that different instructors all deliver the curriculum in the same manner. Without a training program in place, you may find that training classes tend to morph and change over the course of their duration. Consider an engineer training program in which fire fighter A is free to change the curriculum as he or she pleases. As a result, fire fighter A's approach in the year 2012 is markedly different in the year 2014, despite the fact that the lesson plan course/class itself has not changed much. A training program can help to ensure the core

requirements are met when delivered from one year to the next.

- Create a level of transparency
  - Creating a formal training program and ensuring it is accessible to all members provides all participants with a standard set of guidelines and expectations on what they need to do to successfully complete the program. After a new member is added to the department, he or she should begin career planning. If the new member knows exactly what will be required from him or her in each of the programs, he or she can schedule the appropriate time off to ensure class attendance and participation, address family and personal concerns, and other related matters. If this information is not provided or is inconsistent, then members who are participating in the training program may run into problems with attendance and their ability to commit to the course. Some training divisions may serve more than one department, and schedules and job-related demands can vary from one department to another. A formal training program ensures that every participant knows what is expected to finish the program and receive full credit.
- Provide for legal compliance
  - Training programs can help to ensure your department stays within legal mandates, such as the Occupational Safety and Health Administration (OSHA) respiratory compliance training. By establishing a training program, you are sharing with the department, and anyone who looks into your processes and procedures, that you have a plan in place to address the requirements imposed by outside organizations such as OSHA, the National Fire Protection Association (NFPA), and the like. A formal program can also provide documentation on the content that was delivered.
- Provide training without interfering with the daily operation of the company
  - The day-to-day operations of the modern fire company can be very demanding for the company officer. Between company-level inspections (or preplanning), fire safety education, and call volume, the company officer is tasked with the challenge of managing time and resources for the fire company. Having a training program in place ensures that the company officer knows what is expected on a day-to-day basis and allows him or her to plan so that daily operating needs are met and the company's schedule is maintained.

What Is a Training Program?

Training is typically thought of as ongoing technical skills development. This is in contrast to education, which is typically thought of as the acquisition of new knowledge. Your training division may provide both.

Training programs describe the vision for and process of a program of study. They also lay the ground rules of a program of study for the given department. A single training program can also be used to provide a framework for the training function overall. There may be different training programs for different disciplines within the same department.

For example, a department may utilize one training program to describe how they will meet their yearly self-contained breathing apparatus (SCBA) training. The training program may provide information such as the month it will be done, what members within the organization must take part, and how it will be delivered. The training program should also describe the method of evaluation to measure the results of the training.

Some departments publish the training programs as a part of the standard operating procedures (SOPs) collection within the department. They may be treated as an individual SOP and updated on a schedule just as an SOP would. An individual SOP that covers the overall training function should be provided; this would not address a particular area of study, but rather describe the general training function. For example, a department may have a training SOP that describes how many hours each member of the department must earn in a calendar year, how many multicompany drills will be performed, the structure of the training division, etc. The same department may have individual training programs that address how they will handle new engineer training, SCBA training, etc. Training programs can mirror the career ladders or requirements set forth for each position by the department.

Some departments choose to utilize committees in the creation of their training programs, schedules, and SOPs. The training officer is usually the chairperson of the committee if such a process is used for training. Some of the benefits of using a training committee are:

- They allow members who are closely connected with the needs of the department to have a say in the training functions.

- They create a buy-in from line members, in the sense that they are less likely to complain about something if they had a hand in creating it.
- They help the training officer with developing the direction and needs of the training area.
- They assist in spreading the workload.

Some of the drawbacks of using a training committee may be:

- They can create requirements that the training division is unable to meet. A common expression that can be applied to training divisions is, "Nothing is impossible to the man who doesn't have to do it."
- Another saying that can be applied to training divisions is, "Too many hands in the soup bowl makes a poor soup." In other words, it can become hard to develop a clear course of action if the committee is not managed properly.

Components of the Training Program

In a well-organized document, a training program should lay out everything that one needs to understand about the particular subject being covered, based upon the needs of the organization. The training program document should provide clear direction as to all of the requirements, both for the line fire fighter and the instructor charged with its delivery. These requirements may be set forth by the authority having jurisdiction (AHJ), the NFPA, Insurance Services Organization (ISO), or other agencies. This allows the training personnel to customize the document according to their requirements, desires, budget, etc. One department may choose to simply comply with legal mandates, while another may choose to exceed the minimum mandates and create additional training. The following is a list of what items should be included in a written training program:

- Author
- Revision date
- Goal, mission, or vision
- Instructional/delivery method
- Allowable training locations
- Equipment needs
- Timetable of delivery
- Passing requirements
- Attendance requirements
- Who is responsible for delivery
- Evaluation method(s)

SOP 6.01 Training
Position Certification

Personnel in the following positions shall be considered "certified" when all applicable initial training requirements have been met and associated promotional testing has been passed. Training requirements can be found in SOP 7.03 and/or when a vacancy memo is published.

Apparatus Operator

Fire Officer (Lieutenant & and higher ranking Officer)

Fire Inspector

Fire Investigator

Fire & Life Safety Educator

To maintain the certification, members are required to meet the annual training requirements as outlined in this policy, and maintain the promotional requirements that have been outlined for the respective position.

FIGURE 3-1 An example of a written training program included in an SOP. This is an excerpt from an SOP that includes a training program that outlines an AHJ certification program to help meet ISO requirements.

Courtesy of Hall County Fire Services, Hall County Georgia.

- Acceptable alternatives
- Documentation

Some of the above items may not need to be in a training program because they are addressed in another form of documentation. For example, a department may have an overall training SOP in place that includes the statement: "In order to receive credit for a given class, the fire fighter must be in attendance and shall not miss more than 10 percent of the class hour requirements." Since the attendance requirements are already explained in the SOP, they may not need to be included in each individual training program.

Regardless of the contents of the training program, each different training program should be consistent with the others within the same organization. **FIGURE 3-1** provides an example of a written training program used as a part of an SOP.

The Process of Developing a Training Program

There are several different methods that can be used to create the training program. Three examples are:

- NFPA Five-Step Process
  1. Assess needs
  2. Establish objectives
  3. Develop the training program
  4. Deliver the training
  5. Evaluate the impact

- MFRI (Maryland Fire Rescue Institute) Four-Phase Instructional Design Process
  1. Assessment
  2. Program development
  3. Pilot test
  4. Evaluation
- AIDE process
  1. Administration
  2. Implementation
  3. Delivery
  4. Evaluation

More information on each of these methods can be found by researching currently available texts.

If you look these different processes over carefully, you will see that they share some common criteria:

1. Needs assessment
2. Develop/Produce content
3. Implement/Deliver
4. Measure/Evaluate

As you will see in this section, we use the four items listed above in our discussion of the process of developing a training program for a department. However, before we discuss that process, let us take a closer look at one of the items of importance listed above, needs assessment. Roger Kaufman, the "father of needs assessment," is a prominent figure in the field of educational technology and has developed a model for defining the "needs" of a particular program, system, or process. According to Kaufman, the presence of a gap or a discrepancy in a set of desired results or outcomes indicates a "need." For example, say a fire department wanted to assess the department's proficiency in donning and doffing SCBA. In this scenario, the desired result would be that 100 percent of the personnel in the department know how to properly don or doff SCBA. To determine what the actual results are, the training officer of the department tests the entire department on this skill. Upon completion of the testing session, the training officer determines that only 70 percent of the department know how to perform this skill. The discrepancy between the desired result and the actual result indicates a need to provide additional training on this skill. Now that the training officer has identified a need for more training, he or she can begin developing a program to ensure that the department achieves the desired result. The Kaufman model is included because it provides a straightforward method of identifying training needs. The process also includes justification to conduct the training, which has become increasingly important in the current economic climate.

Assess the Needs of Organization

Creation of a training program should always begin with its purpose, which is typically based on a particular need of the organization. Some of the organizational training needs may be very obvious, such as a recruit fire fighter's initial training, which may be spelled out in a state/provincial mandate. Other areas may require more research and effort by the training officer. Common sources used to identify needs or gaps in a training program include:

- **Hazard evaluation**
  - Hazard evaluation is the review of all occupancies within the response jurisdiction of the organization to determine the presence of hazards in the area. One department may identify large fuel-tank farms in its jurisdiction and determine that a need for flammable liquids fire control training exists. Meanwhile, another department may not have any such industry within its response area, and so the need for such training would not be present.
- **Legal mandates**
  - Legal mandates can come from federal, state/provincial, or local levels of government. A federal-level mandate could be a Presidential directive addressing the National Incident Management System (NIMS). A state/provincial-level mandate may require an annual live fire component for fire fighter recertification. On the local level, a mandate could be a department's decision to adopt NFPA 1403, *Standard on Live Fire Training Evolutions*. Each of these mandates could be addressed with the use of a training program.
- **Community needs**
  - The community itself can be a source of training programs. Jurisdictions that include large sections of forest may create the need for a wilderness search and rescue program, whereas organizations situated in more urban areas would have no need for such training.
- **Risk analysis**
  - The level of risk that an organization decides they are willing to accept is another area that can be used to guide the development of training programs. For various reasons, some

organizations may only be prepared to offer a supporting role in the area of technical rescue. Meanwhile, there may be other organizations that conduct technician-level training in common technical rescue areas. Each would have far different training programs and expectations in the technical rescue area. For example, a department that would be considered a "bedroom" community may have little need for technical rescue training and therefore may elect to only provide an awareness level training to their department members. That department may rely on a nearby agency to provide for the next level of technical rescue if needed. This would be identified during a risk-based analysis of the response district. Risk analysis is similar to the hazard evaluation described above.
  - The ISO could fall under this category as well, since it helps determine the amount of risk the department will accept or can afford. The various ISO requirements can be difficult to track. The use of training programs can make this task easier. By utilizing training programs, you lay out the plan that you will use to meet various training-related requirements, such as officer training, or facility use requirements.
- **Employee needs**
  - The line fire fighter can be an excellent source for needed training programs. A questionnaire; promotional assessment; in-person interview; or an on-scene, direct field observation could be used to determine what skills are not performed frequently enough or what skills need to be reinforced via training programs. These assessments can be used to provide guidance on what skills need to be included as a part of a training program. The evaluation method that is chosen can be a source of needs and its worth merits mentioning.

Regardless of the source of the identified need, the program should address a true need for training. Look at the use of seat belts while responding in apparatus. Ongoing training and even discipline has been used to increase compliance. Apparatus manufacturers are now using systems to verify and even prevent a truck from moving without all occupied seats being belted in. While training has helped and increased awareness of this need, engineering is making seat belt use a requirement.

During this needs identification phase, the frequency at which the training will be offered is decided. The identified training program may be offered once to the entire department, or it may become an annual training requirement for particular ranks or positions within the agency. For example, consider one part of a multi-company drill (MCD) in which each company arrives on the scene of a simulated single-vehicle accident that requires extrication, patient care, extinguishment, and other vehicle incident-related tasks. During the drills, the evaluators note that most of the companies fail to use their emergency vehicles to block the work area to protect personnel while working at the accident scene. An apparent gap has been noted, but what exactly is the gap? The gap is the difference between the knowledge of the student as compared to the expectations of the student in regards to the particular skill or knowledge. Upon further investigation via in-person interviews, it is found that most of the vehicle operators who parked their apparatus incorrectly did so as a result of at least one of two main reasons:

- Peer pressure from the law enforcement community has led them to believe they must keep a lane of traffic "open" at all times
- A desire for access to the rig so as to avoid having to carry tools farther than necessary

Looking more closely at the involved parties, the evaluators find that the fire department operates with several different law enforcement agencies that do not have the same concept of emergency responder roadway safety as they do. In the end, the gap becomes a twofold issue:

- A lack of education and understanding of the seriousness regarding highway safety by the responders when responding to roadway incidents
- Multiple policies from all of the involved agencies causing confusion for responders—they simply do not know what to do to please everyone

After determining the gap, a solution is created. A mandatory monthly training module is created that highlights the importance of roadway safety and creates relevance for apparatus operators. A formal SOP is developed to outline the expected and specific desired behaviors when responding to roadway incidents. To create consistency across various agencies, a task force is created that involves all local law enforcement agencies, EMS, and fire departments.

■ Develop the Content and Curriculum for the Training Program

Some training programs may identify particular classes in which the content is commercially available. Numerous sources of content can be found, and some organizations may choose to purchase such content to save them from the time and effort of developing the content themselves. Some publishers will provide a series of instructor resources if the book is adopted by the organization. Although departments may benefit from developing their own curriculum, which allows them to tailor the program to address the particular needs of the organization, developing a curriculum can be very time consuming. Furthermore, any curriculum development should be completed by a qualified instructor. Organizations that lack the appropriately trained personnel may be more inclined to purchase commercially available training programs.

The use of subject matter experts (SME) from within the department can help to increase buy-in from line personnel and lend legitimacy to the program. SMEs could be involved in:

- Reviewing content after it is first produced
- Customizing commercially available content
- Validating and reviewing the test questions associated with a particular program
- Establishing the practical application, or hands-on components; testing requirements; and documentation
- Serving as evaluators during the practical testing

After the course content has been created, you must decide whether outside training courses, such as a state/provincial or national fire academy program, or those of another fire department, will be accepted. The reality for many departments is that they cannot afford to offer every training class they want for their members. Even when classes are offered, some department members may not be able to attend the course when it is offered by the agency. One alternative is to accept another agency's training class in place of your own. Only after the curriculum is decided upon can a decision be made as to whether your department offers the training or if it will outsource to another agency's training. Some considerations affecting the decision may include the objectives, required hours, and hands-on training component(s). Typically, this decision is made by the training officer and on a case-by-case basis.

Consideration should also be given to the delivery method. Will the program be taught both in the classroom and on the training center grounds, or only in the classroom? Some departments have begun to utilize the Internet or closed computer networks to conduct training when appropriate. A blended learning environment can work as well. A blended learning class is one in which a computer program is used in conjunction with a classroom learning experience to achieve learning outcomes. During content production, this approach should be considered. The use of online classes can help alleviate scheduling problems regarding the use of training assets, such as classroom spaces or props, and save time for the training division. If a task level skill is included as part of the course content, skill sheets may be needed to ensure the skill is mastered and to demonstrate what is expected of the student. Any such skill sheets should be written as well and included in the course curriculum.

The last step before formally implementing the program is to conduct a pilot version of the curriculum with veteran personnel, SMEs, and department leadership. The pilot version should be performed in its entirety with all props and activities included. Doing so allows the training officer to receive feedback from key personnel on content, relevance, flow, delivery, appropriate knowledge level, and time frame. Once the pilot version is completed, the training officer should review the feedback, make the necessary adjustments, and prepare the program for formal implementation. Conducting such a review provides the time and space the training officer would need to practice and work through any personal issues before the first formal delivery.

■ Implement the Training Program

After the program content has been developed, the program can be implemented. During this step in the process, support from line personnel and the chief of the department are critical to the success of the program. It does little good to produce a training program and have no participation. Often the requirements set forth in a training program are also included in a career ladder or are considered mandatory. Consideration must be given to enforcement of the requirements during implementation. How will the members of the training division enforce participation? Often, this task is shared among the command staff of line personnel in conjunction with training division members. Enforcement may be a non-issue for programs that members are required to take to become eligible to test for a promotion.

In some departments, a member fills the role of professional standards officer. This position is responsible for ensuring eligibility for promotional testing. Enforcement can be made readily apparent at the individual level by the use of a self-assessment document that is completed by the member who wishes to test for a promotion. As the member completes the paper, he or she must address the promotional requirements set forth by the department; consequently, he or she would obviously see which classes he or she would need to take to meet the requirements.

Other methods of implementation can take the form of a task book that must be completed in a given amount of time. As the member documents the completion of each task, he or she may be required to obtain a signature from a line officer. One benefit of this approach is that the member participates in the enforcement of each requirement because she or he has to track his or her progress. This way, the member and the line officer share the responsibility of accounting for each requirement set forth in the task book.

■ Evaluate the Program

Measuring the Effectiveness of the Program

If you do not measure what you are administering, how will you know whether you are being effective? A method must be determined to measure the outcomes of the training program. This could be done in many ways, including:

- **Multicompany drills**
  - During the performance of multicompany drills, direct skills observation can be applied to most training programs. Check-off sheets written specifically to capture certain points of individual training programs are a great resource for trained evaluators. Training division staff often serve in this role, but line division personnel, particularly those who are subject matter experts, can be utilized as well.
- **Single company drills**
  - Individual company drills conducted at the station can be utilized to gauge the effectiveness of the training program. Skill sheets may be developed to address certain points covered by individual training programs. The skill sheets could be turned in to the training division to become a part of the training record.
- **Computer-based testing (CBT)**
  - A test can be administered on a computer by using a closed network or the Internet. This frees up time at the training site, as now that time can be spent on training members on hands-on skills. Modern testing software available to the fire service allows a much deeper test question analysis than the standard paper and pencil test method, which typically involved much more time and effort. Budgetary requirements, the lack of technical skills, and limited knowledge often preclude many training officers from utilizing this testing method. However, these concerns can easily be overcome. See the chapter on computer-based training for more information in this area.
- **Written test**
  - A written test may be given as the culminating task of a class. This method of evaluation is what most fire fighters are accustomed to. When determining the passing requirements for these tests, the associated time and monetary print costs should be carefully considered. Many organizations are moving toward a complete/incomplete categorization rather than using the traditional method of using a number, such as 90 percent, as a passing requirement. This categorization method could apply to the final course documentation only, or it could apply to anything associated with a traditional pass/fail criterion such as a written test or a practical skills-based test.
- **Practical skills application or hands-on skills component**
  - Many courses offer some type of hands-on application component, and this is a valuable evaluation method. Consideration should be given to documentation of the hands-on application if it is to serve as your evaluation method. Skill sheets are commonly used to document or verify the skill level being demonstrated by the student. Another consideration is the number of trained evaluators that may be needed during this evaluation exercise. It is best to determine this number before the class begins, as it will eliminate the concern of having to do this during crunch time when class is in session.
- **Task book assignment**
  - Task books are often underutilized. A task book can be used for a single class or a collection of skills that need to be completed during a predetermined timeframe. Some states/

provinces require a task book be completed annually for the fire fighter recertification process. One approach is to build a list of skills in the task book and include all associated skill sheets. The skills that each fire fighter must complete can be organized by rank or by certification. The task book can also be used for fire- and EMS-related skills. In either case, verification of each completed skill is often required in the form of a signature by the task book owner and an officer. When using a task book, it is important to consider notating the job performance requirements (JPRs) and the standard referenced with the actual skill. The standard could be an NFPA standard, state/province agency standard, or one created by the AHJ. Noting the standard can also help reduce the associated liability risk of the material. This notation helps to identify the source of the skill. It will also be helpful when the time comes to update the referenced skill sheets based on standard changes. These changes are normally made by training division staff members or the training officer.

- **Direct observation by trained staff**
  - Using direct observation by trained staff as the only evaluation method is not wrong, but it does have inherent problems. One problem associated with this method is finding a way to properly document each member's progress. Using skill sheets to document the student's performance is a typical solution. During delivery of a course, an instructor can become easily overwhelmed with paperwork, which can fall through the cracks and not get completed. Having an adequate number of appropriately trained evaluators available when using this evaluation method will spread the workload. An emphasis should be placed on *adequately trained* evaluators. Using anyone who is available is not an acceptable practice and could lead to injury and or unnecessary liability exposure. In most medium to large departments, this task is often delegated to an instructor who is subordinate to the training officer. The training officer should always ensure the proper evaluation forms or documentation are used as a part of any training exercise and must remember that he or she is ultimately responsible for the evaluation.
- **Legal mandate reporting**
  - Periodically, agencies may be required under a legal authority to report on their activities to an outside agency such as OSHA or ISO. During these reporting activities, an evaluation opportunity exists. Stating whether a requirement is complete or incomplete may be all that is desired in the report, and this simple task can be considered an evaluation method. Consider the example of the OSHA-required infection control annual in-service training. Department A has the lesson plans and all associated course materials on their internal computer network. The training program states that each company officer is to conduct the training class sometime during the year and document the training activity in their computerized record keeping system. In this scenario, the training division's only concern is confirming whether each fire fighter completed the requirement or not. As the training officer, you would not require a task book, as you would only need confirmation that the requisite training has been completed and validated by an officer.
- **Industry standard best practices, such as NFPA standards or state/provincial-level documents and requirements**
  - NFPA and state/provincial-level requirements or standards can be a great source of evaluation materials for the training officer. One benefit of using such documents is that they may allow you to compare your department to other departments in the nation, state/province, or area. NFPA 1720, *Standard for the Organization and Deployment of Fire Suppression Operations, Emergency Medical Operations and Special Operations to the Public by Volunteer Fire Departments*, includes documents for a volunteer department in the form of check-off sheets that could be utilized to evaluate personnel performance.

Evaluation should be a continuous process that begins right away with a needs assessment. The needs assessment should be used to both evaluate the individual and the entire process. It should be considered a part of the continuous feedback mechanism. Consider the following scenario: During the hands-on evaluation component of an engineer program, a training officer finds that problems are often encountered when students are asked to draft from a static water source. This discrepancy motivates the training officer to look through the curriculum. Upon reviewing the curriculum components covering this skill, the training officer discovers that this skill only gets minimal coverage in the classroom and a brief demonstration during the hands-on component of the program. During the revision cycle, the teaching materials are expanded to explain the theory and application of drafting, and the hands-on component is increased from a brief to a full demonstration. The training officer also implements a new requirement in which all of the students are required to complete at least one drafting operation on their own during the class. The skill is then added to the task book that is part of the class. During the next offering of the class, the problem with drafting is not noticed. The process of assessing needs and developing an action plan to meet those needs has been utilized and has worked as expected.

When using evaluators as a part of your evaluation process, you should always use individuals who are trained in the evaluation process and are familiar with the operations being tested. It cannot be said enough that using evaluators who are not trained can lead to less than desirable results. Some states/provinces offer a formal evaluator course, or you can offer your own course to train them locally. Regardless of how you choose to train your evaluators, they must understand their role in the process. By definition, a strict evaluation exercise is not a learning opportunity, even though the student is learning during the evaluation by demonstrating, or practicing, a skill. This concept can be difficult for line personnel to understand, and you should make this expectation clear before beginning the exercise.

One other point that should be made about the evaluation process is that it should ensure that the actual discrepancy in the member's training has been corrected, not just the symptoms of the problem. For example, a group of recruit fire fighter candidates are taught that they should always exit an immediately dangerous to life or health (IDLH) environment if their SCBA low air alarm goes off. When practicing, however, they always stay in the IDLH beyond the time that the low air alarm goes off. During a hands-on evaluation, a recruit fire fighter stays in the IDLH after the low air alarm goes off and fails the evaluation. Before the retest, simply reminding the fire fighters to leave the IDLH when the low air alarm goes off would be treating the symptoms of the problem; correcting the way they practiced what they were taught by making them leave the IDLH when the low air alarm goes off would be treating the problem itself. This method would likely result in the recruit fire fighter passing the skills evaluation in the future. It can be difficult to determine whether this has been accomplished, especially for inexperienced training officers. This potential problem can be partially addressed early on in the planning process by describing in detail what the correct procedure or evolution should look like. This way, the training officer has a prearranged list of expectations to guide him or her in determining whether the members can proficiently perform the skill. This early planning will help the training officer later on when he or she examines the evaluation data to confirm whether the identified problem has been corrected. Another way to avoid this trap is to ensure that enough qualified personnel are available to conduct the evaluation method effectively.

Utilizing Resources

There are various documents and organizations that the training officer should consider, as these sources can provide a wealth of information to aid in the development of training programs, evaluation, etc. Many of these are available 24/7 on the Internet. Other opportunities, such as fire-related association meetings or industry trade shows, may require the training officer to attend in person. Juggling the demands of participating in various organizations and continuing education can be challenging. The modern training officer has a very busy schedule, so time management is an absolute necessity. Regardless, every effort should be made to be involved in available organizations to take advantage of the many benefits they offer. Networking through conferences and training-related groups can be a very useful resource for the training officer and it is highly recommended that he or she participate in these activities. Available resources include:

- National Fire Protection Association (NFPA)
- National Board on Fire Service Professional Qualifications (NPQ)
- The International Fire Service Accreditation Congress (IFSAC)
- Occupational Safety and Health Administration (OSHA)
- Federal National Incident Management System (NIMS)
- International Society of Fire Service Instructors (ISFSI)
- Georgia Firefighters Standards & Training Council (GFSTC)

- Florida State Fire Chiefs Association
- Also, see the Web-based Resource List in the appendix for additional resources.

This list is not meant to be all-inclusive, but rather is intended to demonstrate that a wealth of information is available to the training officer.

Pursue Continuous Improvement

All of the different steps and processes discussed in this chapter should be a part of the continuous training cycle. In the case of a previously established training program, the process can resemble the lifecycle of an SOP. An example of such a process is illustrated below:

- Needs assessment
  - Ensure that personnel needs are being met by the existing program
- Develop/Produce the content
  - Check and update content as required
- Implement/Deliver
  - Implement and change delivery in accordance with the above
- Continuous quality improvement
  - Conduct ongoing measurement and evaluation to ensure the training program meets its intended goals and/or vision

As you can see, the cycle is never-ending. The continuous pursuit of improvement is a constantly evolving process that allows you to meet the ever-changing needs of the all-hazards emergency response.

Continuous Quality Improvement

We have discussed measuring the effectiveness of a given program; however, once a program is established, we, as training officers, have to evaluate the entire process involved with the training program. This is part of the continuous quality improvement process. Evaluating training programs is one of the most important aspects of being a training officer. The programs that you oversee need to be up-to-date with current JPRs within NFPA standards, OSHA regulations, best practices, and current department policies. Performing these tasks will ensure that your program is providing line personnel with the best information possible to take back to the field and implement.

Processes

Two processes that can be used to evaluate programs are formative and summative evaluations. The formative evaluation process is typically conducted for the purpose of improving the program's performance by identifying its strengths and weaknesses during the class. Situations may arise that require the training officer to adjust the program in order to meet the needs of the class. In addition, a major incident may occur somewhere in the country that could provide a valuable lesson for the class.

The summative evaluation process measures the program's achievements through testing or the completion of evaluation forms to determine the strengths and weaknesses of the program. These types of evaluations take place at the end of a course or during observations in the field.

Both of these processes should be used to measure the success of a program. Understanding the strengths and weaknesses of a program will only make it better and ensure that students receive the greatest possible benefit. Items that seem trivial to you may make all the difference in the world to the student. Be open to new ideas and methods of conducting your program. We will discuss methods that can be used to collect information and data concerning your program, including interviews, documentation reviews, observations, focus groups, and case studies, as well as the advantages and challenges associated with each method.

Methods Used for Collecting Data During Evaluations

Questionnaires, Surveys, and Checklists

Questionnaires, surveys, and checklists are used to collect a lot of information in a quick, nonthreatening way. Advantages to questionnaires, surveys, and checklists are that they can be completed anonymously; are inexpensive to administer; can be easy to compare and analyze, especially if electronically administered; can be given to a very large group of people; typically produce a lot of data; and many templates are freely available. Challenges to questionnaires, surveys, and checklists include reliance on participants to provide constructive feedback and that improper wording can bias a client's responses. These forms of data collections are seen to be impersonal and incapable of capturing the entire story; they also may require the use of sampling experts for reliable results.

Interviews

Interviews are used when you want to fully understand someone's impressions or experiences, or learn more about their answers to questionnaires. Interviews can provide a full range and depth of information and help to develop a relationship with the people involved.

Interviews can also be flexible to meet the needs of all parties. The challenges of interviews include that they require an increased amount of time to complete and the results may be hard to analyze and compare, requiring the assistance of an expert; the associated costs of interviews can be high; and the interviewer may bias the client's responses.

Documentation Review

A documentation review is used when you want an impression of how a program operates without interrupting the program, and it involves reviewing applications, finances, memos, minutes, and skill sheets. The advantages of documentation review are that they provide a comprehensive overview and historical information. They also do not interrupt the program or the personnel's routine in the program. All of the information may already exist and information bias is reduced. The challenges of documentation review are that it is time-consuming due to the large amount of documentation that must be reviewed, and it requires preplanning to ensure that information is complete. Before starting, be very specific about what you are looking for—this is not a flexible means to get data, because the data are restricted to what already exists.

Observation

Observation is used to gather accurate information about how a program actually operates, particularly regarding methods and processes. Observations provide a unique opportunity to view the operations of a program as they are actually occurring, and can be adapted to events as they occur. Observations pose unique challenges, however, as it can be difficult to interpret seen behaviors, especially by untrained observers, and to categorize and quantify observations. Observations may be influenced by the behaviors of the program participants when they know they are being observed. Observations can also be expensive, both for the personnel training and during the actual observation process. Consistency can be difficult to maintain between multiple observation sessions and/or multiple observers.

Focus Groups

Focus groups explore a topic in depth through guided group discussion, e.g., about reactions to an experience or suggestion or understanding common complaints. Focus groups can quickly and reliably provide common impressions. They can be an efficient way to get an increased range and depth of information in a short amount of time and can convey key information about programs. The challenge of focus groups is that the responses can be difficult to analyze and may require a good facilitator for safety and efficient use of limited time. It is also often difficult to successfully schedule groups of people and facilities together.

Case Studies

Case studies help to fully understand or depict the students' experiences in a program and it is possible to conduct a comprehensive examination through a cross comparison of cases. The advantages of case studies include that they fully depict the students' experience in program input, process, and results, and they are a powerful means to portray the program to outsiders. The challenges of case studies are that they are usually quite time-consuming to collect, organize, and describe, and they normally represent the depth of information, rather than breadth, although a well-prepared case study can provide both.

Areas to Consider During the Program Evaluation

■ Budget

As a training officer, you will likely be charged with the task of reviewing last year's budget in order to determine whether you are on track to meet or exceed this year's budget. In performing this task, you can forecast future expenditures when repairing props, ordering new student manuals, or creating new programs. In some cases, the governing board (Board of Commissioners, City Manager, etc.) will ask for budget reductions. Understanding a program budget can assist with determining which programs offer the most in return. The programs that have a high cost, but are not necessarily mandated, may be the programs that should be reduced.

■ Scheduling

As a training officer, you will need to determine how many classes were held last year and compare this number to how many students actually completed the training. Given the many obligations of the training officer, time is very valuable. Ask yourself the question, "Is it possible to conduct the class three times instead of four times and still fill all of the classes?" To save yourself some time, it is a good idea to designate one person to schedule all of the classes with the goal that the classes do not overlap, if possible. By ensuring that classes do not overlap, additional instructors and back-up instructors can be available to fill a class, if needed. Also, be sure to allow yourself time and personnel to evaluate these programs. This step should be a part of your yearly schedule.

■ Audio and Visual Aids

While evaluating your programs, you may find that you need to update your library and audio/visual aids as well. A relevant library is an excellent resource for people to learn on their own time, but it needs to be available for everyone to use. Additionally, it needs to be kept current and organized. Your audio/visual aids need to be maintained properly. Make sure that all of your programs are properly backed up and stored, so if you should suffer computer failure, you will not have to rebuild all your records and training programs. A library is also an item that can be considered in an ISO rating evaluation. Below is a list of items that you may consider adding to your library.

- Fire fighter I and II videos or DVDs
- Specialized videos specific to your local community
- Industry publications, such as educational manuals or videos
- NFPA publications, such as the NFPA Fire Codes
- Fire fighter I and II Training Curriculum
- Other training curricula to fit the needs of your department
- Back and current issues of trade magazines
- Lesson plans for classes taught at your department

Credibility

■ Providing Quality and Meeting a Standard

As you develop a training program, you will need to consider the requirements of various organizations, such as the OSHA; state/province and national standards, such as those from the NFPA; and third parties, such as the International Fire Service Accreditation Congress (IFSAC). The consideration of these is not only necessary to ensure compliance with standards and federal mandates, but they also help you develop a more effective and thorough training program. Most if not all of the requirements posed by these organizations and third parties are based on past experiences in the field. They can guide you in your development of what the training program should entail.

For some, the role of a training officer is the first leadership position above that of a fire fighter that they have ever held. It can sometimes be a challenging transition as the training officer steps out of the role of being a fire fighter. As a fire fighter, the training officer was only accountable for his or her own skill level and development. In the new role, he or she is now directly responsible for the skill levels of other fire fighters. The training officer role is a position of leadership. Being responsible and holding oneself accountable for the skill level and development of other line personnel comes with the territory of a training officer.

Teaching to the standard also makes the process of developing a training program easier for officers. It provides an end goal, so you are not constantly trying to reinvent the wheel or having to adapt each time the next "big idea" comes around.

Training programs could include, but are not limited to, the following items:

- Standardized lesson plans, including relevant handouts, skills sheets, etc.
- Instruction by certified/qualified instructors
- Evaluation instrument(s) to be used
- Record keeping
- Acceptable equivalencies
- Budgeting
- Local resources
- Prerequisite training requirements
- Prop/facility requirements
- Relevant SOP references
- Promotional requirements

Standardized Lesson Plans

When creating lesson plans, it is best to standardize the format. By doing so, it makes it easier for you to consistently develop effective lessons. Such standardization also makes teaching the class easier when different instructors are involved. The instructor who is tasked with teaching from any particular lesson plan is not always the same instructor who wrote the plan. The lesson plan is the tool that allows the course author to ensure the class is taught the way it was intended, and this should always be kept in mind during lesson plan development. We will not spend much time on this task, aside from introducing the steps, as each step is described in more detail in the following chapters. The National Fire Academy uses a four-step format to developing lesson plans. This format is widely used by instructors.

1. **Preparation**
   - What is the objective?
   - How much time will you need to prepare?
   - How much research is necessary?
   - Do you need any equipment or props?

- Is the instructor prepared to teach the class?
- Who is the audience, and how can I target them so the lesson is relevant?
- Are the lesson plans complete, videos and pictures ready?

2. **Presentation**
   - Does the audience have any special needs?
   - What is the time format?
   - What are the delivery methods?
   - Were the objectives stated?
   - Was the content covered adequately?
   - Have all objectives been covered?
3. **Application**
   - What tasks will you use to show whether students can apply what they have learned in a realistic fashion?
   - Should you have the students participate in a scenario-type drill?
   - Did the instructor ensure a safe environment?
   - Remember to use demonstration techniques and allow for student practice after the demonstration.
4. **Evaluation**
   - Will there be a written or a practical exam, or both?
   - How will you measure the competency of the students in performing the skill (i.e., improved job performance, skill sheet, trained observation, etc.)?
   - Should you use criterion-based or norm-based testing?
   - What objectives will be tested?

A sample lesson plan is included in the appendix.

Instruction by Certified/Qualified Instructors

As you are developing a training program, it is essential that you consider the capabilities and skills of your instructors. The instructors in your training division may be certified or qualified to teach the material, and it is important to understand the difference between the two types. Certified instructors have successfully passed an accredited program, and in doing so, have demonstrated competency against a standard. A qualified instructor, on the other hand, is an individual who has been deemed qualified by a person who possesses jurisdictional authority, such as the fire chief of a department. Certification levels vary from state to state (or province to province), and some states/provinces only offer limited certifications. However, the NFPA recognizes three levels of instructor certification, Instructor Level 1, 2, and 3. We will discuss the differences between certified and qualified instructors in more detail in the chapter, *Guiding Requirements*. Regardless of whether the instructors are certified or qualified, they must be able to perform the task they are teaching to a certain level of competency.

You may not have any certified instructors in your ranks when you begin developing the training program, but one of your many tasks as training officers will be to locate or develop qualified instructors. From there you should develop them into certified instructors. This process takes a tremendous amount effort on the part of the training officer to correctly mentor developing instructors. However, if done correctly, these instructors will be your core group for many years to come and will have the chance to further their development as instructors as they become more experienced. When conducted properly, this should be considered a part of succession planning as well.

Do not be afraid to take a chance on young fire fighters who demonstrate a desire to learn and advance their career. Most of us, at some point, were given a chance; be sure to return the favor. However, it is prudent to be careful here. It is a natural human tendency for training officers to take the quick path and force developing instructors through certification as quickly as possible, although in doing so, you only short yourself, your program, the department, and, most importantly, your fellow fire fighters. Instructors who are not adequately trained will not possess the required depth of knowledge to teach the material properly and safely. This is unprofessional and an unnecessary liability exposure that could also lead to an injury or death in the training environment. Certification should be carefully considered and only issued when earned appropriately.
Tips for mentoring include:

- Discuss the importance of being prepared as an instructor and expanding your knowledge level every chance you can.
- Explain your expectations as the training officer to ensure that you and the instructor are on the same page.
- Spend time reviewing techniques and providing constructive criticism. Be positive; no one ever got it all right the first time.
- Allow the instructor to watch you instruct and encourage him or her to ask questions.
- Use the team teaching format and eventually allow the instructor to teach all aspects of the program.

- Provide feedback.
- Challenge the instructor to personalize the program to fit his or her style and to find ways to develop a better program.

Record Keeping

According to NFPA 1401, *Recommended Practice for Fire Service Training Reports and Records*, section 1.4.10, a record is "A permanent account of known or recorded facts that is utilized to recall or relate past events or acts of an organization or the individuals therein." Training officers are required to maintain records of training activities. You should consult your local legal counsel to determine your agency's length of record keeping. Most jurisdictions consider educational records to be protected under the educational act and not subject to open records requests of public records. The standard definition of a public record is that it includes the original and all copies of any paper, correspondence, form, book, photograph, photostat, film, microfilm, sound recording, map, drawing, or other document regardless of its physical form or characteristics. Additionally, the public record has been made or received as the result of a public business transaction by a state/province or a political subdivision (i.e., the fire department).

It is very important to maintain records on training activities. There are many different types of records that can be maintained by the training officer, but at minimum, all training activities should be documented. Training records should be kept apart from personnel records and should not include any medical data. To further emphasize the importance of proper record keeping and not falsifying documents, OHSA has the ability to fine individuals up to $10,000 and/or 6 months in jail.

Methods of Recording There are many different ways of retaining records. The methods you choose will depend on the size of your department, the resources at your disposal, and the types of records you need to retain. Many agencies have begun to utilize electronic recordkeeping practices, others choose to keep printed paper records that have been computer-generated, and of course some may still utilize handwritten records. Regardless of how you retain the records, federal, state/province, and local laws mandate that you document all training conducted in your department. Some examples of documented training include: multicompany drills, in-service company-level training, and recruit class training. As the training officer, you are responsible for deciding which format is used, but regardless of what that format is, it should be readily accessible to everyone. Original copies of records should be stored in a centralized location, such as the training division office or an in-house records archive process. Depending on your departmental policy, you may also want to have a soft copy of your training records stored on your computer. There are several questions you should consider as you decide which method of record retention you should use. When selecting a recordkeeping system, consider the following:

- How does the organization track training?
- What is the method of tracking?
- Who is responsible for record keeping?
- What information from a training session is recorded?
- What types of reports are generated from the recorded information, and how will I be required to report or present the information?
- What other information is kept?

Length of Record Retention Training officers also need to manage the length of time in which records should be retained on file. At minimum, you should keep all training records of personnel for at least 1 full year after termination of employment, though it is strongly recommended that you archive all training records for at least 7 years post-employment. Medical records and personnel records should be kept separately from any training records. Policies and regulations regarding record retention vary from department to department and state to state (or province to province). Therefore, it is also important to review your department's policy and any local or state/province laws that may require additional retention time. The length of time that you decide to keep records should be made after consulting your agency's legal representative and local (AHJ) policy. Several agencies and agreements that dictate which records must be retained and for how long include but are not limited to:

- State/province
- Contractual agreements in the case of a contracting a privately held fire department
- Grant or outside contracts
  - Wildland or mutual-aid organizations that mandate specific record keeping
- Your department
  - May specify additional requirements for record keeping, such as patient contact forms, individual fire department investigation forms, etc.
- NFPA
- OSHA

LESSONS LEARNED OF TRAINING

Sample recordkeeping forms are included in the Appendices.

Here are some examples of how long typical records may be kept:

Training reports and records: 1 year after the employee leaves the department

911 Communication radio logs: 3 years after the incident

Personnel-related files: 7 years after the employee leaves the department

Vehicular accident report: 7 years after accident or 7 years after conclusion of any associated litigation

Departmental incident reports: 7 years after run or 7 years after conclusion of any associated litigation

Fire prevention inspection records: 7 years after inspection date

Medical history: 30 years after the employee leaves the department

Bloodborne pathogen exposure reports and records: 30 years after the employee leaves the department

Information to Include in the Records Keep in mind when putting together the training record requirements that you not only track members' drill attendance, but also their certification. Some certifications, such as those for Fire Investigator, may have expiration dates and requirements to recertify. These need to be tracked on a constant basis. There should be an identified routing/filing method to ensure that all records are properly recorded and filed. Additionally, some agencies may require that all members use a full signature when signing in for training, no initials are allowed. This has definite implications on your sign in roster. Some agencies have begun to utilize electronic signatures in their records and reports. This should be considered carefully only after consulting with legal counsel and the AHJ policy.

Your training records should answer the five Ws:

- **Who:** Who taught the class (the credentials of the instructor should be in your files), and who was in attendance (students), including some form of signature (handwritten or electronic)
- **What:** Whether the training was hands-on or classroom, documentation of the written and/or manipulative evaluation to ensure the class objectives was met, and listing the equipment used during the class
- **Where:** If the class was taught in house or obtained elsewhere (location), including all if there were multiple locations
- **When:** Date, time, and length of the class
- **Why:** What was the objective of the training; whether the training was a night drill, multi-company drill, or a mutual aid drill

Good recordkeeping practices are one of the most important duties and responsibilities of the training officer. Always keep in mind that those records may be requested by outside agencies for use in the courtroom, for recertification, and many other applications. Ensuring that all of the training-related records are complete and accurate is a worthy investment of your time.

Total Quality Management

Training officers should strive to employ some measure of quality control over their training programs to ensure that the content remains consistent with newer standards, technologies, and operational needs. Several methods of quality control are available such as using a quality circle, Taguchi methods, business process engineering, or total quality management (TQM). While all methods have their strong points, the most inclusive for fire department training program use is TQM because it is an all-member involvement process. Common in private industry, TQM is adaptable for use in the public sector with its use of performance monitoring, member participation, public feedback, and the pursuit of continuous improvement. The principles of TQM are:

1. Striving to meet the customer's needs (in this case, the members of the agency).
2. Being able to do things right the first time, free from defects (mastering skills).
3. Striving for continuous improvement in service delivery.
4. Constant seeking of improvement.
5. A measurable result of all training programs.

In order for the system to work, the training officer must be a coach, advocate, and mentor. There must be a belief in the ability of members of the agency to solve problems. If this belief is in place and instilled, then the ability to challenge members and allow for growth while remaining effective is supported. Along with this, there is a need to relay to the members of the agency that the people who do the work are the best at solving issues that arise in relation to it. The belief is that if all members of the department feel a responsibility and ownership of the training program, it will thrive and be more successful because of their levels of involvement. As deficiencies in operations are encountered and new missions are taken on, the training officer should be willing to seek solutions and input from the members themselves. Since the solutions can come from the membership, the belief that all members are accountable for the quality of operations that they carry out is reinforced. The training officer may very well be the perfect messenger for this system. By mentoring and coaching, the training officer can create training and help members develop solutions to needs as they arise.

In order for a training program to successfully utilize TQM, there must be support from the chief of the department. Without the backing from upper levels, a training officer would be hard pressed to enforce and maintain a quality training program. Also, the training program should be based off of a purpose-driven mission statement. This could be nothing more than the simple, raw goal of the program. It need not be wordy or fancy, just a statement of an operational excellence. A great example of a mission statement is "It is the mission of the XYZ Fire Department Training Division to provide the highest quality training, instruction, and evaluation to our members."

Another key to success is proper planning for the goals of the training program, which allows the training officer to lay the groundwork of instructional development (SMART goals). The program's goals should reflect the needs of the department and be based on the purpose of the mission statement. These goals should also be the drive of the training officer to remain focused on the needs of the department.

Once the foundations are in place, a set of measurements must be implemented in order to gauge the quality or effectiveness of the program. This may be done with aptitude tests, skill drills, evaluations, and after-action reviews of incidents. Other data, such as a reduction in injuries, decreased fire loss, or information that reflects any improved or quality performance, can also provide a basis for judging the program. The training officer also has the ability to use his or her influence as a coach to empower members of the department. The members, once empowered, will often police the program themselves, seeking a high standard due to pride and ownership in the results. The sense of empowerment may also serve to forge a greater ability for the members of the department to function as a team. They could very well serve to build "espirit de corps" and lead to more productive planning of advanced programs, due to a common drive and motivation.

Training officers must also be wary of becoming blind to the fact that, over time, the needs and goals of the program may need to change. Various reasons may drive the need for change, whether from changes in laws, NFPA standards, new target hazards in the response district, threats to homeland security, or new missions of operation for the department. While current initiatives are being monitored, any needed improvements should be sought as well. Another improvement, which may be slightly more difficult because of budget constraints, is to provide dedicated resources to the training program. This would include adequate and applicable equipment and instructors well versed in their respective subjects.

Statistician and author W. Edwards Deming developed 14 points for quality management that reflect the basis for total quality management:

1. **Create constancy of purpose for improvement.** Training programs should be based on the goals and mission statement, thus remaining consistent.
2. **Adopt a new philosophy.** As change is needed, be willing to adapt to it and improve from it.
3. **Cease dependence on mass inspection.** Individual testing and evaluation allows the program to be tweaked to the benefit of each member.
4. **Do not award business on price alone.** Before buying a service, training materials, or awarding a contract, make sure the product meets the needs of the training program.
5. **Work continually on the system of production and service.** A training officer must never become totally satisfied or complacent with the program. It must continually be evaluated and adjusted as needed to meet needs that arise.
6. **Institute modern methods of training.** Training officers cannot be afraid to try new things and use technology to their advantage.
7. **Institute modern methods of supervision of workers.** Part of evaluating the training program would also include evaluating those conducting it. By staying abreast of new philosophies, principles, and theories of supervision, empowering behaviors could be created.
8. **Drive out fear.** A training program should never be used to "weed out" members or exploit weaknesses.
9. **Break down barriers between departments.** A training program should include plans to train with other agencies and departments.
10. **Eliminate slogans, exhortations, and targets for the work force.** These well-intentioned goals of perfection may be too lofty to ever

achieve. Instead, a coaching approach to improve where needed may prove more effective, by introducing progressively challenging goals.

11. **Eliminate numerical quotas.** Most training programs are based around a predetermined number of hours. While this may still be needed to meet various requirements, the program should be results-based. The predetermined hours should not be the gold standard.
12. **Remove barriers preventing pride of workmanship.** By encouraging empowerment and promoting self-management of the training program, pride can become a driving force in the program.
13. **Institute a vigorous program of education and retraining.** Training officers must be willing to learn new techniques and embrace new technologies. These should be instituted into the training program as soon as feasible.
14. **Take action to accomplish the transformation.** If change or improvement needs to be made, the training officer can often forge the way to transformation.

Deming also believed that all managers (training officers) should have a "System of Profound Knowledge." This system is comprised of four things important to the training officer that have to be congruently utilized. First, there must be an appreciation of the system and how all aspects must be met for it to become effective. In the case of fire departments, this may pertain to understanding the relationship of all department services, divisions, hierarchy, and operations within local government. Maintaining the structure of the organization is key to all aspects interacting effectively. Secondly, the training officer must have knowledge of the existence of variation. This means that problems cannot always be fixed with a blanket solution; rather, tailored solutions to meet the situation or individual are needed and that the responsibility for overseeing this is one of the responsibilities of managing the program. A prime example would be that a particular gallon per minute flow or hose load will not always be ideal in every situation, and an open mindset would allow a company to tailor their training and tactics to fit their most common scenarios. Third is the theory of knowledge, which dictates that there are limits to what can be known and that some things will be unpredictable. Often times, generic training can lull fire fighters into a false sense of security. The world of firefighting contains many variables. Instead of trying to solve all problems before they occur (which would be impossible since no two fires or incidents are alike), the empowerment of members can lead to improved problem-solving and critical thinking capabilities. With this, the problems associated with the variables (i.e., heavier than expected fire load, delayed response of additional companies, water supply issues, or multiple victims) can be worked through in a smoother fashion. The fourth and final component is knowledge of psychology and the concepts of human nature and motivations. From this comes the ability to coach and encourage the individual members and focus any counseling directly at their needs. As this is accomplished, students improve, become more empowered, and better function in their assigned role.

The concept of TQM should be understood by a training officer, as it can make beneficial contributions to a training program. The training officer should use this as a tool to seek new ways to improve and streamline a training program, with the end goal of providing the best training possible to the members of the department. Finally, with this understanding, a training officer should not be afraid to change and empower members to have pride and ownership in the program (**TABLE 3-1**).

TABLE 3-1 TQM Transforming a Training Program/Academy

From	To
Motivation through fear and loyalty	Motivation through a shared vision
Attitude: "It's their problem"	"Ownership of every problem"
Attitude: "That's the way it has always been done."	"Continuous improvement"
Decisions based on assumptions and judgment calls	Decisions based on data and facts
Everything begins and ends with upper management	Everything begins and ends with those receiving services from the department
Crisis management and recovery	Doing it right the first time

Key Points

- Training programs should provide consistency, uniformity, transparency, and legal compliance.
- A training program describes the vision for and process of a program of study. It should also lay the ground rules of a program of study for the given department.
- When developing a training program, there are several models that can be followed, such as the AIDE, NFPA, or MFRI models. Regardless of which model you choose to use as a guide, the basic steps are the same with each:
  - Needs assessment
  - Develop/Produce the content
  - Implement/Deliver
  - Measure/Evaluate
- Creation of a training program should always begin with its purpose, which is typically based on a particular need of the organization.
- Training officers developing content for a training program should consult with the SMEs from the department. This will allow them to obtain a buy-in from line personnel and provide legitimacy to the program.
- Evaluation should be a continuous process that begins right away with a needs assessment. The needs assessment should be used to evaluate both the individual and the entire process. Evaluation will also assist in measuring the effectiveness of the training program.
- As a mentor, discuss the importance of being prepared as an instructor and increasing your knowledge level every chance you can.
- The standard definition of a *public record* is the original and all copies of any paper, correspondence, form, book, photograph, photostat, film, microfilm, sound recording, map, drawing, or other document regardless of its physical form or characteristics. Additionally, it has been made or received as the result of a public business transaction by a state/province or a political subdivision (i.e., the fire department).
- Information about training could be requested by outside agencies if something happens in the department that reflects on the quality of training that has been conducted.
- Common in the private industry, total quality management (TQM) is adaptable for use in the public sector due to its use of performance monitoring, member participation, public feedback, and the pursuit of continuous improvement.

References

Buckman, J.M. III. *Chief Fire Officer's Desk Reference*. Burlington, MA: Jones & Bartlett Learning; 2006.

Deming, W.E. *Out of the Crisis*. Cambridge, MA: MIT Press; 1982.

Edwards, S.T. *Fire Service Personnel Management, Third Edition*. Upper Saddle River, NJ: Brady/Prentice Hall, Pearson; 2010.

International Association of Fire Chiefs and National Fire Protection Association. *Chief Officer: Principles and Practice*. Burlington, MA: Jones & Bartlett Learning; 2012.

International Association of Fire Chiefs and National Fire Protection Association. *Fire Officer: Principles and Practice, Second Edition*. Burlington, MA: Jones & Bartlett Learning; 2010.

National Fire Protection Association (NFPA) 1401, *Recommended Practice for Fire Service Training Reports and Records*. 2012. http://www.nfpa.org/codes-and-standards/document-information-pages?mode=code&code=1401. Accessed July 18, 2014.

National Fire Protection Association (NFPA) 1403, *Standard on Live Fire Training Evolutions*. 2012. http://www.nfpa.org/codes-and-standards/document-information-pages?mode=code&code=1403. Accessed July 18, 2014.

National Fire Protection Association (NFPA) 1720, *Standard for the Organization and Deployment of Fire Suppression Operations, Emergency Medical Operations and Special Operations to the Public by Volunteer Fire Departments*. 2014. http://www.nfpa.org/codes-and-standards/document-information-pages?mode=code&code=1720. Accessed September 7, 2014.

National Fire Protection Association (NFPA). Document Information Pages (list of NFPA codes & standards). http://www.nfpa.org/aboutthecodes/list_of_codes_and_standards.asp. Accessed July 18, 2014.

Occupational Safety and Health Administration (OSHA). OSHA Online. https://www.osha.gov/dte/outreach/index.html. Accessed July 18, 2014.

Wyoming Department of Fire Prevention & Electrical Safety. *Manual for Training Officers*. December 2011. http://wyofire.state.wy.us/pdf/trainthetrainemanual.pdf. Accessed July 28, 2014.

CHAPTER

4 Guiding Requirements

Brian P. Kazmierzak, Nita Ham, Chad Cobb, Rich Rodewald, and Brian Ward

Training Mandates and Requirements

This chapter will discuss the training requirements put forth by the Occupational Safety and Health Administration (OSHA) and the Insurance Service Organization (ISO), so that you as a training officer can make informed decisions about which classes and how much training in specific areas need to be a part of your program.

Your Department's Standard Operating Procedures

As a training officer, you have a tremendous amount of influence on the amount of training offered for your department, from the basics of recruit school to the more specialized chief officer level. Using the needs determination phase discussed in the *Evaluating and Creating Your Own Program*, determine the needs of your department and institute standard operating procedures (SOPs) based around the additional levels of training. This will allow the training officer to implement several new levels of training for the officer and chief positions (see the appendix for outlines of similar programs such as Acting Officer and Acting Captain).

It is also important that the department sets a standard for its training officers, instructors, and adjunct instructors. These are the individuals who will assist you on a routine basis and will represent you in your absence. Training officers should establish SOPs at a competent level and select assistants wisely. All levels of instructor should have some minimum qualifications and/or certifications to ensure a certain level of competency as subject matter experts. These various titles are generally tied to promotions in most departments; for example, in many jurisdictions, attaining a level of Instructor I is required before acceptance is granted to sit for the lieutenant's promotional test.

Generally speaking, the various levels of instructor qualifications are based on each department's needs. However, in most areas there are three categories of training staff: adjunct instructor, instructor, and training officer.

Adjunct Instructor: The adjunct instructor is generally certified as an Instructor I through International Fire Service Accreditation Congress (IFSAC) or National Professional Qualifications (NPQ) and is qualified as an adjunct by the authority having jurisdiction (AHJ) concerning specific topics. For example, one department may allow an adjunct to assist with live fire, while a similar department may only allow him or her to assist as a basic fire fighter skills adjunct.

Instructor: The instructor typically has an Instructor I and/or II NPQ/IFSAC certification and is qualified by the AHJ on various topics. The instructor also may be solely responsible for several aspects of a curriculum or designated as the instructor-in-charge for station skill evaluations.

Training Officer: The training officer has oversight and is responsible for specific programs or training divisions and is certified as an Instructor II and/or III through NPQ/IFSAC. Training Officers are appointed individuals and typically are

TABLE 4-1 Sample SOPS for Promotions Within Fire and Emergency Services

Fire Fighter I	Fire Fighter II	Fire Fighter III	Driver/Engineer
NPQ FF I NPQ FF II NPQ Haz Mat Awareness & OPS Emergency Response to Terrorism Safety and Survival	1 year as FFI Med Unit Cert. ICS100, 200, 700, & 800 Relief Driver after 1 year as FFII	3 years as FFII Relief Driver School Acting Officer School	1 year as FFIII Relief Driver School Acting Officer School
Lieutenant	**Captain**	**Battalion Chief**	
3 years as FFIII or 2 years as Driver Management I and II (NFA Leadership I and II) NPQ Instructor I Public Education (Fire and Life Safety Educator) Building Construction EMT-I	2 years as Lt. Administration I (NFA Leadership III) Fire Investigation Fire Prevention Acting Capt. Program Incident Safety Officer	2 years as Capt. Incident Safety Officer Acting B.C. Program	

2014 Associate's Degree for Battalion Chief
2016 Bachelor's Degree for Battalion Chief
Source: Courtesy of Brian Ward.

assigned to a fire academy, though they may work on a shift but maintain training officer duties.

TABLE 4-1 shows an example of an SOP for promotions within Fire and Emergency Services.

National Fire Protection Association Standards

The National Fire Protection Association (NFPA) publishes a collection of consensus standards that is produced by a group of fire service and industry peers. Although these are generally not mandates, they can carry the power of law in court and need to be treated as such.

It is *highly* recommended that the training officer obtain a copy of NFPA standards for continuing reference. You should review and become familiar with those that are pertinent to the training division. Realize that people frequently will say that something does or does not follow NFPA standards, and part of your job will be to know which is correct. It is important to decipher the standards and ask questions. There is a certain amount of interpretation and cross-referencing that needs to be done when reading some standards to fully understand the intentions. Regardless, if you are unsure of the intent or purpose, lean towards the side of safety. Other sources for clarification include your command staff, local departments, your state's/province's fire academy, or NFPA itself.

It is also important to understand how NFPA standards are constructed. Each document starts with a standard, and within each standard are job performance requirements (JPRs). Job performance requirements are the tasks or knowledge levels that must be addressed in order to be in compliance with NFPA standards. They are generally broken down into subcategories and are grouped together by a common topic. Another item of note is that it is a common practice for NFPA to reference several other standards within a specific standard; for example, NFPA 1403, *Standard on Live Fire Training Evolutions*, references NFPA 1500, *Standard on Fire Department Occupational Safety and Health Program;* NFPA 1001,

Standard for Fire Fighter Professional Qualifications; and others.

■ Validating Training Against NFPA Standards

As it relates to training and certification testing, state/province testing agencies are responsible for validating the test that they proctor. As the staff go through each question in the test bank, they should match the question to a JPR in the NFPA standard for that specific topic. In addition, training officers should match JPRs to the content of their lesson plans in order to ensure that all required information is covered (**FIGURE 4-1**). This is the process that should be used whenever a new promotional class or department curriculum is

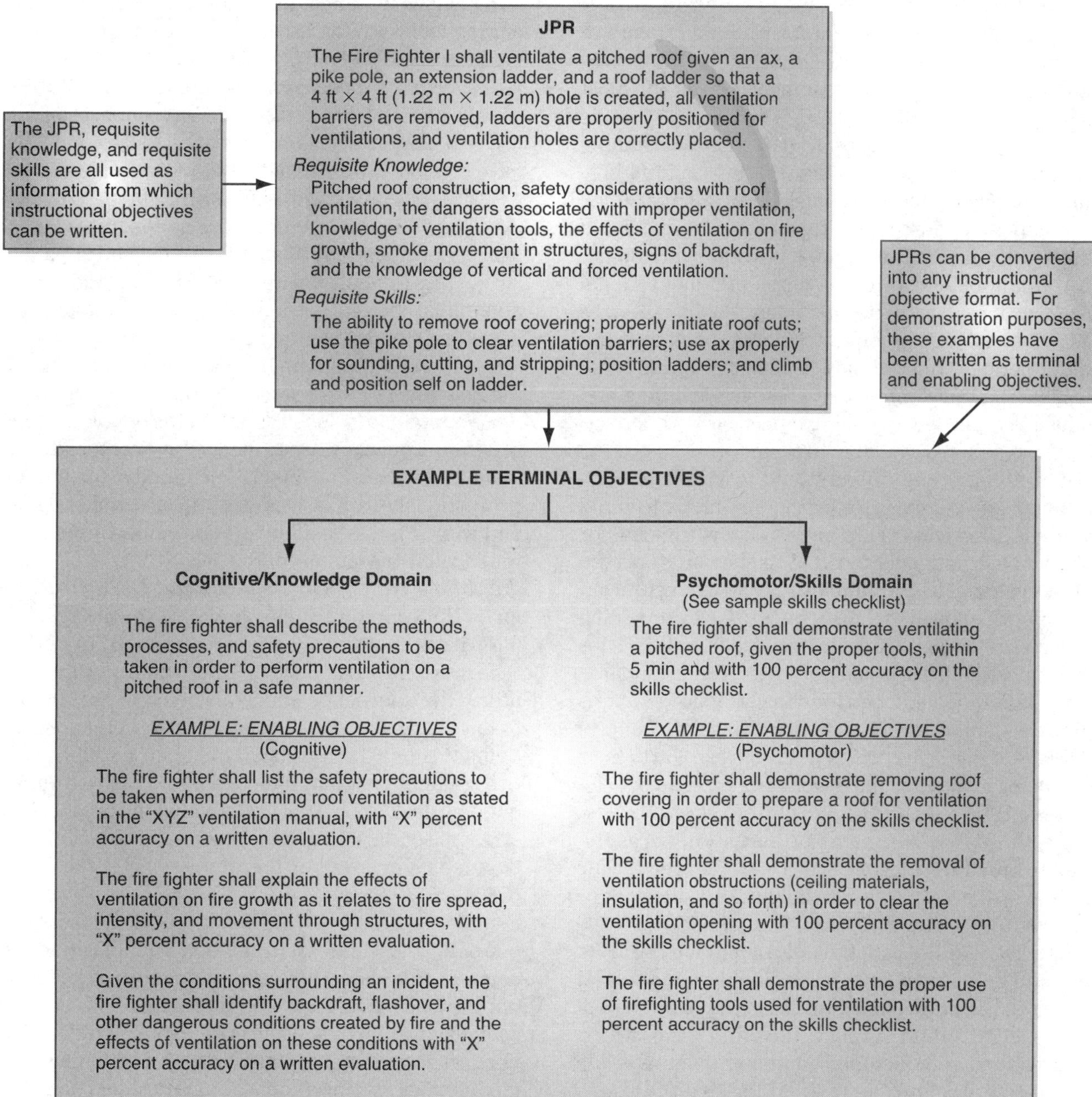

FIGURE 4-1 Converting JPR into instructional objectives.

developed and implemented. The outlines in the appendix detailing the Acting Officer, Acting Captain, and Acting Battalion Chief were constructed in this manner. The equivalents to these classes are Fire Officer I, II, and III. The only difference is that these classes were department-specific and were not nationally accredited through programs such as NPQ and IFSAC.

■ Applicable Standards for Training Officers

NFPA standards are essential for training officers to understand and incorporate into their training program management and development. One of the most challenging duties of fire service training officers is developing and delivering up-to-date, relevant training programs. The fundamental goal of any training program is to ensure fire personnel are properly trained and competent enough to mitigate incidents safely. In addition, it is crucial for training officers to document the training provided to fire personnel due to legal considerations. Just as in emergency medical services (EMS) patient care reports, if it is not documented, it did not happen. Accordingly, training program management and development can be overwhelming and confusing, especially if training officers are not familiar with the resources available to them.

Unaware training officers tend to utilize old lesson plans and outdated reference materials to build programs. This can lead to gaps in knowledge and/or skills not being taught because NFPA standards are constantly evolving. NFPA standards give training officers a framework to build their programs upon by providing guidelines regarding requisite knowledge and skills for various positions or ranks. Moreover, NFPA standards can serve as the foundation for all safety considerations within the training exercise. The following are examples of NFPA standards that training officers can add to their toolbox for managing and developing programs:

NFPA 1001, *Standard for Fire Fighter Professional Qualifications*, may be better known as the standard for Fire Fighter I and Fire Fighter II. For training officers, this should be the starting point of your training for new recruits. The standard is written as a list of JPRs, with the goal of making the JPRs clear and concise to help determine if an individual possesses the knowledge and skills required to perform as a fire fighter. Along with defining the task to be performed and the tools, equipment, or materials needed to complete the task, JPRs give the training officer evaluation parameters and/or desired performance outcomes. NFPA 1001 is also used as an evaluation tool for experienced fire fighters to test their retention of the basic knowledge and skills needed to operate safely on the fireground. Although the title of this standard and several other standards cited in this chapter use the term "Professional Qualifications," these standards apply to all fire fighters, career and volunteer.

Having qualified personnel driving and operating your department's fire apparatus is equally as important as having well-trained personnel advancing hose lines, cutting ventilation holes, and commanding your emergency scenes. NFPA 1002, *Standard for Fire Apparatus Driver/Operator Professional Qualifications*, should be used by training officers for initial training of driver/operators and as an evaluation tool for those who have previously been qualified as a driver/operator. Like NFPA 1001, it is written as JPRs. The standard features different JPRs for different types of apparatus, including pumpers, aerials, tiller trucks, wildland apparatus, aircraft fire fighting apparatus, and mobile water supply apparatus; therefore, training officers can tailor their use of the standard to the needs of their organization.

A training officer can have a tremendous effect on the long-term success of the organization when *career development* is included as part of the ongoing training program. Too many departments leave career advancement up to the individual to "self-study" for the next promotional exam. A training officer should be ready and willing to assist all levels and ranks of personnel reach their maximum potential. NFPA 1021, *Standard for Fire Officer Professional Qualifications*, gives the training officer guidance on the requisite knowledge and skills that personnel need to perform the duties of a fire officer. The training officer can utilize the knowledge and skills listed in the standard to develop classes and/or activities to help those participants hone their skills. Remember, a well-rounded fire training program includes initial instruction and ongoing evaluation of all ranks within the organization.

As one of the four original professional qualification documents created for the fire service in the 1970s, NFPA 1041, *Standard for Fire Service Instructor Professional Qualifications*, is one of the oldest, yet one of the most relevant professional qualification standards used today. The standard includes three levels of instructor qualification (Instructor I, II, and III), each having increasing levels of responsibility and job performance requirements. Training officers should identify their organization's need for trained personnel at each of the different instructor levels and understand the capabilities and limitations of each. Among other duties and responsibilities, those personnel qualified at the Instructor I level can *teach* from a prepared lesson plan and adapt lesson plans to the unique requirements of the students and the authority having jurisdiction. Instructor II-qualified personnel possess the knowledge and skills required to *create* lesson plans, *write* objectives, *develop* course materials, and *design* a course evaluation plan. Instructor III-qualified personnel are trained to *plan, develop,* and *implement* comprehensive programs and curricula. These personnel can conduct needs analysis, design recordkeeping and scheduling systems, and develop training goals and implementation strategies. The role of the training officer is to ensure all instructors within their department are trained to the appropriate Instructor level to successfully perform the tasks they are assigned.

Training records have many important functions in the management of a fire department. They provide proof of compliance with statutory requirements, assist with budget development, and supply basic information for effective training program management. NFPA 1401, *Recommended Practice for Fire Service Training Reports and Records*, outlines a systematic approach for training officers to develop, record, and manage these vital records. It explains the "who, what, when, where, and why" of appropriate documentation. The different *types* of training documents are discussed and explained. In addition, the standard presents valuable information on the legal aspects of fire service recordkeeping. NFPA 1401 demonstrates that proper training documentation is much more than participants writing their name on a sign-in sheet, and it explains the elements needed to make your documentation complete.

While technically not a standard, NFPA 1402, *Guide to Building Fire Service Training Centers*, provides training officers with guidance on the design and construction of facilities for fire service training. It includes useful information on planning the site, infrastructure considerations, and the construction process. Training officers can find information on a wide variety of building types and mobile props, including drill towers, live fire training structures, smoke buildings, driver training courses, and trailers for self-contained breathing apparatus (SCBA) maze training.

NFPA 1403, *Standard on Live Fire Training Evolutions*, was developed after the tragic deaths of two Boulder, Colorado fire fighters during a live fire training exercise in 1982. Prior to that, there were no nationally accepted guidelines on how to conduct live fire exercises. Since its introduction in 1986, NFPA 1403 has provided training officers with a solid blueprint for safe, effective practices during live fire training evolutions. The standard addresses critical safety issues that must be taken into account by the training officer prior to conducting any live fire exercise, including:

- Student training prerequisites
- The presence of a safety officer
- The role of the instructor-in-charge
- The roles other instructors play
- Personal protective equipment
- Communications
- The need for on-scene EMS
- Water supply requirements
- Acceptable fuel types and loads
- Pre-burn safety briefings
- Fire fighter rehab

NFPA 1403 gives the training officer guidance on how to conduct live fire training in acquired structures, gas-fired facilities, non-gas-fired facilities, and exterior live fire training props. Finally, training officers can utilize the reports and checklists found in NFPA 1403 for proper documentation of the training event. All training officers who conduct live fire training should have a thorough knowledge of NFPA 1403 and follow its guidelines to ensure the safety and health of their fire fighters.

NFPA 1404, *Standard for Fire Service Respiratory Protection Training*, addresses the minimum training requirements for the respiratory protection program found in NFPA 1500. All fire departments should have a written respiratory protection program, and it falls upon the training officer to ensure proper training and documentation is completed. NFPA 1404 guides the training officer as to what breathing apparatus skills should be taught to new recruits and outlines the requirements for retraining and evaluating incumbent fire fighters. The standard also defines different types of respirators and methods of respiratory protection, enabling the training officer to train on the specific items that apply to his or her organization. A recent addition to the standard calls for students to be taught the concept of individual air management, which is intended to increase an individual's awareness of his or her personal air consumption rate while working in SCBA. NFPA 1404 covers a wide range of information that training officers can easily implement into their training program.

From a training officer's perspective, few training topics are as important to the success of the organization as those concerning fire fighter safety and survival. Prior to the creation of NFPA 1407, *Training Fire Service Rapid Intervention Crews (RIC)*, it was left up to the individual jurisdictions how to train fire fighters the skills needed to rescue a downed comrade. With its release, NFPA 1407 has standardized the specific skills and evolutions needed to develop a solid rapid intervention program. In addition to giving the training officer guidance on RIC policies and procedures, the standard establishes instructor requirements and student prerequisites. It gives the training officer a standard set of survival skills to include in the training program and defines the required performance of successful completion of RIC training evolutions.

Originally adopted in 1966, NFPA 1410, *Training for Initial Emergency Scene Operations*, began as a set of basic hose line evolutions to test the speed and skill of engine and ladder companies. The standard has evolved over the years to include chapters on the operational performance of truck companies and rapid intervention crews. Training officers can utilize NFPA 1410 to design and organize company drills, evaluate the performance of those participating companies, and properly document compliance with the standard. NFPA 1410 offers flexibility, as the evolutions found in the standard can be adapted to local conditions by the authority having jurisdiction. The training officer can make NFPA 1410 the basis for many different types of drills, including:

- Recruit training
- Driver/operator training
- Live fire drills
- Laddering drills
- Defensive operations training
- RIC evolutions

NFPA 1451, *Fire Service Vehicle Operations Training Program*, establishes the minimum requirements for a fire service vehicle operations training program. For those departments that do not have a formal driver/operator training program, this standard should serve as a starting point and reference for training officers to develop such a program. NFPA 1451 also serves as a method to meet all of the requirements of NFPA 1500, 1002, and 1003 (*Standard for Airport Fire Fighter Professional Qualifications*) with regards to training and education of fire department driver/operators. The standard addresses training frequency, education requirements, instructor qualifications, driver safety, and training record documentation. Given that historically between 20 and 25 percent of all fire fighter deaths annually are related to motor vehicle accidents, the training officer should ensure all department driver/operators are trained to this standard.

A major tenet of any fire training program should be to establish and maintain a strong safety culture within the department. Training officers can make or break that safety culture by the attitude they take and the example they set, both in the classroom and on the training grounds. That is why a thorough understanding of NFPA 1500 is critical to a training officer's success. NFPA 1500 is *the* comprehensive safety and health document for the fire service. Originally published in 1987, the standard was developed as the first consensus standard for occupational safety and health for the fire service and was intended to reduce the staggering number of fire fighter injuries and deaths suffered in the 1970s and early 1980s. The standard has evolved and expanded over the years to address the ever-changing dangers we face as a fire service, placing new emphasis on traffic incidents, establishing control zones, and fire fighter fitness and health.

While the training officer should be familiar with all aspects of NFPA 1500, Chapter 5, "Training, Education, and Professional Development," deals with the specific role training plays in the safety and health of fire fighters. Training officers can use Chapter 5 as the basis for establishing member qualifications, as ancillary NFPA professional qualification standards are listed for a variety of member duties. In addition, NFPA 1500, Chapter 5, gives training officers direction on minimum training requirements for their department members by offering guidelines for frequency of training and topics to be covered.

The other chapters of NFPA 1500 can be used by the training officer to further develop the training program by identifying the safe practices outlined in each and then incorporating those practices into the drill tower and classroom exercises.

In many fire departments, the training officer also serves as the department health and safety officer and/or incident safety officer. When this is the case, the importance of the training officer being familiar with NFPA 1521, *Standard for Fire Department Safety Officer Professional Qualifications*, is clear. For larger departments, where the roles are filled by different personnel, the value of this standard to the training officer may not be as obvious, but it is no less important. NFPA 1521 contains separate chapters for health and safety officer (HSO) and incident safety officer (ISO).

In NFPA 1521 Chapter 5, "Functions of the Health and Safety Officer," there are numerous references to the HSO ensuring that fire department members are trained in safety procedures. The training officer must be able to work with the HSO to develop an overall fire training program that includes topics important to fire fighter health and safety.

The duties of the ISO are discussed in NFPA 1521, Chapter 6, "Functions of the Incident Safety Officer." A thorough understanding of Chapter 6 will assist the training officer in making safety decisions at training exercises. Finally, NFPA 1521 provides training officers with guidance on how to prepare and maintain department members who may be designated as the health and safety and/or incident safety officer by covering the assignments, duties, and responsibilities of both positions.

NFPA 1584, *Standard on the Rehabilitation Process for Members During Emergency Operations and Training Exercises*, establishes a rehabilitation (rehab) process for fire department members working within an incident management system, during both emergency scene operations and training exercises. With the greater awareness of the importance of fire fighter health and safety, it is appropriate that more attention be given to the rehab process. Training officers should utilize NFPA 1584 to plan the rehab process for all scheduled training sessions whenever those exercises pose the risk of participants exceeding a safe level of physical or mental endurance. The standard explains the specific components of a complete rehab system and includes many useful forms, guidelines on vital signs, and temperature/heat index charts. Training officers should be advocates for the rehabilitation process both in training exercises and on emergency scenes.

For a training officer, *safety* on the training grounds should be the driving force behind the design and delivery of all hands-on drills. As most drills require participants to don at least a portion (if not all) of their personal protective ensemble (PPE), a thorough understanding of NFPA 1971, *Standard on Protective Ensembles for Structural Fire Fighting and Proximity Fire Fighting*, is essential for training officers. We know gear should be inspected prior to engaging in any training and/or emergency scene activity. Typically, this is done by the individual user at the start of a shift for career fire fighters or on a meeting/training night for volunteer fire fighters. Even though PPE is presumed to be inspected prior to attending a training session, it is the responsibility of the training officer to ensure fire fighters are wearing gear that is compliant, properly fitted, and in good condition before participating. This is especially true when conducting live fire training evolutions.

Changes in NFPA 1971 may also require training officers to provide updated training on the proper care, use, and inspection of PPE. Such was the case after the 2007 revision to NFPA 1971 added the drag rescue device (DRD) to fire fighter turnout coats. This change necessitated training on how to use the DRD in an emergency, how the DRD should be cleaned and cared for, and how to reload the DRD for its next use.

NFPA Compliance

Compliance with NFPA standards is not federally mandated; they are only mandated if the state/province or local jurisdiction has adopted them. However, NFPA standards are regularly recognized in legal proceedings as an industry norm and thus used against fire departments and individuals who do not follow or meet their requirements. Training officers can protect themselves and their department from litigation by knowing the standards that apply to them, working towards compliance with those standards, and maintaining complete documentation of compliance.

Even if an NFPA standard is not officially adopted, this does not preclude a fire department from modeling their programs after certain NFPA standards. Training officers can utilize NFPA standards to help *solve the puzzle* of training program development and management. They can be used to develop new training programs, update existing programs, and fill in the knowledge and skill gaps of fire department personnel of all ranks. Because NFPA covers such a wide variety of topics in their standards, training officers will find them an invaluable resource when developing their department's training program.

Occupational Safety and Health Administration

In 1970, under the direction of President Nixon, Congress enacted the Occupational Safety and Health Act, which created the Occupational Safety and Health Administration (OSHA). Its mission is to help employers and employees reduce on-the-job injuries, illnesses, and deaths through prevention training, regulation implementation, and auditing.

OSHA directs the national compliance initiatives in occupational safety and health throughout all states. However, some states have chosen to implement their own OSHA State Plan; the plans for these states, at a minimum, must match or exceed the federal plan. Through the methods described below, OSHA helps businesses protect their workers and reduce the number of workplace deaths, injuries, and illnesses. When employees stay safe and healthy, companies can reduce workers' compensation, insurance costs, and medical expenses; decrease payout for return-to-work programs; reduce the incidence of faulty products; and lower costs for job accommodations for injured workers. Indirectly, additional benefits such as increased productivity, lower training costs due to fewer replacement workers, and decreased costs for overtime have also been attributed to OSHA's research and guidance.

Since 1970, workplace deaths have been cut by more than 60 percent and occupational injuries and illnesses have declined 40 percent. In addition, OSHA has conducted almost 39,000 inspections, issued over 85,000 citations for safety violations, and has assisted businesses with its Consultation Program by making over 31,000 visits to employers.

OSHA promotes workplace safety and health by:

- Implementing new (or improved) safety and health management systems.
- Completing worksite inspections. Companies failing to adhere to OSHA Regulations may be cited and/or fined.
- Promoting cooperative programs including Voluntary Protection Programs, OSHA Strategic Partnerships, and other industry alliances.
- Establishing specific rights and responsibilities of employees and employers.
- Supporting innovation in dealing with workplace hazards.
- Establishing recordkeeping and reporting requirements for employers.
- Developing training programs for occupational safety and health personnel.
- Partnering with states that operate their own occupational safety and health programs.
- Supporting the OSHA Consultation Program.

One important point that commonly causes confusion is the relationship between OSHA and NFPA. Training officers should understand that NFPA sets guidelines and "standards," while OSHA sets regulations. NFPA does not have a regulatory component, although these are considered consensus standards and may be held against an individual and/or a fire department in a court of law. On the other hand, as a branch of the executive government, OSHA sets minimum requirements and has the regulatory component to enforce them.

OSHA establishes the minimum standards for the safety of public and private employees; however, many industrial facilities exceed these standards. Training officers should consider taking a class from OSHA in order to understand the thought process of compliance officers and what they are allowed to do. One interesting concept is the "General Duty Clause," which states that employers have a duty to provide an environment free of recognized hazards. The concept behind this clause is that there may not be a standard for all topics, but employers should know about and be able to recognize safety and training concerns. If the compliance officer feels that there are hazards present that are obvious and should have been recognized by the department, they can use the "General Duty Clause" to cite the department. It is also important to remember that OSHA is not in the business of telling us how to fight fire or what strategy and tactics to use; its job is to ensure we are prepared, trained, and safe.

If you review the Code of Federal Regulations (CFRs), similar to the JPRs in NFPA standards, OSHA references different schools, accreditation agencies, and the International Society of Fire Service Instructors (ISFSI) as training agencies. What OSHA does not want to see is someone leading training who is not qualified or certified, and it is equally as important to keep good documentation. This documentation should reflect all of the vital items required, including instructor qualifications, attendance reports, dates, times, topics taught, curriculum, and a few others discussed later.

Note: It is important to consult with your local and state OSHA representatives to determine the applicability of OSHA mandates and if your state is an OSHA state.

OSHA CFRs

The following are examples of OSHA CFRs[1] that training officers should refer to when developing classes and determining how much area-specific training needs to be included.

1910.38 Emergency Action Plans

- *Kept in writing and available for employees.*
- *Provide procedures for those performing rescue or medical duties.*
- *Refresher training provided when plan changes or upon employment.*

[1]Reproduced from: OSHA Law & Regulations. The Code of Federal Regulations. Occupational Safety & Health Administration.

Continued

1910.39 Fire Prevention Plans

- *Same as above for training purposes. The only difference is the type of plan, which dictates the type (Fire Prevention) of information provided.*

1910.67 Vehicle Mounted Elevating and Rotating Work Platforms

- *(b)(3) The requirements of this section do not apply to firefighting equipment or to the vehicles upon which aerial devices are mounted, except with respect to the requirement that a vehicle be a stable support for the aerial device.*

1910.95 Occupational Noise Exposure

- *(5) Baseline within six months of an employee's first exposure.*
- *(6) Testing completed annually and compared against baseline.*

1910.120(q) Emergency Response to Hazardous Substance Releases

- *(6) Initial Training*
- *Before responders are permitted to take part in an actual emergency in accordance with the below information:*
- *(8) Refresher Training*
- *Shall receive annual training of sufficient content and duration to maintain their competencies, or shall demonstrate competency in those areas at least yearly.*

1910.120 Appendix E Training Curriculum Guidelines

- *Used for training conducted in accordance with 1910.120(q). See Appendix J in this book.*

1910.132 Personal Protective Equipment

- *(f)(1) Employer shall provide training detailing when, how, what, limitations and the proper care/maintenance of PPE.*
- *(f)(3) Retraining - Changes in workplace, changes in type of PPE, or inadequacies in an affected employee's knowledge or use of assigned PPE.*

1910.134 Respiratory Protection

- *(f) Fit testing and initial training required prior to working in exposed area. Fit test shall be completed annually thereafter. 1910.134 Appendix A explains the mandatory Fit Test procedures.*
- *(k) Training and Information*
- *Shall be completed annually and demonstrate the following competencies:*
  - *(i) Why the respirator is necessary and how improper fit, usage, or maintenance can compromise the protective effect of the respirator.*
  - *(ii) What the limitations and capabilities of the respirator are.*
  - *(iii) How to use the respirator effectively in emergency situations, including situations in which the respirator malfunctions.*
  - *(iv) How to inspect, put on and remove, use, and check the seals of the respirator.*
  - *(v) What the procedures are for maintenance and storage of the respirator.*
  - *(vi) How to recognize medical signs and symptoms that may limit or prevent the effective use of respirators.*
  - *(vii) The general requirements of this section.*

Continued

1910.146 Permit Required Confined Spaces – Emergency Services

- *(k)(2)(iii) At least one member of the rescue team shall be trained in first aid and CPR.*
- *(k)(2)(iv) Minimum of one simulated rescue operation in which they remove dummies, manikins or actual persons from the actual permit spaces or from representative permit spaces once every 12 months.*

1910.151 First Aid

- *In the initial training - Employees shall be adequately trained to provide first aid.*

1910.155 Fire Protection

- *1910.156(c) Initial training is required before starting fire fighter duties.*
- *1910.155 Refresher training is to occur as required.*

1910.157 Portable Fire Extinguishers

- *Initial training is required upon initial employment*
- *Refresher training is required at least annually*

1910.1000 Air Contaminants (Toxic and Hazardous Substances)

- *There is not a specific training requirement for this standard, however may prove useful for Hazardous Material Teams and/or employee awareness concerning Permissible Exposure Limit (PEL), Time Weighted Average (TWA) and Parts Per Million (PPM).*

1910.1030 Bloodborne Pathogens

- *(g)(2) Information and Training: At the time of initial assignment to tasks where occupational exposure may take place.*
- *Annually thereafter.*
- *(h)(2) Training Records*
- *Include date, contents, name and qualification of trainer, name and titles of attendees, and shall be maintained for 3 years.*

Insurance Service Organization Requirements

The Insurance Service Organization (ISO) requires specific types of training to take place on an annual basis:

- Eight single-company drills per year on any topic, which are at least 3 hours in length.
- Four multicompany drills per year, which are at least 3 hours in length.
- Two night drills per year, which are at least 3 hours in length.

Note: If you conduct a night drill, it can count for the night drill and either the single company drill or the multicompany drill.

In order to receive the maximum allowable points, fire department personnel should complete the following training:

- 20 hours of training per member per month.
- 2 days of officer training per year.
- 4 half-day training sessions for drivers and vehicle operators.
- 40 hours of training for each new driver or apparatus operator.
- One half day of hazardous materials training per member per year.
- 240 hours of recruit training for all new personnel.

Training officers should be familiar with additional items that may allow your organization to be granted points to your rating if they are in possession, or within 10 miles of your fire department and a Memorandum of Understanding is in place:

- Drill tower
- Fire building (including smoke room)
- Combustible liquid pit (meeting all applicable standards)
- Library and training manuals
- Slide and movie projectors

- Hydrant cutaway
- Training area (this may include streets or open areas when no other training facilities are provided)

The ISO established the Fire Suppression Rating Schedule for insuring purposes and based half of the allowable points on the fire department specifically—training, response, and maintenance. It is a recommended list of training and is not intended to analyze all aspects of a training program.

Note: ISO is currently reviewing its requirements to bring them more in line with those of NFPA. Training officers are encouraged to stay abreast of this ongoing change. Also, consult your local ISO agent for specific details and questions related to your area.

Federal Requirements

Currently, in order to receive monies through the National Assistance to Firefighters Grant Program (AFG), departments must conduct approved training in the National Incident Management System (NIMS). These programs include the NIMS series 100, 200, 300, 400, 700, and 800, according to respective rank levels. Additional training information at the federal level can be found through one of the following courses described on the Federal Emergency Management Agency's (FEMA's) National Preparedness Directorate National Training and Education website:

- **Center for Domestic Preparedness (CDP)**
- **Emergency Management Institute (EMI)**
- **National Training and Education Division (NTED).**

Grant Requirements

When attempting to secure grant funding, there are several items to keep in mind concerning training. First, prepare for any training needs prior to submitting the grant; this shows good faith. Second, follow through on training after receiving the grant. For example, if a department is requesting grant funding for an apparatus, the department should have driver/operator training in place prior to requesting funding. However, driver/operator training should be included with the request for the apparatus. Generally, when submitting a grant request, choose what level of training that you have already conducted or the level that you wish to conduct with the assistance of the grant. Multiple FEMA grants are available and the application criteria should be reviewed prior to submission to ensure compliance, as the procedures can be altered from year to year. Reproduced here are several key quotes found in the AFG Rules and Procedures for 2012 concerning NIMS compliance and training programs. Additional information can be found at http://www.fema.gov/firegrants/

AFG: Assistance to Firefighters Grant
SAFER: Staffing for Additional Fire and Emergency Response Grants
FP&S: Fire Prevention and Safety Grants
SCG: Station Construction Grants

Assistance to Firefighters Grant

Assistance to Firefighters Grant (AFG) - 2012 Frequently Asked Questions (FAQs):[2]

Do fire departments and EMS organizations have to comply with the requirements of the National Incident Management System (NIMS)?

Yes. In accordance with regulations, all 2011 federal grant awardees must agree to adopt the National Incident Management System (NIMS), must have an understanding of its principles and policies, and must comply with their state, local, or county EMA authority. Responders who have already been trained in ICS do not need retraining if their previous training is consistent with DHS standards. In order for us to document compliance, grantees will be required to certify their recognition of NIMS/ICS as part of their grant closeout process.

DHS offers training for those fire departments that have not implemented ICS or are unfamiliar with the principles and policies of the NIMS. For information regarding ICS and NIMS training,

[2]Reproduced from: Federal Emergency Management Agency (FEMA). Assistance to Firefighters Grant Program: 2012 Frequently Asked Questions, Washington, DC: 2013.

Continued

visit the NIMS Web site at http://www.fema.gov/nims. For more information about ICS training, contact the DHS Centralized Scheduling and Information Desk (CSID) at 1-800-368-6498, or visit the Web site of the DHS Emergency Management Institute (EMI) and the National Fire Academy (NFA) at http://www.usfa.fema.gov. Your state emergency management training office may also be a source of information in this regard. Recipients of AFG funds should also participate in any community-wide NIMS efforts that may be initiated to implement NIMS in accordance with the guidance that was provided to each State Governor on September 8, 2004.

Our department does not have a driver/operator training program in place. Can we have the training program in place after the grant award?

Applicants requesting vehicles that do not have driver/operators trained to current NFPA 1002 or equivalent and are not planning to have a program in place by the time the vehicle is delivered will not receive a vehicle award. If the program is not in place by the end of the period of performance, the grant may be revoked. The funding to train drivers/operators can be requested under additional funding.

Why are training flashover simulators an ineligible expenditure but the manufactured burn trailers are not?

Manufactured burn trailers allow firefighters to train in a controlled and environmentally sound atmosphere. They are gas-fueled and computer-controlled training systems that realistically replicate fire emergencies. Flashovers can be simulated in a manufactured burn trailer. The Guide to building Fire Service Training Centers, NFPA 1402, provide guidelines for fire departments on purchasing and the use of these trailers. There are currently no National Fire Protection Association (NFPA) standards covering mobile live-fire training simulators.

When excess funds are used for training, what type of training is allowable?

All training, justified by the grantee, including Firefighter l and ll certifications, EMS, Driver/Operator, Fire Officer I-IV, Hazmat response, alternative fuel firefighting Incident Command, supervision and safety, CBRNE awareness, performance, planning and management, and train the trainer classes are allowable.

When requesting breathing apparatus, how many SCBAs should I request for our firefighters?

*The number of breathing apparatus needed for entry into environments immediately dangerous to life or health is directly linked to the number of firefighters who are trained and certified to conduct fire suppression or entry operations. Firefighters who are not **trained and certified** to conduct operations in these environments should not be issued breathing apparatus. For many departments, this equates to the number of firefighters reasonably expected to show up at an event, i.e., the shift on duty or the number of seated positions in your fleet of vehicles.*

Can a fire department buy EMS equipment or obtain EMS training?

Fire departments can apply for EMS training and EMS equipment within the training and equipment activities under the Operations and Firefighter Safety program area. Ambulances, however, are not eligible within this program area. All vehicles, regardless of function, must be requested in the Vehicle Acquisition program area.

Local Requirements

Fire service leadership, including the training officer, should understand the policies and procedures for their local jurisdiction. The training officer should also be an influence in establishing the minimum standards and core competencies of all fire department personnel. Each policy and/or procedure should have a training aspect, whether it is classroom, video, presentation, or documentation with examples. The training officer needs to ensure that each individual understands the requirements of the policy and/or procedures. Changing or altering these requirements may require approval from the fire chief.

Certification, Accreditation, and Credentialing

Important definitions to know and understand are:

- **Credentialing:** This involves establishing qualifications for licensing professionals or organizations through assessment of their experience, skills, and knowledge.
- **Accreditation:** Schools, colleges, and similar organizations are required to meet formal requirements, mandated by a governing agency, that guide the curriculum, facilities, and other components of the educational process.
- **Qualification**: Individuals wishing to function in a certain position are assessed for their knowledge and skills to determine if they are competent to meet the needs of that function.
- **Certification:** Individuals complete a course of study and/or successfully pass an examination that measures a competency level.
- **Reciprocity:** A mutual exchange of certifications among states/provinces or organizations. Reciprocity will vary from state to state (province or province). Some states/provinces will accept reciprocity if a fire fighter possesses an IFSAC or NPQ certification, while others will not.

Important acronyms to know and understand include:

- **AHJ:** Authority having jurisdiction
- **CPSE:** Center for Public Safety Excellence
- **CFAI:** Commission on Fire Accreditation International
- **JPR:** Job performance requirement
- **KSA:** Knowledge, skills, and abilities
- **ISFSI:** International Society of Fire Service Instructors
- **IFSAC:** International Fire Service Accreditation Congress
- **ISO:** Insurance Services Office
- **NPQ:** National Board on Fire Service Professional Qualifications
- **NREMT:** National Registry of Emergency Medical Technicians
- **OSHA:** Occupational Safety and Health Administration

■ Becoming Credentialed, Accredited, and Certified

"I don't get no respect," a phrase made famous by Rodney Dangerfield, summarizes how many fire and EMS professionals feel. One of the most significant ways to gain respect in one's chosen career is to seek and gain certification. When you look at the world today, most professions require some type of professional credentialing or certification, and many industries are now requiring accreditation, such as the International Organization for Standardization (ISO) or Six Sigma. In addition, an educational institution's accreditation (or lack thereof) is a good indicator of its educational quality. Even the National Fire Academy has chosen not to accept certain universities' degrees as part of their Executive Fire Officer Program, due to the university issuing the degree not being properly accredited. It is important to research these schools and their level of accreditation prior to enrollment. Most reputable schools are either nationally or regionally accredited through the government or a third party such as Southeastern Association of Credentialed Schools (SACS).

Since the inception of firefighting as an occupation, there have been minimum skills, such as techniques and tactics, that are required to function as a fire fighter. The NFPA was developed to establish minimum standards for firefighting such as NFPA 1001 for introductory fire fighters. The International Fire Service Accreditation Congress (IFSAC) and the National Professional Qualifications (NPQ) issue certifications based on individuals meeting the minimum standards established by the NFPA.

The majority of states/provinces do not develop their own standards; instead, they will adopt standards from organizations such as the NFPA or they will adopt JPRs from within a standard. This grants the AHJ the right to decide how to train its fire fighters and what the standard procedures will be. Most professions mandate that their employees to be trained to a certain level and demonstrate a certain amount of competency prior to engaging in any activities. Because of the inconsistency of state/province standards, fire fighters can perform one of the most dangerous jobs with varied degrees of training and competency. However, as training officers, we morally cannot allow this to happen. Developing national training standards for emergency operations, whether it is in regards to medical, fire, or technical skills, would help ensure that all fire fighters are trained to a minimum standard. It would also improve the interoperability between organizations.

Reciprocity

One of the biggest issues facing fire fighters across the country is the transportability or reciprocity of certifications. If a fire fighter in New York moves to California, one would think that their certifications would transfer, but this is not always the case. The International Fire Service Accreditation Congress (IFSAC) and the ProBoard are the two main agencies working to expand the reciprocity of certifications; however, not all states/provinces are members of these organizations, which limits their effectiveness.

Center for Public Safety Excellence/Commission on Professional Credentialing

The mission of the Commission on Professional Credentialing (CPC), part of the Center for Public Safety Excellence (CPSE), is to assist in the professional development of emergency service personnel by providing guidance for career planning through participation in the Professional Designation Program, which includes several credentialing levels including Chief Fire Officer (CFO), Chief Training Officer (CTO), Chief EMS Officer (CEMSO), Fire Marshal (FM) and Fire Officer (FO). An eleven member commission is responsible for establishing professional standards, awarding designations, and providing oversight to reviewers. The application reviews the candidate's educational background; participation in emergency services at the local, state/province, and national levels; and active involvement in the community.

ISFSI Live Fire Instructor Credentialing Program

The ISFSI Live Fire Instructor Credentialing Program is the first of its kind concerning NFPA 1403, *Standard on Live Fire Training Evolutions*, and is the only national live fire credentialing program in the country.

ISO and Fire Fighter Training Requirements

One of the toughest training standards to achieve in the fire service is that set forth by the Insurance Services Office (ISO). Most departments are not usually able to meet the personnel training requirements because of the number of hours in the different subjects that are required. While ISO is not a certification agency, it does tie to NFPA standards and in some ways is similar to an accrediting agency, as it holds agencies to specific guidelines to achieve different classes of ISO ratings. Review the appendix for a sample ISO training schedule and for a guideline addressing hours and topics required by ISO.

Functions of the IFSAC

According to the IFSAC[3] website, the IFSAC has several functions, which include:

- *The IFSAC accredits both public fire service certification programs and higher education fire-related degree programs. The IFSAC Certificate Assembly provides accreditation to entities that certify the competency of and issue certificates to individuals who pass examinations based on the National Fire Protection Association fire service professional qualifications and other standards approved by the Assembly.*
- *The IFSAC Degree Assembly accredits fire science or related academic programs at colleges and universities. Accreditation is available for both two-year associate degree programs and four-year bachelor degree programs. An IFSAC accredited program indicates that course offerings, institutional support, and qualified faculty are working together in completing their educational mission.*
- *The IFSAC Certificate Assembly provides accreditation to entities that certify the competency of and issue certificates to individuals who pass examinations based on the National Fire Protection Association fire service professional qualifications and other standards approved by the Assembly. IFSAC is one of the agency's that issues certifications based on the NFPA's professional standards. Some states/provinces only recognize IFSAC or NPQ certifications for fire fighters. Without an IFSAC or NPQ certification, the department or agency has no way of ensuring fire fighters meets minimum qualifications.*

[3]Reproduced from: International Fire Service Accreditation Congress (IFSAC). Certification. IFSAC.org. Accessed July 22, 2014.

The ProBoard

The ProBoard, also known as National Professional Qualification, accredits organizations that use NFPA's professional qualification standards. The ProBoard[4] website describes their requirements to become eligible for accreditation, which includes an application and a detailed self-study document.

- *The application package is then reviewed by the members of the Committee on Accreditation (COA) for completeness and compliance with the accreditation criteria of the ProBoard.*
- *The next step is a site visit by a team of COA members, usually two, who perform an extensive on-site review of the organization's testing and certification processes. The site visit team prepares a report and presents it to the COA. The COA decides if accreditation is granted.*
- *An entity that is accredited by the ProBoard makes certification available to its members, which are then eligible to be entered into the ProBoard's certification registry and receive a certificate from the accredited entity or apply for a certificate from the ProBoard, affirming that they were certified by a ProBoard-accredited entity and that they are entered in the ProBoard's certification registry.*
- *In addition, the ProBoard encourages reciprocity among certifying entities, which helps assure that ProBoard certifications will be recognized by departments as members seek advancement or should they seek to transfer within the fire service.*
- *The cost of national registration including a ProBoard Certificate ranges from $5.50 to $15.00 depending on the certificate distribution plan selected by the accredited agency.*

[4]Reproduced from: The ProBoard. *NPQ certification.* theproboard.org. Accessed July 22, 2014.

Crosswalk of National Fire Academy Courses and NFPA Standards

Since its inception in the early 1970s, the NFPA professional committee has identified the competencies necessary to perform the various positions within the fire service. Since early 1980, these standards have also been the cornerstone for state/province certification programs.

The CPSE

The CPSE[5] website describes the benefits of achieving a designation as the following:

- *Demonstrating that an individual has developed a strategy for continued career improvement and development.*
- *Highlighting the CFO's professional and personal growth and providing an individual source of pride.*
- *Showcasing superior leadership characteristics.*
- *Developing a commitment to continuing education, training, and skill proficiency.*
- *Establishing critical core competencies for personnel who serve in senior officer positions.*
- *Affirming an organization's dedication to proficiency and the proactive delivery of services to the community.*
- *Improving the hiring process by enabling authorities to quickly identify individuals with superior skills, knowledge, and abilities.*

[5]Reproduced from: Center for Public Safety Excellence. http://publicsafetyexcellence.org/professional-credentialing/chief-fire-officer/the-benefits.aspx. Accessed December 3, 2014.

ISFSI Live Fire Instructor Credentialing Program

The ISFSI Live Fire Instructor Credentialing Program:[6]

- *Is designed to deliver a comprehensive live-fire training program in accordance with NFPA standards 1142, 1402, 1403, and 1500.*
- *Both lecture and practical is used, allowing participants to apply knowledge of the above standards to live fire training activities.*
- *After successful completion of online work, skills evaluation, task book and post-test, students will be eligible for an ISFSI Live Fire Instructor Credential.*
- *The Live Fire Instructor Credential program currently credentials for instructors involved in fixed facilities; however, other credentials, such as Instructor in charge of live fire training, is under development.*
- *The program provides certification for 3 years and requires a minimum of 4 hours of live fire/1403 continuing education/upgrade training from ISFSI and a minimum of one documented burn in each credentialed discipline per year to maintain the credential.*
- *To take this course, you must already be certified as an Instructor I or higher.*
- *After successful completion of this course, students will receive an ISFSI Accredited, "Live Fire Instructor" Credential in the applicable area of Live Fire Training.*

[6]Reproduced from: International Society of Fire Service Instructors (ISFSI). *Live Fire Credential*. http://isfsi.org/links/live-fire-credential/. Accessed August 4, 2014.

As professional qualification programs such as IFSAC and NPQ have grown, fire departments and state/province agencies have started to ensure that curricula and certification exams are aligned with the NFPA standards. Furthermore, some states/provinces have a formal process in place for approving training programs before an individual is eligible to take the certification exam. One example of this process is the use of a crosswalk, such as the one provided below showing the correlation between the National Fire Academy (NFA) course, "Challenges for Local Training Officers," and NFPA standards. Using the crosswalk is the preferred method to ensure that a curriculum or exam is aligned with the NFPA standards. This will also serve as documentation that specific knowledge and skills were addressed in the case of simple questions or an investigation.

Guides and Recommendations for the EMS Training Officer

The most recent data from the U.S. Fire Administration's National Fire Incident Reporting System (NFIRS) demonstrate that fewer than 10 percent of runs made by fire departments involve fire, and more than half (55 percent) of fire department responses are categorized as EMS or rescue. Therefore, the EMS training officer has become a necessary member of the training staff within fire departments.

Whether the fire department functions as a first responder for their local 911 provider, or has an EMS transport component within, EMS personnel are expected to meet (or exceed) industry standards for patient care. To better understand these standards, we should first look briefly at the evolution of prehospital emergency medical care in the United States.

The theory of accessing, extracting, providing limited treatment for, and transporting injured individuals to a location where further medical care could be provided can be traced back to the Civil War. The first known hospital-based ambulance service was established at Commercial Hospital in Cincinnati, Ohio in 1865. Other hospitals, such as Grady Hospital in Georgia and Bellevue Hospital in New York, soon established ambulance services. These early ambulance services utilized horse drawn carriages, and carried medical equipment and supplies such as splints, stomach pumps, morphine, and brandy. The first volunteer rescue squad was organized by Julian Stanley Wise in Roanoke, Virginia in 1928. Between 1928 and 1965,

FIGURE 4-2 A rescue squad.

rescue squads (both volunteer and those incorporated into fire departments) sprang up around the country (**FIGURE 4-2**).

In 1966, the National Academy of Sciences published *Accidental Death and Disability: The Neglected Disease of Modern Society* (often referred to as the "White Paper"), which signaled the beginning of modern-day EMS. This paper provided concerning statistics related to the number of Americans who were killed, injured, and permanently impaired due to accidents—specifically, motor vehicle crashes. The financial impact on the United States was also detailed in the report, and, for the first time, politicians and civilians alike could see the toll taken by this unaddressed epidemic.

The EMS White Paper acknowledged that most people (including police, fire department personnel, and those employed by ambulance services) did not have basic first aid training. Ambulance services and the care they provided varied widely. Approximately 50 percent of the ambulance services in 1965 were operated by funeral homes, simply because hearses were better equipped to carry stretchers. Routinely, no care was provided to the patient en route, and many patients rode alone in the back of the hearse to the hospital because there was no space for an attendant to provide care.

Congress passed the Highway Safety Act of 1966, and from this Act, the Department of Transportation (DOT) was created. The DOT improved EMS systems by designing and implementing standards for the training of all EMS providers. In 1971, the DOT published the first EMS national curriculum.

The findings of the White Paper sparked universal attention and eventually resulted in adding the Emergency Medical Services (EMS) Systems Act of 1973 to the Public Health Service Act. This included the creation of 15 essential components of an EMS system.

On February 1, 1977, after a complaint was filed by the American Red Cross related to the common use of an Omaha orange cross on a square background, the newly designed six-barred cross was registered with the Commissioner of Patents and Trademarks in the name of the National Highway Traffic Safety Administration. Each of the bars of the blue *Star of Life* represents the six system functions of emergency medical services (**FIGURE 4-3**). Those functions are:

1. Detection
2. Reporting
3. Response
4. On Scene Care
5. Care in Transit
6. Transfer to Definitive Care

Based on the EMS Systems Act of 1973, federal and state agencies began developing the system that we know today. The EMS industry remained in a state of evolution, and in 1996, the National Highway Traffic Administration, other federal agencies, and a group of stakeholders developed the *EMS Agenda for the Future*. The *Agenda* recognized 14 EMS attributes

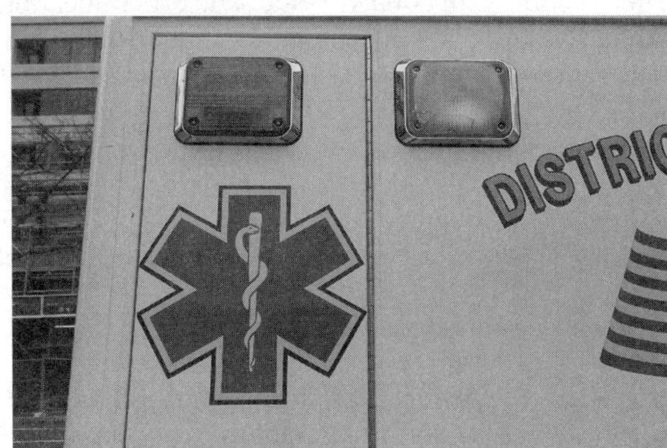

FIGURE 4-3 The Star of Life.

necessary in the continued development of the EMS system. These attributes were built upon the 15 components outlined in the *White Paper*. The attributes outlined in the *Agenda* are:

1. Integration of Health Services
2. EMS Research
3. Legislation and Regulation
4. System Finance
5. Human Resources
6. Medical Direction
7. Education Systems
8. Public Education
9. Prevention
10. Public Access
11. Communications Systems
12. Clinical Care
13. Information Systems
14. Evaluation

The *National EMS Scope of Practice Model* (*Scope of Practice*) was published in 2007. This document defines the levels of EMS personnel and delineates the practices and minimum competencies for each level. The *Scope of Practice* does not have regulatory authority, but provides guidance to states/provinces for four described levels of EMS personnel licensure: Emergency Medical Responder (EMR), Emergency Medical Technician (EMT), Advanced Emergency Medical Technician (AEMT), and Paramedic.

The National Emergency Medical Services Education Standards, published in 2007, define the competencies, clinical behaviors, and judgments that must be demonstrated by entry-level personnel to meet practice guidelines defined in the *National EMS Scope of Practice Model.* In 2009, the National EMS Education Standards Gap Analysis Template was published. This document was designed to assist in the transition and implementation of the National EMS Education Standards for states who were preparing for the transition.

The National EMS Education Standards are a set of comprehensive objectives set for each level of certification or licensure. This allows program coordinators to determine the curriculum and length of the training program based on successful completion of these objectives. Most program coordinators work in conjunction with their state/province regulatory agency and sponsoring educational facility to determine the most reasonable program length based on the minimum set of objectives.

The EMS training officer should be familiar with these national documents, though most states/provinces align with the *Scope of Practice*, at least to some degree. EMS training officers should be familiar with the scope of practice and all rules and regulations for the state/province in which they practice. Educational standards and the scope of practice for each state/province will provide the foundation for patient care protocols and clinical expectations. Other recommendations made within the National EMS Scope of Practice are relative to accreditation of EMS Education Programs and credentialing of EMS personnel.

The National Registry of EMTs also supports the direction recommended in the *EMS Agenda for the Future* and *Scope of Practice*. Currently, certification through the National Registry of EMTs is a requirement for certification or licensure in many states. Effective January 1, 2013, paramedic students who desire certification testing through the National Registry of EMTs are required to have graduated from a nationally accredited paramedic program.

Agencies that provide medical care to the community should have, and work closely with, a medical director. It is strongly recommended that the medical director be actively involved in the development of local protocols and standing orders, as well as provide oversight for a credentialing process for the EMS personnel working under his or her authority.

Currently, credentialing of EMS personnel is a recommendation, but not a commonly established process. As medical directors move toward adopting all aspects of the *Scope of Practice*, it is recognized that no standard process has been published, and most agencies find themselves designing a process of their own.

The purpose of local credentialing is to verify personnel qualifications and objectively evaluate the individual's knowledge and skill level to ensure he/she meets the minimum acceptable level of performance for his/her level of certification or licensure. This is usually obtained through valid written and practical evaluations to document competency and may incorporate agency-specific testing on policies and protocols. The medical director will determine how frequently the evaluations should be done, and many services require personnel to successfully complete core competency evaluations on an annual basis.

EMS training officers are frequently required to work with the local medical director to manage all aspects of oversight related to medical care provided by that department. The training officer may be required to verify the certification (or licensure) of new applicants; design, coordinate, and teach new employee orientation programs; manage in-house continuing education

FIGURE 4-4 A training officer.

programs for employees; establish an internal quality assurance process; design appropriate remediation pathways for personnel; participate in employee recognition programs for successful performance; or make disciplinary recommendations when necessary (**FIGURE 4-4**).

Whether the training officer has very broad or very limited responsibilities, it is imperative that he or she be thoroughly familiar with the national documents outlined in this chapter. In additional to a thorough working knowledge of those documents, the Training Officer must be familiar with his/her own state/provincial laws, rules, regulations, and scope of practice. Medical directors may also have the authority to outline additional local mandates, so the training officer must incorporate those expectations into an evaluation process.

The inclusion of EMS responders within the fire department has certainly proved its value during the last four decades. Whether the fire department includes EMS transport capabilities, or simply provides first responder medical support, those who are ill or injured in the community are receiving valuable medical attention from prehospital personnel in a timely manner. As the expectations of the public continue to increase, the role of the training officer becomes even more critical. The EMS industry has set high standards for responders, and it is the responsibility of the training officer to ensure those standards are met, or exceeded, on every EMS call.

Key Points

- As a training officer, you have a tremendous amount of influence on the level of training within your department, ranging from recruit school to the chief officer level.
- Review and become familiar with NFPA standards that are pertinent to the training officer's position. In addition, study the correlation between NFPA, ISO, and OSHA standards.
- The NFPA provides guidelines and is considered a consensus standard; the standards are developed by industry specialists, researchers, and fire service personnel.
- As a branch of the executive government, OSHA sets regulations and has the ability to enforce them through citations and/or fines.
- ISO provides recommendations on fire protection across several categories of residential or corporate buildings. In return, ISO makes recommendations on the actual insurance cost.
- Training officers should review all aspects of the local, state/province, and federal regulations and laws that oversee their operations.
- Training officers should keep in mind that many of the grants that fire departments are receiving are coming with caveats that require the department to perform some task.
- Review the need for accreditation, certification, and credentialing in your department; as the training officer, it is your duty to choose the appropriate level for your operations.

References

Center for Public Safety Excellence. http://www.publicsafetyexcellence.org. Accessed July 22, 2014.

Federal Emergency Management Agency (FEMA). *Assistance to Firefighters Grant Program: 2012 Frequently Asked Questions*. Washington, DC: 2013.

Federal Emergency Management Agency (FEMA). *Crosswalk*. http://www.usfa.fema.gov. Accessed July 22, 2014.

Federal Emergency Management Agency (FEMA). *Fire Grants*. http://www.fema.gov/firegrants/. Accessed July 22, 2014.

Federal Emergency Management Agency (FEMA). *Frequently Asked Questions*. http://www.fema.gov/2012-frequently-asked-questions. Accessed July 22, 2014.

International Fire Service Accreditation Congress (IFSAC). *Certification*. IFSAC.org. Accessed July 22, 2014.

International Society of Fire Service Instructors (ISFSI). *Live Fire Credential*. http://isfsi.org/links/live-fire-credential/. Accessed August 4, 2014.

National Fire Protection Association (NFPA). *Standards*. NFPA.org. Accessed July 22, 2014.

National Fire Protection Association (NFPA) 1001, *Standard for Fire Fighter Professional Qualifications*.

2013. http://www.nfpa.org/codes-and-standards/document-information-pages?mode=code&code=1001. Accessed December 9, 2013.

National Fire Protection Association (NFPA) 1002, *Standard for Fire Apparatus Driver/Operator Professional Qualifications*. 2014. http://www.nfpa.org/codes-and-standards/document-information-pages?mode=code&code=1002. Accessed September 7, 2014.

National Fire Protection Association (NFPA) 1003, *Standard for Airport Fire Fighter Professional Qualifications*. 2010. http://www.nfpa.org/codes-and-standards/document-information-pages?mode=code&code=1003. Accessed September 7, 2014.

National Fire Protection Association (NFPA) 1021, *Standard for Fire Officer Professional Qualifications*. 2014. http://www.nfpa.org/codes-and-standards/document-information-pages?mode=code&code=1021. Accessed September 7, 2014.

National Fire Protection Association (NFPA) 1041, *Standard for Fire Service Instructor Professional Qualifications*. 2012. http://www.nfpa.org/codes-and-standards/document-information-pages?mode=code&code=1041. Accessed September 7, 2014.

National Fire Protection Association (NFPA) 1401, *Recommended Practice for Fire Service Training Reports and Records*. 2012. http://www.nfpa.org/codes-and-standards/document-information-pages?mode=code&code=1401. Accessed July 18, 2014.

National Fire Protection Association (NFPA) 1402, *Guide to Building Fire Service Training Centers*. 2012. http://www.nfpa.org/codes-and-standards/document-information-pages?mode=code&code=1402. Accessed September 7, 2014.

National Fire Protection Association (NFPA) 1403, *Standard of Live Fire Training*. 2012. http://www.nfpa.org/codes-and-standards/document-information-pages?mode=code&code=1403. Accessed December 9, 2013.

National Fire Protection Association (NFPA) 1404, *Standard for Fire Service Respiratory Protection Training*. 2013. http://www.nfpa.org/codes-and-standards/document-information-pages?mode=code&code=1404. Accessed September 7, 2014.

National Fire Protection Association (NFPA) 1407, *Training Fire Service Rapid Intervention Crews (RIC)*. 2015. http://www.nfpa.org/codes-and-standards/document-information-pages?mode=code&code=1407. Accessed September 7, 2014.

National Fire Protection Association (NFPA) 1410, *Training for Initial Emergency Scene Operations*. 2010. http://www.nfpa.org/codes-and-standards/document-information-pages?mode=code&code=1410. Accessed September 7, 2014.

National Fire Protection Association (NFPA) 1451, *Fire Service Vehicle Operations Training Program*. 2013. http://www.nfpa.org/codes-and-standards/document-information-pages?mode=code&code=1451. Accessed September 7, 2014.

National Fire Protection Association (NFPA) 1500, *Standard on Fire Department Occupational Safety and Health Program*. 2013. http://www.nfpa.org/codes-and-standards/document-information-pages?mode=code&code=1500. Accessed September 7, 2014.

National Fire Protection Association (NFPA) 1521, *Standard for Fire Department Safety Officer Professional Qualifications*. 2015. http://www.nfpa.org/codes-and-standards/document-information-pages?mode=code&code=1521. Accessed September 7, 2014.

National Fire Protection Association (NFPA) 1584, *Standard on the Rehabilitation Process for Members During Emergency Operations and Training Exercises*. 2008. http://www.nfpa.org/codes-and-standards/document-information-pages?mode=code&code=1584. Accessed September 7, 2014.

National Fire Protection Association (NFPA) 1971, *Standard on Protective Ensembles for Structural Fire Fighting and Proximity Fire Fighting*. 2013. http://www.nfpa.org/codes-and-standards/document-information-pages?mode=code&code=1971. Accessed September 7, 2014.

Sheridan, J. A $37 billion waste, *Industry Week*, September 4, 1989.

The ProBoard. *NPQ certification*. the proboard.org. Accessed July 22, 2014.

United States Department of Labor, Occupational Safety & Health Administration (OSHA). https://www.osha.gov/pls/oshaweb/owadisp.show_document?p_table=standards&p_id=9696. Accessed August 4, 2014.

Wyoming Department of Fire Prevention & Electrical Safety. *Manual for Training Officers*. December 2011. http://wyofire.state.wy.us/pdf/trainthetrainemanual.pdf. Accessed July 28, 2014.

CHAPTER

5 Strategic Planning for the Training Division

Brian Ward and Leigh Hubbard

Introduction and the Purpose of Strategic Planning

Strategic planning in the fire service is of the utmost importance, regardless if you are a career, combination, or volunteer department. Even the industrial side of emergency services must develop a meaningful strategic plan in order to succeed in the long term. As Abraham Lincoln once said, "If I was given 6 hours to cut down a tree, I would spend the first 4 sharpening the axe." Putting the time, effort, and resources into developing a strategic plan that dictates the direction, measureable goals, and objectives of your training division will more than pay off in the long run. Everyone stands to benefit from a well-structured, high-functioning training division.

The easiest way to perform strategic planning is simply to determine your long-term goals for department training. Strategic plans come in various lengths and have various topics, in accordance to the specific needs of your department. These topics may include reaching a certain accreditation level, increasing training levels, lowering the Insurance Service Organization (ISO) rating, or the construction of new stations. All of these achievements result in the acquisition of new recruits and promotional classes. As a training officer, you should have a strategic plan that aligns the training division with the intent of the overall organization. Each division will bring objectives to the table and goals to consider. This is a good time to start the planning process of prioritizing objectives and selecting goals to accomplish.

Strategic planning is a cyclical process (**FIGURE 5-1**). The steps include:

1. Establish a vision.
2. Create goals.
3. Examine internal factors.
4. Examine external factors.
5. Formulate a strategy.
6. Reevaluate.

The strategic planning cycle is a continuous cycle that requires updating in order to receive the greatest benefit. The most difficult part is starting this process; once the cycle is established, it becomes a routine maintenance item. However, with that being said, do not become complacent and settle for the status quo. Always try to think outside the box and consider how you can create value with a strategic plan.

Establishing a Common Vision for Your Training Division

The first step in strategic planning is to develop a vision for the department as a whole. Establishing a vision will allow you to know the direction of your department; it provides a picture of what the end result should look like. More importantly, it describes a state of being that your department wishes to achieve. Employees feel a sense of pride and are motivated by the simple fact that leadership is looking for ways to better the department. Improvements typically include advancing safety, updating equipment, and being fiscally responsible. As the

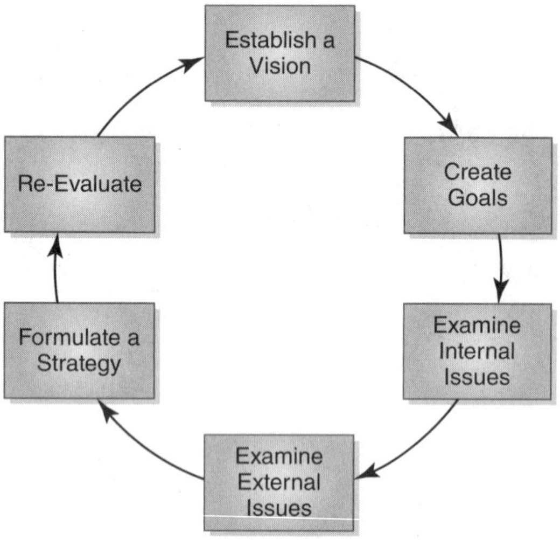

FIGURE 5-1 The steps of the strategic planning process.
Courtesy of Brian Ward.

training officer, knowing this information can assist you in guiding the department towards that vision. The improvement could involve a skill set or it could involve a culture change, but both would be equally important to the morale and effectiveness of the department.

While the training division drives most of the other divisions' initiatives, the other divisions also play an intricate role in the development of the overall training strategic plan. It is an absolute necessity for the department to work together to develop a common vision. Without collaboration, confusion, duplication, and wasted money and efforts may occur instead. As the fire department develops the overall vision, each division within the department should develop their own vision that aligns with that of the department as a whole. It is also important to understand the difference between vision and mission statements. Vision statements describe what the organization would like to become, while the mission statement describes the purpose or reason for the organization's existence.

There is a certain amount of interpretation to the underlying meaning of any vision, therefore it is important to take a deeper look and ensure that everyone is on the same page. In fact, since the vision of a fire department is so subjective, it is vitally important that the leaders communicate and discuss the vision with everyone in the department. A fire department may have weekly meetings with all department supervisors so that everyone understands how to use the vision to drive their decisions on a daily basis.

In any discussion, especially those involving philosophy and personal beliefs, questions will be asked. To prepare for this, we should think through the process ourselves. Using the example vision below, some questions that may come up are: *How do you define the best protection? How do you achieve the best protection? When do you know if you are there?* Having these discussions allows everyone to take ownership and develop an understanding of the expectations coming from above.

How you explain these concepts to your officers will have a direct impact on how they apply them. In this particular quote, the phrase "the best protection" implies that no matter how good you are, there is always room for improvement. There are many ways we can strive to achieve "better protection of our citizens." We can train in different skill sets, expand and increase our knowledge, make sure we are fit for duty,

Mission Versus Vision

An example of a department's vision could be, "To respond to all emergency calls within 5 minutes, while following all applicable laws, in order to provide the best protection possible to our citizens." The training division could provide training on driving, applicable laws, all skill sets for protecting citizens, quick dress, and many more topics from this vision statement.

The department may then choose to develop a mission statement such as, "We will deliver the highest quality service to all citizens," and the training department's mission statement could be, "Provide the highest quality training to our employees to best protect our citizens." This will personalize the mission for everyone involved and will also allow training officers and instructors to understand their role in the overall plan of the department and inculcate a sense of ownership.

Mission statements—describe the purpose or reason for the organization's existence.
Vision statements—describe what the organization would like to become.

Case Study

The United States Marine doctrine, *Warfighting*,[1] describes the process that the U.S. Marines use in order to carry out a task. Through the use of intent of command, the U.S. Marine Corps provides the background of each task at hand: what is the task, why the task is performed, and when it will be performed. By explaining the intent of the command, the troops can make their own decisions regardless of the situation. In the book's foreword, C.C. Krulak, General, U.S. Marine Corps, explains how this text is not a manual, but rather describes the U.S. Marine Corps' philosophy for "how [they] fight and how [they] prepare to fight" and that "the thoughts contained [within the text] are not merely guidance for action in combat but a way of thinking." The one item that the U.S Marine Corps does not delve into in this text is how the task will be performed. As General Krulak indicated, "This book contains no specific techniques or procedures for conduct. Rather, it provides broad guidance in the form of concepts and values. It requires judgment in application."

In a essence, the intent of command is a vision of a better state. Although this text discusses how the intent of command is used in the context of the U.S. Marine Corps, the intent of command is a very important concept that can be applied to anyone who works in emergency response situations, because what it allows us to do is evaluate the situation we are facing and make a decision. It also allows us to step outside of the status quo and achieve superior results. Just as not every fireground is the same, neither is the battlefield. However no matter what we are up against, the end result we desire is the same: to be successful and have everyone go home.

[1]Data from: U.S. Marine Corps Staff. *Warfighting*. North Charleston, NC: CreateSpace Independent Publishing Platform; 2014, page 3.

and learn to be respectful and cordial to the public we serve. A vision should not give you an end point—it is simply a direction to a better place. In actuality, the end point is where goals come into play. The goals that we set do have an end point, are measureable, and will guide us toward reaching our vision.

Creating SMART Goals for Your Training Division

Once you have a vision for your training division, you need to establish goals that will help you achieve this vision. Creating SMART goals is one method that you and your training division can use to establish goals. There are two things you should consider when creating SMART goals:

1. You should look at your personal training goals and think about what you want to achieve.
2. You should review your department's training goals and consider how you will achieve them. The department goals and your personal goals may not always align, but you should strive to bring them closer together.

Regarding the first item, you will need to determine your career goals as a training officer and be encouraged to look beyond your position. What do you want the department and, yourself as a training officer specifically, to accomplish in the next 1, 2, 3, 5, and 10 years? Even if you are unsure what these goals should be, it is generally a good idea to start thinking about them now (**FIGURE 5-2**). Developing these goals is good for you as an individual and for your department.

You will need to determine the goals of the department. Where do you and the administration see the department in a 1- to 10-year time frame? In order to be prepared for the future, you must start looking at what the needs may be for your community and organization.

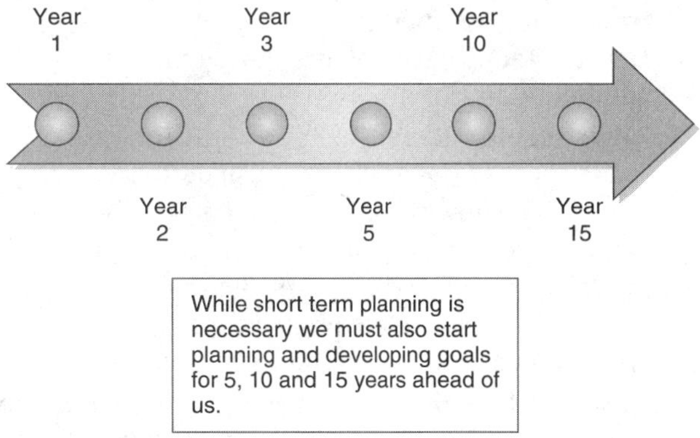

FIGURE 5-2 While short-term planning is necessary, we must also start planning and developing goals for the next 5, 10, and 15 years.

Courtesy of Brian Ward.

In addition, if your department wishes to receive revenue in the form of grants or Special Purpose Local Option Sales Tax (SPLOST)—generally, a one cent sales tax on purchased goods within a jurisdiction—it is a good idea to have a firm understanding of how this money will influence your department. This money could help increase training levels and allow for training on new equipment and equipment maintenance, but what is the succession plan for remaining competent on the equipment's use or skill sets after the initial training? These are all questions that strategic planning can assist in answering if performed correctly. Regardless of which part you are working on—career goals or department goals—a clear understanding of how to develop and achieve goals is needed.

One of the most important traits of any successful individual or department is their ability to establish and meet goals. Individuals or departments who wish to be successful will determine how they wish to improve and what it would take to make those improvements. Once you reach your first set of goals, always establish new ones, with the ultimate goal being that you never become stagnant in whatever you are doing.

Some individuals become discouraged very quickly, because they set goals that are not immediately obtainable. Throughout your career, you should set both short- and long-term goals. Doing so will provide you with a sense of accomplishment in the short term, but it will also allow you to continue working towards the "big picture." One method of ensuring that you remain focused on your goals is to write them down using the SMART format discussed below. This will allow you to work toward your goals while remaining flexible with the timeframe.

SMART:

Specific– Be specific when stating your goal. Define exactly what you wish to accomplish, which will help you stay focused.

Measurable – Identify some avenue for measuring the success of the goal. Establish definitive benchmarks that show your progress. These benchmarks could be time-based or grade-based, or they may utilize some other form of measurement. The key is to ensure that your measurement reflects the specific task you are trying to complete.

Attainable – Ensure that the goal is reachable. Pace yourself and be *reasonable* with your goals. Research and explore possibilities that will assist you in gaining the knowledge and abilities you will need to obtain each goal.

Realistic – Set realistic goals. Establish a plan or a series of steps that will allow you to accomplish a satisfying goal. Do not set goals that are easy, but rather, set ones that require some work and dedication. One of the reasons for working toward something or setting goals is that we all must step out of our comfort zone from time to time. This allows us to grow as humans and to learn. The key to doing this effectively is to not set your sights so high that you cannot possibly reach them. Remember, it takes time, work, and effort to obtain each goal. If the goal was easy, everyone would do it.

Timely – Establish an endpoint or timeframe for accomplishing your goal. This part may also be closely tied to your measurement benchmark. However, it is important to remember that just because you set a date, this does not mean the date cannot be modified, especially if it is a long-term goal. It is OK if dates have to be altered. Simply establish a second date and remain focused on meeting it. Do not allow yourself to procrastinate.

Although there are other ways of setting goals, the SMART method provides one of the simplest and most straightforward ways of putting your intentions down on paper. While it is acceptable to keep track of personal goals without writing them out, if you are developing a strategic plan for your department, it is highly recommended that you document this process in order to show others (i.e., administration, city managers, council, etc.) what the plans are.

Examining Internal and External Factors

While we will discuss internal and external factors separately, we feel it is important to first show a model that could be used to examine internal and external factors simultaneously. The SWOT (strengths, weaknesses, opportunities, and threats) analysis allows you to compare the external and internal factors within the four categories (**FIGURE 5-3**) and determine the areas in which the fire department is doing well and the areas that require improvement. In the more traditional sense, this model can also be used by the fire department administration to review their current state of affairs.

■ Examining Internal Factors

Examining factors within your department involves identifying your department's strengths and weaknesses.

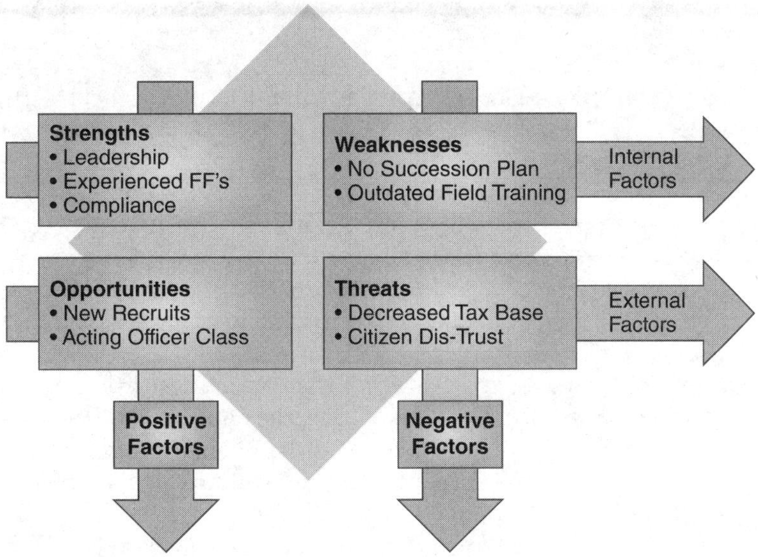

FIGURE 5-3 A diagram of the SWOT analysis model.
Courtesy of Brian Ward.

The strengths of your department identify areas that you should plan to build upon. Areas of strength may include ISO ratings, department accreditation, or recruit training levels. Determining your department's strengths can help you to leverage actions against its weaknesses. Whether you are facing the fire department administration or the governing board, or you are asking for grant money, being able to provide proof that you are a responsible division or department will assist you in your request. Another reason it is important to know your department's strengths is that it will help you determine when your department has achieved its goals.

The weaknesses within your department may include factors such as understaffing, injuries, or inadequate water supply. Once these areas are identified, they should be prioritized by frequency and severity, while also considering which factors could have the greatest impact on the department. Safety and health topics should always be at the top of the list of priorities. Once these have been addressed, you should transition your focus into other areas that will improve your department. Areas of weakness will make up a vast majority of the goals that you create.

Examining External Factors

Every fire department faces external factors, or situations, that present either opportunities or threats to your department. In general, the opportunities presented to your department could become part of its list of strengths if it plans for them. Opportunities that would strengthen your department include increasing the department's ISO rating, increasing the knowledge of your instructors beyond what the authority having jurisdiction (AHJ) recognizes, or the acquisition of SPLOST funds. The opportunities could also fall into the weakness category because they have not been made a priority or represent a knowledge/skill that is lacking. For example, your department may already be strong in the area of continuing education for your instructors; however, the chance to improve may be right there in front of you, such as a free workshop or class. In most cases, these opportunities are not the kind that you can produce internally—they will require assistance from grants or an outside participant such as the ProBoard and ISO.

Threats can also be presented to your department from an external source. Threats typically represent items such as a decreased tax base, decreased citizen approval, and/or a lack of current technology/equipment. Some threats we cannot control, such as decreased tax base due to a slow economy. However, we can effect change for some, such as the public's perception of our department. Whether you agree or disagree with what the public is thinking or saying, they ultimately write our checks. Establishing a good relationship with the public is vital to receiving the support that we need to operate safely and effectively. Right or wrong, politicians will react in line with the public's perception of us, so it is best if the public's perception is in our favor.

Case Study

In one Metro Atlanta department, it was identified that the training levels for their recruits and acting officers was well above average. Recruits attended a 32-week fire and EMS recruit school and graduated with various certifications, including Emergency Medical Technician–Intermediate (EMT-I), Firefighter I (FFI), and Firefighter II (FFII) in accordance with National Board on Fire Service Professional Qualifications, in the area of Hazardous Materials Awareness and Operations, as well as several other certifications. Not only did recruits have excellent opportunities to get trained and certified in fire service areas, acting officers also had to undergo extra training before they could ride seat. Before assuming any officer-related responsibilities, these individuals attended a 120-hour course taught by field subject matter experts (SMEs) that included practicals and a precepting or ride along phase.

Although the department demonstrated high levels of training in these areas, the training unfortunately stopped at the developing officer level. With this knowledge, the department leveraged the success of these two programs and created an acting captain and acting battalion chief program. Now, all officers, including developing officers, are prepared to perform at a level above their official position. Personnel in both of these classes leave with additional knowledge and practical exercises based on the respective position. The reason these programs were such a success is because they were included as a part of the department's strategic plan. To build training and maintain programs of this scale, it requires the department to allocate sufficient numbers of personnel to each training level. With the tight budgets that departments face, it would be impossible for the fire department to reallocate personnel or increase head count without a plan in place.

One way we can keep the threats of changing technology and equipment at bay is to stay on top of the NFPA standards as well as any literature published regarding equipment changes. Since these changes can affect our operations, it is necessary that we make sure we are informed of them prior to making any purchases. It is irresponsible to purchase a piece of equipment if there is information published that warns us the equipment is on the verge of becoming obsolete or noncompliant.

As you examine both the internal and the external factors facing your department, identify which factors could have the greatest positive impact on your department. Research these factors and consider what "good" looks like. Questions to consider include:

- What can we achieve now, with minimum effort?
  - These particular items are the "low hanging fruit." It should not take a tremendous amount of effort to achieve anything that falls under this category.
- Which item has the greatest overall impact?
  - These items may or may not take a tremendous amount of additional effort and time to complete; however, the investment on return may deliver results that compensate for the effort.
- Where is our biggest threat?
  - These items may or may not have the greatest overall impact, but they typically represent the areas in which we are most vulnerable and should be addressed before items posing smaller threats.

Formulating the Strategy for the Training Division

The next step to developing a strategic plan is to consider your goals for the training division and the goals of the fire department, incorportae the impact of all internal and external factors, and formulate a strategy on how you will tackle each item. As you begin to understand the impact of each factor, it will become easier to prioritize these factors and determine how to address them. The key to formulating a strategy for the training division is to not try and develop it without support from others. In addition, if the funds exist, inviting an outside agency or a neighboring department to review your plan may be beneficial. Within this plan, include the information that you have collected through the process of creating goals and identifying factors. A timeline showing your benchmarks and completion dates for each goal should also be developed; this will help you measure the success of your SMART goals. A sequence of steps on how you are going to execute

Case Study

The Impact of Citizen Approval

The Gwinnett County Fire and Emergency Services department realized the impact of citizen approval roughly 20 years ago. During this period, the department succeeded in building a well-respected relationship with their citizens. They made efforts to treat each person respectfully and by the Golden Rule. They also created a community risk-based division (similar to most prevention divisions) that spoke to and created interactions with these citizens. In fact, the relationship they built with their community was so strong that it allowed them to be the recipients of the Special Purpose Local Options Sales Tax (SPLOST) for the last 17 years.

The community's approval of their department also became the driving force behind many developments that enhanced and augmented the department and its services, including numerous stations being built and apparatus being purchased. With these purchases and additions, the department could focus its internal resources toward other areas in need of improvement, which only made their department even greater.

As you can imagine, this type of relationship can take a long time to build. It took many years for Gwinnett County Fire and Emergency Services department to build a relationship of mutual respect and cooperation with its community. The department understood that one of its greatest resources was its community and that many potential opportunities could arise if community–department relations were strong. As a result of this one decision, the department was able to grow and improve in ways that benefited not only the department, but the community it served as well. It is important to remember, however, that such a relationship can be lost very quickly.

this plan should also be developed as you formulate your timeline.

Constructing a strategic plan that will be reviewed and used by others is similar to preparing lessons, in the sense that both situations require you to identify your audience. You will need to ask yourself: *Who will be reviewing this plan?* The education level or level of understanding of these individuals may vary; in addition, intangibles specific to you and/or your department may not be obvious to external review boards. Generally speaking, governing boards have individuals with financial, public administration, and business degrees, which makes them experts in their respective fields, but they may not understand the motivating factors behind each plan and decision. Being able to predict how these individuals will perceive this plan is very important when you want them to make a decision that is in your favor.

Keeping these important points in mind, there are several tools and tactics can be used to ensure the development of a strategic plan is a success. As a training officer, learn to use a variety of visuals to demonstrate the quantitative benefits of your plan. Remember: A picture is worth a thousand words.

Tips for Creating an Executive Summary

- Generally, the summary should be 1–10 pages long, or 10 percent of the original document.
- Summarize your research.
- Accuracy is key—make sure all the numbers and statistics are correct.
- Begin with a concise statement concerning your conclusion.
- Make a recommendation.
- Justify your recommendation.
- Attach a cover sheet to the executive summary that outlines the specific points addressed in your summary. If your audience chooses to learn more, they can read the specific parts of interest to them.

Being able to use numbers to show the financial impact of every decision and how the training division and the fire department will benefit from this plan will have a direct impact on their decision. One method to achieve this is to research outside departments, best practices, and other forms of industry research. Due to the time constraints that all governing boards face, it is also extremely useful to create an executive summary that enables them to understand your research in a succinct and efficient way. Creating an executive summary should be one of the first items that you learn as a training officer, as it has many implications for its use.

■ Re-Evaluate the Plan

As is the case with any plan, the strategic plan must be continuously evaluated and updated. Remember that the creation and development of a strategic plan is a dynamic and cyclical process. As the plan is executed and each item on the agenda is completed, move to the next goal or find ways to improve on the division's accomplishments. If the plan does not go as expected, we have to determine why, then go back through our strategic planning process and make changes based on our needs and the SWOT analysis. The key to this entire process is to never settle simply because a milestone was reached; that achievement only becomes a steppingstone for a bigger and better future.

Another key concept to consider is the speed with which you execute each step of the plan. You should not over plan to the point where you forget to execute. We must find a speed that will allow us to progress through the strategic planning process, while allowing us to make adjustments and alter the plan as we move forward. The greatest plan ever is of no use if we fail to execute it.

Creating an Annual Training Plan for Your Division

One of the benefits of developing a strategic plan is that the process enables you to create an annual training plan that will assist you in executing against your overall plan. The same SMART goal process can be used for establishing your long-term organizational goals. In addition, the training division is usually tasked with ensuring that the fire department has the appropriate knowledge and skills to drive these goals. In conjunction with the organizational goals, there are minimum standards that must be met for compliance, generally on an annual or biannual basis.

The immediate concerns, or mandatory annual training topics, are where most training officers begin the development process. These are the classes that must be covered within a definitive timeframe for compliance or certification. A simple spreadsheet is a sufficient way to track this data. You can also break a spreadsheet down by months, and each a tab can represent a different year. With once click of a mouse you can have an annual or even a 5-year training plan laid out. There are software programs available, such as Target Solutions, that may assist you in developing your training plan and provide a few additional tools that can be used for tracking completion and updating of content. Target Solutions offers a complete online learning and tracking platform that can assist with a department training plan.

If you decide to use the basic spreadsheet to structure your plan, the first step to take is to divide all of the annual training topics evenly across 12 months. Attempt to reduce the number of mandatory training topics for December, for obvious reasons—the crews will appreciate this and it helps you with last minute catch-ups. It is inevitable that at least one fire fighter will need to catch up on certain training topics due to various reasons, such as military duty, Family Medical Leave Act (FMLA), sick leave, vacation, and other types of extended leave. However, once the 12-month calendar is developed, copy it over for the next 5 years (these topics generally will not change from year to year). These are your base training requirements and can be altered from month to month according to the initiatives that are rolling out in your organization. This part of the plan should be aligned with the goals of your organization.

The second step is to identify the biannual training topics and to plug them in at the appropriate time on the rotating years. This is still part of your baseline training requirements, but they are on a different reporting schedule. Topics which fall under this category include CPR and state certifications, such as Emergency Medical Technician (EMT) and Paramedic certification.

The third step is to identify the different levels of certifications within your organization and when these certifications need to be renewed. Once these types of certifications are identified, place them in the appropriate month/year section of your spreadsheet and fill in the title listings in the same manner that was used for the annual training plan. This will complete the majority of your strategic training plan.

The final step of the process is to review with your supervisors where the department needs to be at various points in the future. Determine what steps are necessary to reach these goals and where training will play a part. Once your goals are identified and the training division determines its level of participation, research the best possible way to deliver this training. One question you should ask with this and all levels of training: Is there a more efficient, yet still effective, method to deliver this training? Options you may want to consider exploring include computer-based training (CBT), instructor-led classes, self-study, or contract (for more information on these options, see the chapter *Alternative Learning*). After determining the answers to these questions, identify in your training plan the methods with which the topics will be delivered with respect to the appropriate timeframes for that specific initiative. For example, the chief says that this time next year everyone will be trained in Advanced Safety and Survival. If this is the department's main initiative, training in this area should be where your focus is for the time being.

After completing these steps you will have a multi-year training plan developed and a clear direction to travel with execution dates. The best part of having this plan mapped out for you is that it just takes one click to see what topics are to be covered this month and what topics are coming up next month. Furthermore, once your certification levels and goals are determined, this plan will assist you in staying on target and executing against the allotted time frames.

Establishing the base training plan will assist you and your department in determining where you wish to be with regard to accreditation and ISO requirements (see the chapter *Guiding Requirements* for additional information on ISO requirements). For example, you will have a clear understanding of where your fire department stands in relation to the number of hours and topics required by ISO and the various forms of fire department accreditation.

Project Management for Training Officers

As a training officer develops their strategic plan, there will be projects that need to be led and managed by appropriate personnel. While some of these items may seem very similar to strategic planning and creating an annual plan, they present different challenges. The keys to being a successful project manager include having a genuine passion for your work, a clear vision, good communication skills, being a disciplined and accountable leader, and creating a well-structured environment with clearly defined direction. Whether you are a new project manager or a seasoned veteran, the rules of project management have evolved over the years as the industry has experienced changes in technology, communications, and customer expectations for immediate results.

This section on project management will provide you with the direction and resources to successfully tackle your project. While each project is unique, the processes used to manage each project are similar. A successful project manager will ensure that the project life cycle is well defined in regards to initiating, planning, executing, controlling, and closing of the project. For the training officer, project management could include several topics and initiatives, including:

- Burn building construction
- Development of a Fire Fighter I and II curriculum
- Fire Academy infrastructure development
- Upgrading response services based on new target hazards
- Apparatus design
- Major in-service training for new standards or best practices

Leadership and the Team

The key to running a successful project is ensuring proper leadership. Most of the qualities of a good leader are timeless and apply regardless of the specifics of the project. Ensuring that the project manager possesses these key traits and is driven by them will ensure that your team will have a sound foundation from the beginning of the project.

- Lives the vision
- Driven by integrity
- Good communicator
- Delegator
- Builds the team
- Problem solver

The successful project manager is defined as having 20 percent hard skills (i.e., technical and procedure-based knowledge) and 80 percent soft skills (i.e., attitude and behavior). The visual of the project manager who makes quick judgment, bangs on the table, intervenes in daily activities without warrant, and is known for cracking the whip will suffer the consequences of a dysfunctional team. Team members are more likely to waste time, show less interest in project delivery, and deliver sub-par work when managed by this type of personality. However, a project manager who behaves as a member of the team, collaborates on deliverables, and supports individual team members will receive the benefit of higher levels of team productivity.

The role of the project manager is to mobilize a team that is innovative and to translate those innovations into logical processes. Instilling structure and discipline alleviates areas of ambiguity and ensures accountability by all team members. A team that is built on trust will weather the storm more efficiently. Building a team that shares the same value or common vision and is committed to the project success will decrease the amount of intervention required amongst team members during the project life cycle. Control the assignments of team members to ensure that you are taking into account behavioral factors that might impact overall success. Individual temperament and moral fiber should be considered along with skills when working within the team. Ensure that the team is constructed with recognized value of each participant to the project and continually foster a collaborative team spirit——the focus should be on the team achieving the goal, not on individual successes.

Initial communication with team members will ensure that all expectations of the project lifecycle are defined. Key information that should be shared with the team members includes:

- Project objectives
- Visualization of project result
- Impact of project completion to the organization
- Project approaches
- Expectations regarding team culture and environment
- Individual versus team performance

Consider the use of team-building events prior to project implementation, as well as at different points during the project. These events provide the project manager with the opportunity to display the appropriate attitude and gratitude, steer the process, say the right things, share knowledge, and deliver critical messages for success. Recognize and reward individuals, teams, and mechanisms for their success throughout the project. When rewarding an individual, be sure to acknowledge his/her contribution to the success of the team.

Balancing the complaints of "too many meetings" to "not enough communication" can be a challenge, but a successful project manager will be mindful and work to address both. Consider using the following tools to enhance these communication outlets:

- Web chat services
- Circulars
- Team newsletter
- Project newsletter
- Project website
- Webinars

■ Project Scope

The project scope is defined as what the project is supposed to accomplish within the allotted time and money. This statement serves as the agreement between the project team and the customer (department, citizens, students). It specifically relates the objectives for the work from the individual with overall responsibility, such as the Fire Chief or government administrator, and includes the project deliverables, costs, gains, and measurable successes involved in the project.

A *scope statement* that is well-written will assist the project manager in his/her ability to make knowledgeable decisions during the project life cycle. Early research by a project manager helps to avoid obstacles during the project, and should those obstacles arise, the manager will be more adaptable. A change in scope at any point during the project process should have a matching change in the budget (in either time and/or resources). If you are unable to address a change to the budget, a change in the scope should NOT occur. Include the following in your project scope statement:

- **Project justification.** Identifies the problem that is to be resolved or the benefit of the project completion, and will include both qualitative and quantitative measures. It should be clear and precise, and should be the basis by which all future decisions are based.
- **Project product.** Serves as a summary of your project description. It will define potential solutions to the problem, and will include the work that is required to either solve the problem or achieve the overall benefit. In addition, it should include any potential work that might fall outside of the project.
- **Project deliverables.** Summarizes the high-level deliverables that ensure project success.

- **Project objectives.** Includes all quantifiable criteria of the project, such as cost, measurable quality, and schedule. Unless it can be quantifiably defined in a mutually acceptable fashion, avoid customer satisfaction as a project objective.

Beware of creeps in the scope—what may appear as a minor change in the direction of the project and manageable with current resources may actually manifest in great burdens over time. Be sure to record all instances of change, regardless of the size or impact, to ensure that the budget and schedule are appropriately aligned with those changes.

Costs

In most cases, the greatest challenge for the project manager will be managing a budget that has been set by the customer, which is typically the case in fire department operations. In fact, the project manager rarely has any say in the creation of the budget. The first evaluation tool of the success or failure of a project manager is his/her ability to complete the project within budget. Below are a few key tips for delivering a project on budget:

- **Earn trust and respect.** As early as possible in the project, take time to develop the trust and respect of the customer. This is best accomplished by honest and frequent communications.
- **Revise it.** You may not have created the original budget, but you can certainly revise it. Communicate with your customer about insufficient funds related to specific details of the project. Doing this early in the project life cycle is critical.
- **Financial planning.** Detail the expenses and when they are most likely to occur during the project life cycle. Present the plan to the customer and attain their buy-in.
- **Manage the plan.** Continuously mange the financial plan and the anticipated expenditures anticipated. Investigate and evaluate over-budget items and address the issues at hand. Identify, record, and track all project costs. Use expense forms for team members to record all expenses as they occur. Formality and documentation is a must!
- **Customer communication.** It is better to tell your customer that you are trending over-budget during the life cycle than it is to do it a week before project delivery. Communicate often with customers to gain buy-in and potential help. If you need more funding, ask for it!
- **Rule of thumb.** Your job is to deliver a project that is on time, on cost, and of quality. However, do not compromise the quality or time constraints just to complete the project under budget.

Be mindful that every project task has an associated cost. That cost may be labor hours or a tangible good, but there is a cost. In the simplest of terms, a budget consists of estimated costs, contingency funds, and allowances, plus any profit (profits are generally only available for EMS-based generated funds, the Inspections Division, or some form of subscription). The role of the project manager is to keep the actual costs at or below the estimated cost, to minimize any use of the allowance and contingency funds, and to maximize the profit.

Developing a budget can be tricky when estimating costs. While some products and services are relatively consistent in price, variables in cost to high-ticketed items can make or break even the most thoughtful budget. Ensure that the budget includes an allowance for those estimates that are well over the budgeted amount.

Resources, Scheduling, and Time Management

Resources, loosely defined as people, equipment, facilities, and funding, are identified by the project management team during the planning phase of the project. A lack of resources can inhibit the timely completion of a project. Scheduling, availability, and optimization of resources are key to the success of a project management team. The creation of an efficient resource schedule will minimize the project duration and maximize the use of resources that are available.

Within any project, the greatest resource will be your team. An engaged team will be capable of circumventing most obstacles that are thrown at it. Additional resource allocation can be challenging, so it is important to maximize available resources without overindulging in the budget. Project managers must effectively lead the people (employees, vendors) and manage the equipment and materials. Do not discount funding as a viable resource.

Be sure to create a detailed plan for each task. This will include a calculation of man-hours and budgetary needs. A detailed schedule and budget will ensure that you are on task and capable of meeting deadlines.

Note: Tasks that require a great deal of staff hours usually cause the most problems with meeting deadlines and staying within budget.

To create a project schedule:

- Identify the tasks.
- Identify a timeline for each task.
- Identify the resources that will be required to complete the task.
- Identify the order in which tasks should be completed.

Projects often exceed budget when managers do not allocate resources for project start-up, instructional preparation, training of team members, and final project review. Ensure that your overall plan, schedule, and budget include allowances for these items.

Effective time management is crucial to providing a successful project delivery. Task duration, dependencies, and critical path management are all variables of time management that need to be considered by the manager. Some tips to ensure that you are on track include:

- **Create a plan and stick to it.** When your team has clearly defined goals with benchmarks for deliverables, your job as a project manager is easier. When there is clarity for your team on the expectations, you will spend less time dealing with confused team members.
- **Identify and focus on the activities that produce results for the project.** By focusing on those activities that matter to the overall project delivery, your time is not wasted on mundane activity.
- **Avoid team meetings that focus on status updates.** These meetings rarely have value for the overall team and waste time. Consider hosting team meetings that focus on project issues and risks or opportunities for improvement. The team is more valuable in brainstorming creative ideas and solutions than it is in recounting their weekly activity.
- **Avoid micromanagement.** You have selected the team, now let them do their job. By delving into the detail of their work, you are impeding the process. Your job is to steer the vessel; leave the day-to-day work to your capable team.

While time management seems to be the most basic of skills, it is often the most underaddressed. Project managers who have difficulty in managing their time will struggle in the management of their teams. A successful project delivery comes easier to those managers who themselves have mastered the art of time management, as they are more likely to stay within their project budget.

There are a variety of project management software systems available. Some are web-based, some are available for a single user, others are collaborative, and there are those that are integrated. Choose the technology system that suits your project. Ensure that it offers functionality for your specific project needs. Keep in mind that the use of a software system may have extensive learning curves. Those projects that are smaller in nature may actually be hindered by such time-constraining software packages.

■ Risk

Minimizing the risk of project threats allows a project manager to deliver on time, on budget, and within the confines of the scope. In addition, you are providing a more comfortable working environment for yourself and your team. Risk management is best defined as using the techniques of analysis and measurement to identify, classify, and manage risk. While there are numerous ways to address risk management, the following rules of thumb will help you get started:

- **Communicate.** The redundancy of the word "communicate" within the project life cycle is powerful. Often, a team member identifies the threat before the project manager ever observes it coming. When the project manager has an open communication policy, the team member feels comfortable addressing the issue of a threat before a crisis occurs. Make project risk management a fundamental part of your agenda during team meetings; it will keep these issues at bay.
- **Identify the risks during project planning.** This is easier said than done. Predicting the future is not the easiest of tasks, but keeping an open mind during the planning process will help identify traps that inevitably lie ahead. Combine several different types of methodology to help in this process.
- **Embrace it.** Risk is unavoidable in project management, so including risk management as part of the planning process will help avoid the pitfalls. Ignorance is not bliss.
- **Prioritize the risk.** Your main focus of attention on risks should be on those that will cause

the largest loss. Create a criterion for which you will judge risk—for example, the effects the risk could have or the likelihood of the risk occuring.

- **Analyze risk.** Define how the risk may impact project performance as it relates to cost, duration, and customer requirements.

There are plenty of data available on project risk management and how to effectively avert threats. The best rule of thumb, however, is to spend time in the planning phase of the project identifying any potential risks to help avoid setbacks.

Quality and Customer Satisfaction

The discounting of customer satisfaction has been likened to one of the seven deadly sins. No seasoned project manager will testify that evaluating the satisfaction of their customer at the conclusion of the project while neglecting their feedback during the project process is acceptable. Reviewing the project plan with the customer on an ongoing basis is fundamental to ensuring that there are no surprises post-project. Ensuring that the project team is satisfying the requirements of the project and that deliverables are being met is vital to the overall customer satisfaction. A primary tool to engage customer satisfaction is strong communication, customer service, and problem resolution in a timely manner. By fulfilling these simple soft skills, a project delivery team is more apt to fulfill the expectations of the client. No project manager has ever been accused of *overcommunicating*. At the project conclusion, you should already have a good idea if the customer and stakeholders needs have been successfully met.

Some potential tools to use in the ongoing evaluation of your customer's satisfaction include interviewing and/or questionnaires. Creating a measurable form of customer satisfaction will guide you through the project process, and ensure that you are using the most appropriate form of communication with the customer based on his/her personality traits. Ask the customer how and when they would like to receive project updates. Frequent conferencing with the customer and your team regarding project deliverables, resource allocation, fiduciary responsibility, and time management will create a plausible project delivery. Some questions to consider including in your communications are:

- Does the client/customer have suggestions for improving the project progress and efficiency?
- What types of improvement to communications would the client/customer like to see?
- Is the client/customer satisfied with the deliverable dates as related to the final project plan?
- Was the client/customer satisfied with their level of involvement?
- Was the information presented to the client/customer clear and concise? Was it enough information for them to report to their executive managers?
- What is the client/customer's overall level of satisfaction with the project to date?
- What is the client/customer's view on the quality of work to date?
- What suggestions does the client/customer have for the project team?

While feedback from your customer is crucial to the success of the final project, the feedback from the team members is equally important. Ensure that the team is engaged in providing constructive solutions for improving the project delivery, and using the same methods and questions listed above, you can gain valuable insight for improvement based on the team feedback.

As a training officer, the above strategies and tactics can provide you with a basic understanding of project management and the direction to successfully tackle your project.

Key Points

- As a training officer, you should have a strategic plan that aligns the goals of training division with the goals of the overall organization.
- Strategic planning is a cyclical process. The steps of the process are: (1) establishing a vision, (2) creating goals, (3) examining internal factors, (4) examining external factors, (5) formulating a strategy, and (6) reevaluating.
- One of the keys to establishing these or any set of goals is to use the SMART (specific, measurable, attainable, realistic, and timely) goal system.
- A SWOT analysis studies the organization's strengths, weaknesses, opportunities, and threats. The training officer should use these items of concern when determining specific goals and objectives in the strategic plan.
- Consider how different people think about and perceive the value of time, resources, and money

and factor these concerns into your strategic plan. Keeping these concerns in mind, write an executive summary to provide a condensed and specific overview of all research you conducted, your recommendations, and a justification of these recommendations.

- Create an annual training plan to assist you in executing against the overall strategic plan you and your department created. Use one of the many formats and tools available to create this plan.
- As you, the training officer, develop your strategic plan, you may encounter projects that will need to be led and managed. Specific strategies and tactics can provide you with a basic understanding of project management. A successful project manager will ensure that the project life cycle is well defined in regards to initiating, planning, executing, controlling, and closing of the project.

References

Richards, C. *Certain to Win*. Bloomington, IN: Xlibris Corporation; 2004.

U.S. Marine Corps Staff. *Warfighting*. North Charleston, NC: CreateSpace Independent Publishing Platform; 2014, page 3.

Wheelen, T.L. and J.D. Hunger. *Concepts in Strategic Management and Business Policy*. Upper Saddle River, NJ: Pearson Prentice Hall; 2009.

CHAPTER 6

The Preparation Step

Kevin Milan

Introduction

The next few chapters will discuss the process of developing a lesson, course, or curriculum. Regardless of how big or small the objective, the same thought process applies when developing any education program. This process includes the preparation, presentation, application, and evaluation phases. The ability of an instructor to understand each category will come with experience in the classroom and the drill ground. In addition, as training officers we need to be well versed in the science behind how adults, especially emergency responders, learn. We will discuss these, and several other items, over the course of the next four chapters.

Preparation

Training today's fire and emergency responders is a daunting task. The sheer volume of content (fire, emergency medical services [EMS], hazardous materials, extrication specialist, weapons of mass destruction [WMD], etc.) that has to be covered in a finite amount of time presents the primary challenge for training officers. Couple this with the intricacy of emergency responses and the training officer's job is complicated even more. With all of these obstacles, it is imperative that training officers determine the most effective means of training their departments to a competent level in the minimum amount of time. Understanding and applying the preparation step will help us make these decisions. In simple terms, every task a training officer performs leading up to the actual class starting is part of the preparation step. Within this step there are several items that will assist any instructor or training officer with delivering an effective class.

In addition, understanding the learner and the way learning occurs is essential for success. Just as each fire or emergency response is unique, so is each student you encounter. Despite this uniqueness, there are more similarities than differences in learners. Understanding the modes of learning and teaching enhances training. A solid foundation and historical perspective of instruction is provided in this chapter. This perspective provides the foundational repertoire for cross-generational teaching. Additionally, this foundation is essential when exploring today's research-based instructional strategies.

Determining and Prioritizing Training Needs

Determining and prioritizing training needs is the crucial first step in the development of effective and efficient training. Without a clear and concise plan, training can become dangerous and ineffective. A purposeful and articulated training plan is essential to support emergency services and creating proactive training.

As you start your needs analysis and attempt to determine your training needs, there are a few items that you should become intimately familiar with in order to create the best training. These items include the task analysis for a specific job, familiarity with national standards, and a solid grasp of your agency's standard operating procedures (SOPs); these will provide the starting point for program development. Each of these elements could easily fill a chapter of this text. Ultimately, the desired outcome of each training episode must be defined and these outcomes must work in concert with the national standards and department SOPs. It is important that each and every member of the emergency response team work towards these same goals, with the training officer as the common link.

As you begin your needs analysis, you are basically reviewing where your training stands and then where does it need to be. Dugan Laird, in his book *Approaches to Training and Development*, describes this gap analysis as the distance between Must do (M) and Is doing (I) is the Deficiency (D). The formula, M – I = D surmises the gap for training.

The gap analysis methodology applies at all levels. It is a methodology used in private and public sectors to determine training needs. The essential and often overlooked step is the determination of a final or summative assessment. Once you have identified your training gap, compare it to your SOPs and ask yourself, "Are we reaching the department's standard?" If the answer is yes, the next question is, "Are we reaching the national standards?" If your department lacks relevant SOPs, look to the standards. The National Registry of Emergency Medical Technicians (NREMT), the National Fire Protection Association (NFPA), and Occupational Safety and Health Administration (OSHA) are just some of the organizations that provide abundant applicable standards. In addition, look to neighboring departments.

After you have determined the gap, you must determine where the students are in relationship to mastery. This is done through a formative assessment, which could entail a written, oral, and/or skills test, or through observation and communication. Once you determine where the students are and where they need to be, appropriate learning episodes are designed to transition them from their current performance to mastery. For those tasks that are not covered by standards, as for thermal imaging cameras (TICs), the best place to determine mastery is through a task analysis. This methodology breaks the complex task into smaller components. Simple steps such as scanning the room and identifying thermal layers and convection currents are essential for information gathering. From the information, higher-level analyses and evaluation decisions are made. Mastery of each of these elements provides the roadmap for developing proficiency with this tool.

Regardless of where you look to determine a mastery level, you have to consider your audience, which could range from a student first being introduced to the tool versus the seasoned veteran simply refreshing their skills. This is a very important point to remember when knowing your audience. New fire fighters need a basic, step-by-step process or building blocks to help them learn new concepts while also challenging them with more in-depth and complex components. For instructors, the law of association, which states that new information can be easily learned when associated to existing knowledge, is an important point to consider when training veterans.

An important key for the emergency services instructor and training program manager to remember is the tendency to teach or design learning in a preferred learning style. For example, hands-on activities may not be the most effective method of learning for all students, so it is important to push instruction into all learning domains. The primary modes for learning are considered to be auditory, visual, and kinesthetic. Many online assessment tools can assist you in determining your preferred learning and teaching style.

It is often stated that emergency service personnel are primarily hands-on or kinesthetic learners. Research does not support this belief however, offering instead that the learning styles of emergency responders parallel the general population. Another takeaway from educational research is the ability to blend learning styles based on circumstance. No learner is purely visual, auditory, or kinesthetic. Strive to blend your instructional programs so that they tap into all learning domains. It is generally a good idea to provide students with the foundational information through the cognitive domain and then transfer to the psychomotor or kinesthetic. Also, it is always a great idea to tie into the affective domain—touch their hearts, values, and beliefs.

The training that you prioritize is based on several factors, including the frequency and severity chart, influence from the chief, and where the greatest gaps are. As you go through these options and weigh the pros and cons, you are prioritizing your work. More times than not, as you conduct your needs analysis your prioritizing is completed for you simply through the process. Otherwise, if there was no gap determined or you considered your department to be performing at a mastery level, there is no need for us to recreate the training you already have. This would not create the greatest value within our department.

Objectives

Once you have determined the end result and your goal(s) for the class, building the objectives will help you specifically define each task that the student will have to successfully complete. Clearly defining the intent of a training episode is accomplished through these objectives. The academic model presents us with formulas for writing objectives. Objectives are often referred to as ABCD (audience, behavior, condition, and degree) objectives or SMART (specific, measurable, achievable, realistic, and timely) objectives.

- **Audience:** Who will be performing the task?
- **Behavior:** What task will they be performing?
- **Condition:** What situation or circumstance will they be faced with?
- **Degree:** How well will they perform the task?

For example, NFPA 1001, *Standard for Fire Fighter Professional Qualifications*, provides the required elements of donning a self-contained breathing apparatus (SCBA). An example of an ABCD objective is:

> The fire fighter (audience) will correctly don an SCBA (behavior) given a department-issued SCBA and full personal protective equipment (condition) within an acceptable timeframe with 100 percent accuracy per the department check sheet (degree).

By referencing standards and SOPs, objectives are relevant and realistic measures of proficiency.

The same task can be expressed as a SMART objective. The specific guidelines outlined in NFPA 1001 provide the specificity in the objective. Measurement comes in the form of compliance with the check sheet. The task is achievable and realistic as it is referenced to the national standard.

Objectives are written at varying cognitive levels. Low-level objectives ask the learner to identify, list, or match elements, while high-level objectives ask the learner to synthesize information, make judgments, or evaluate alternatives. These levels are rooted in Benjamin Bloom's Taxonomy of Learning.[1] The levels build from the simple to the more complex.

1. **Knowledge.** The ability to recall or recognition of information. This is the basis or foundation for all learning.
2. **Comprehension.** The ability to understand or explain the basic information. The student should also be able to interpret or translate a situation based on their competency level.
3. **Application.** The ability to actively engage in a situation and use information, ideas, or skills to make decisions and solve problems appropriately.
4. **Analysis.** The ability to separate components within the information and show relationships between the components.
5. **Synthesis.** The ability to combine several components or ideas to form a new concept or idea. Innovations are developed through this level of learning.
6. **Evaluation.** The ability to judge the component, idea, or situation and weigh it against a criterion. This is the highest level of learning and requires evidence, facts, and research to make the appropriate judgments.

In conjunction with Bloom's Taxonomy, action verbs are presented at each level of complexity. The lowest level objective, knowledge, ask learners to identify, list, or chose items. The highest level, evaluation, synthesizes material and asks learners to evaluate. It is important for training officers to consider all levels when constructing objectives (**FIGURE 6-1**).

Learning Domains

Once you know your objectives, take a minute to decide what learning domain or area they fit into. This decision will help you tailor your training plan to best achieve your department training goals. The three domains of learning are:

- **Affective (Attitude):** based on feelings, beliefs, and/or values.
- **Cognitive (Knowledge):** based on intellectual skills or mental aptitude.
- **Psychomotor (Skills):** based on the ability to transfer the intellectual skills to physical skills and apply the knowledge through a muscular function or motor skill.

[1]Bloom, B.S. *Taxonomy of Educational Objectives*. Boston, MA: Allyn and Bacon, Pearson Education; 1984.

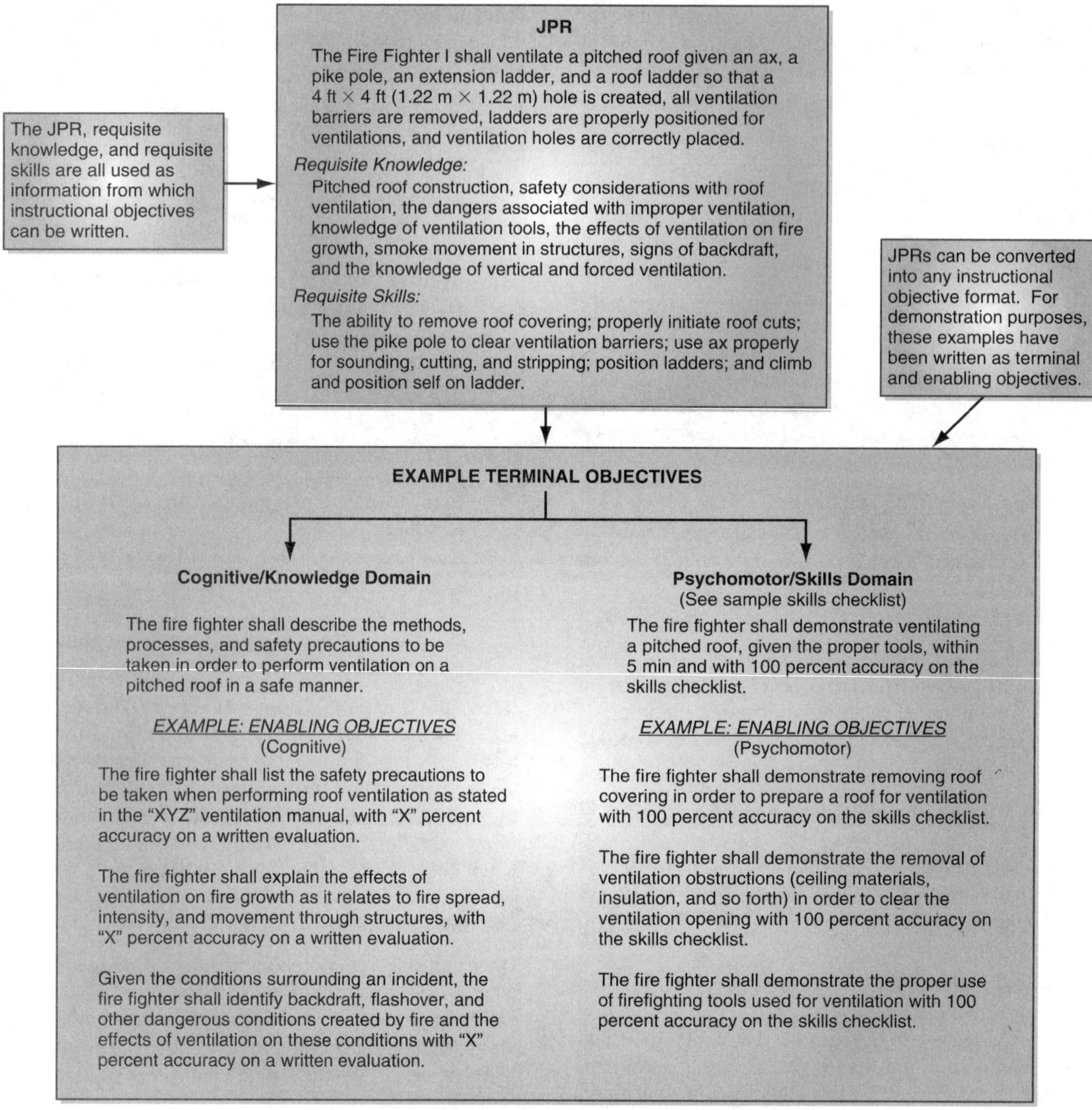

FIGURE 6-1 Illustration of a flowchart demonstrating how to convert a JPR into instructional objectives.

Reproduced with permission from NFPA 1033-2014, Standard for Professional Qualifications for Fire Investigator, © 2013, National Fire Protection Association. This reprinted material is not the complete and official position of the NFPA on the referenced subject, which is represented only by the standard in its entirety.

Fire department training should include all three domains in order for the training to reach its highest potential for success. Although there are no set criteria for how to establish the order of these domains in each course, there is a consensus on how to infuse each of these into the lesson. First we have to reach the fire fighter's heart and beliefs to motivate them for the training they are about to receive. Then, give the fire fighter the knowledge necessary to perform the task while simultaneously keeping them motivated.

Finally, teach them the motor/physical skills based on the knowledge you just relayed to them. The affective domain should overlap each of the other two domains whenever possible. This will allow fire fighters to believe in why they perform a task a particularly way, not just repeat the task or freelance.

A lot of focus in this area is being placed on aggressive firefighting, but smart firefighting combined with that aggression prevents fire fighters from placing themselves in dangerous situations. Also, while everyone knows the dangers of not wearing a seat belt in a rig, not all fire fighters believe in wearing seat belts. Sharing personal stories and showing the devastation that it can cause a family may impact their attitude towards wearing the seat belt. If you are not familiar with the Courage to Be Safe (CTBS) program produced by Everyone Goes Home, I would encourage you to sit in on a class. The CTBS program is built around the affective domain and changing the attitudes of fire fighters concerning self-protection.

Cognitive is the most commonly referenced domain, which is based around knowledge being the foundational level. The higher you go, the more critical thinking skills are involved until you reach the top, where an individual is able to review a situation and make a successful decision (**FIGURE 6-2**). Affective is the most difficult domain to reach or test. The affective domain is based on your heart, values, and beliefs (**FIGURE 6-3**). Psychomotor is the process that will be learned during the hands on drill (**FIGURE 6-4**).

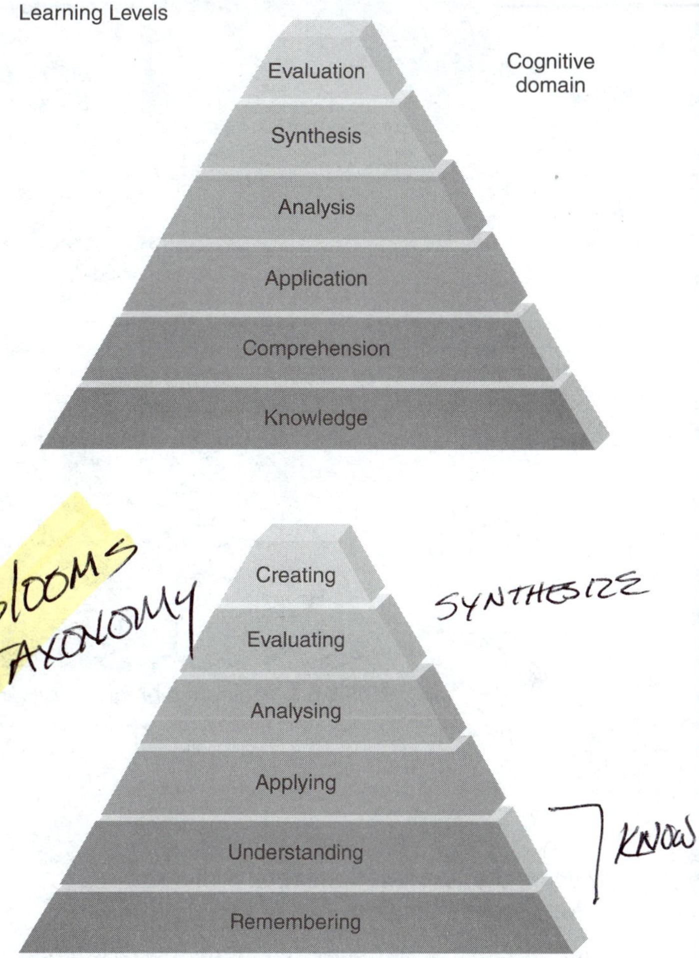

FIGURE 6-2 Illustration of cognitive learning.

Modified from: Anderson, Lorin W.; Krathwohl, David R.; Airasian, Peter W.; Cruikshank, Kathleen A.; Mayer, Richard E.; Pintrich, Paul R.; Raths, James; Wittrock, Merlin C., A Taxonomy for Learning, Teaching, and Assessing: A revision of Bloom's Taxonomy of Educational Objectives, complete edition, 1st,©2001. Printed and Electronically reproduced by permission of Pearson Education, Inc., Upper Saddle River, New Jersey.

Lesson Planning

A training objective will tell you and your class where the lesson is going, but a lesson plan is your map or blueprint, telling you how to get there. A good lesson plan is a reference tool for conducting training. Whether you prepare one yourself, borrow it from your departmental training manuals and instructor's guides, or get it from another fire service organization, a good lesson plan will identify the essentials of the lesson and how to instruct the class (**FIGURE 6-5**). You should be able to find the following elements in a good lesson plan:

- Job title or topic
- Level of instruction/length of instruction
- Instructional materials necessary
- Contingency plan
- References
- Audience/prerequisites
- Preparation or motivation step
- Objectives
- Overview step
- Presentation step
- Application step (safety plan should be built in)
- Summary step
- Assignments
- Testing or evaluation step

A sample lesson plan has been included in the appendix.

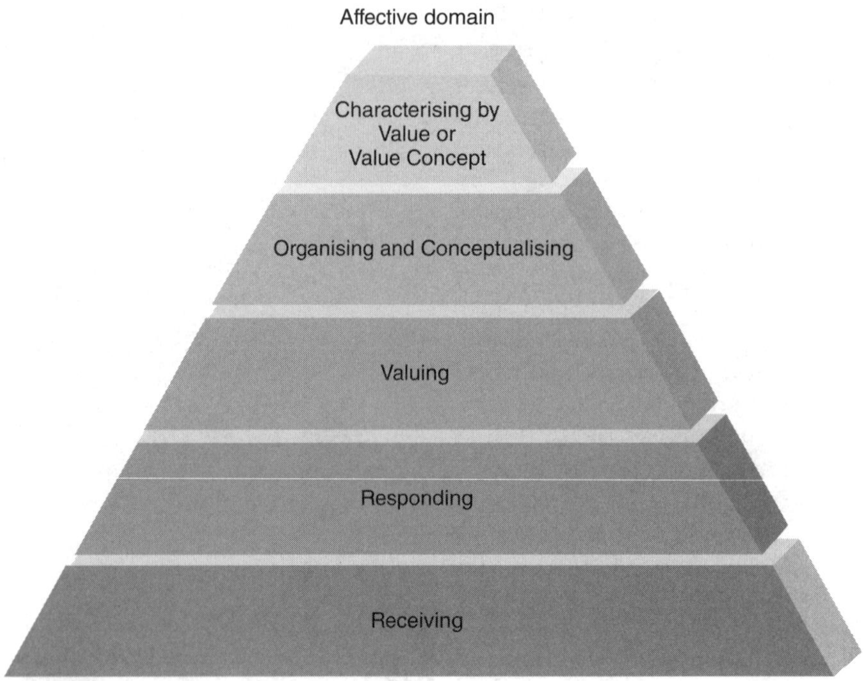

FIGURE 6-3 Illustration of affective learning.

Modified from: Kratwohl D R, Bloom B S and Masia B B (1984) Taxonomy of educational objectives: the classification of educational goals. Logman Group.

Using the Incident Command System for Lesson Planning

There are many templates available for lesson planning; in this chapter, we will utilize the Incident Command System (ICS). The development of an incident action plan takes a formal planning process; this planning process may be applied to training as well. We can begin the process by identifying objectives on an ICS 202 form. Just as with a major incident, there are five things that need to be addressed in a training action plan (TAP) for a major training event. They are:

1. What is to be accomplished?
2. Who is involved?
3. What are the assignments?
4. How are the participants going to communicate?
5. What happens in the event of an injury?

Just as with ICS, the TAP is modular and only the elements needed in the plan are included. For example, a full TAP is prudent for large multicompany drills, whereas individual skills such as ropes and knots can be covered with objectives and a simple lesson plan. Whether or not you use ICS forms for your training, they provide guidance on the essential elements of any lesson plan. The key elements described in a division assignment can articulate methods of instruction. These steps must be delineated in a lesson plan, in any format.

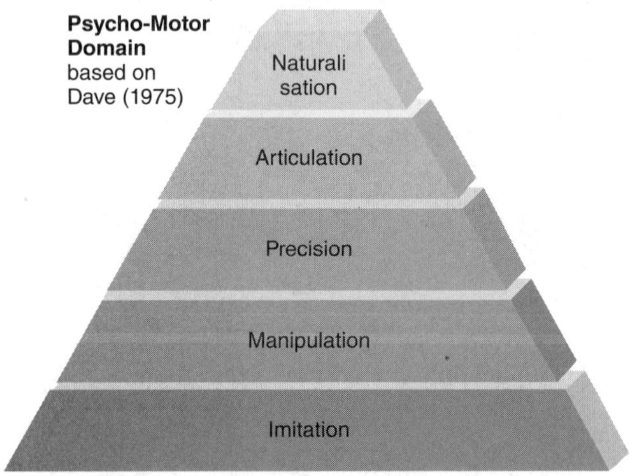

FIGURE 6-4 Illustration of psychomotor learning.

Modified from: Dave R H (1975) in Developing and Writing Behavioural Objectives (R J Armstrong, ed.) Educational Innovators Press.

Instructor Guide
Lesson Plan

Lesson Title: Use of Fire Extinguishers ← **Lesson Title**
Level of Instruction: Firefighter I ← **Level of Instruction**

Method of Instruction: Demonstration
Learning Objective: The student shall demonstrate the ability to extinguish a Class A fire with a stored-pressure water-type fire extinguisher. (NFPA 1001, 5.3.16) ← **Learning Objective**
References: Fundamentals of Firefighter Skills, 2nd Edition, Chapter 7 ← **References**

Time: 50 Minutes

Materials Needed: Portable water extinguishers, Class A combustible burn materials, Skills checklist, suitable area for hands-on demonstration, assigned PPE for skill ← **Instructional Materials Needed**
Slides: 73–78*

Step #1 Lesson Preparation:
- Fire extinguishers are first line of defense on incipient fires
- Civilians use for containment until FD arrives
- Must match extinguisher class with fire class
- FD personnel can use in certain situations, may limit water damage
- Review of fire behavior and fuel classifications
- Discuss types of extinguishers on apparatus

  Demonstrate methods for operation

Step #2 Presentation

A. Fire extinguishers should be simple to operate.
1. An individual with only basic training should be able to use most fire extinguishers safely and effectively.
2. Every portable extinguisher should be labeled with printed operating instructions.
3. There are six basic steps in extinguishing a fire with a portable fire extinguisher. They are:
   a. Locate the fire extinguisher.
   b. Select the proper classification of extinguisher. ← **Lesson Outline**
   c. Transport the extinguisher to the location of the fire.
   d. Activate the extinguisher to release the extinguishing agent.
   e. Apply the extinguishing agent to the fire for maximum effect.
   f. Ensure your personal safety by having an exit route.
4. Although these steps are not complicated, practice and training are essential for effective fire suppression.
5. Tests have shown that the effective use of Class B portable fire extinguishers depends heavily on user training and expertise.
   a. A trained expert can extinguish a fire up to twice as large as a non-expert can, using the same extinguisher.
6. As a fire fighter, you should be able to operate any fire extinguisher that you might be required to use, whether it is carried on your fire apparatus, hanging on the wall of your firehouse, or placed in some other location.

B. Knowing the exact locations of extinguishers can save valuable time in an emergency.
1. Fire fighters should know what types of fire extinguishers are carried on department apparatus and where each type of extinguisher is located.
2. You should also know where fire extinguishers are located in and around the fire station and other work places.
3. You should have at least one fire extinguisher in your home and another in your personal vehicle and you should know exactly where they are located.

Step # 3 Application
Slides 7–10

Ask students to locate closest extinguisher to training area

FIGURE 6-5 Illustration of a lesson plan.

Step # 3 Application cont.

Review rating systems handout—Have students complete work book activity page #389

C. It is important to be able to select the proper extinguisher.
  1. This requires an understanding of the classification and rating system for fire extinguishers.
  2. Knowing the different types of agents, how they work, the ratings of the fire extinguishers carried on your fire apparatus, and which extinguisher is appropriate for a particular fire situation is also important.
  3. Fire fighters should be able to assess a fire quickly, determine if the fire can be controlled by an extinguisher, and identify the appropriate extinguisher.

What happens if wrong type or size extinguisher is used?

     a. Using an extinguisher with an insufficient rating may not completely extinguish the fire, which can place the operator in danger of being burned or otherwise injured.
     b. If the fire is too large for the extinguisher, you will have to consider other options such as obtaining additional extinguishers or making sure that a charged hose line is ready to provide back-up.
  4. Fire fighters should also be able to determine the most appropriate type of fire extinguisher to place in a given area, based on the types of fires that could occur and the hazards that are present.
     a. In some cases, one type of extinguisher might be preferred over another.

D. The best method of transporting a hand-held portable fire extinguisher depends on the size, weight, and design of the extinguisher.
  1. Hand-held portable fire extinguishers can weigh as little as 1 lb to as much as 50 lb.
  2. Extinguishers with a fixed nozzle should be carried in the favored or stronger hand.

Display available types of extinguishers

     a. This enables the operator to depress the trigger and direct the discharge easily.
  3. Extinguishers that have a hose between the trigger and the nozzle should be carried in the weaker or less-favored hand so that the favored hand can grip and aim the nozzle.
  4. Heavier extinguishers may have to be carried as close as possible to the fire and placed upright on the ground.
     a. The operator can depress the trigger with one hand, while holding the nozzle and directing the stream with the other hand.
  5. Transporting a fire extinguisher will be practiced in Skill Drill 7-1.

E. Activating a fire extinguisher to apply the extinguishing agent is a single operation in four steps.
  1. The P-A-S-S acronym is a helpful way to remember these steps:
     a. Pull the safety pin.
     b. Aim the nozzle at the base of the flames.

Have students demonstrate steps using empty extinguisher

     c. Squeeze the trigger to discharge the agent.
     d. Sweep the nozzle across the base of the flames.
  2. Most fire extinguishers have very simple operation systems.
  3. Practice discharging different types of extinguishers in training situations to build confidence in your ability to use them properly and effectively.
  4. When using a fire extinguisher, always approach the fire with an exit behind you.

Complete skills sheet #7–9 for each student

     a. If the fire suddenly expands or the extinguisher fails to control it, you must have a planned escape route.
     b. Never let the fire get between you and a safe exit. After suppressing a fire, do not turn your back on it.
  5. Always watch and be prepared for a rekindle until the fire has been fully overhauled.
  6. As a fire fighter, you should wear your personal protective clothing and use appropriate personal protective equipment (PPE).
  7. If you must enter an enclosed area where an extinguisher has been discharged, wear full PPE and use SCBA.

Review PPE required for extinguisher use

     a. The atmosphere within the enclosed area will probably contain a mixture of combustion products and extinguishing agents.

FIGURE 6-5 *(Continued)*

F. The oxygen content within the space may be dangerously depleted.

Step #4 Evaluation:

1. Each student will properly extinguish a Class A combustible fire using a stored-pressure type water extinguisher. (Skill Sheet x-1)
2. Each student will return extinguisher to service. (Skill Sheet x-2)

Lesson Summary: ← **Lesson Summary**

- Classifications of fire extinguishers
- Ratings of fire extinguishers
- Types of extinguishers and agents
- Operation of each type of fire extinguishers
- Demonstration of Class A fire extinguishment using a stored pressure water extinguisher

Assignments: ← **Assignment(s)**

1. Read Chapter 8 prior to next class.
2. Complete "You are the Firefighter" activity for Chapter 7 and be prepared to discuss your answers.

Step # 3 Application cont.

Discuss hazards of extinguishing agents

FIGURE 6-5 *(Continued)*

We are doing a disservice to our fire fighters by using two separate systems for response (incident action plans, or IAPs) and training (lesson plans built on objective-based learning). ICS forms, action plans, and incident command provide the plans and details needed for both response and training. However, the most important item is to have a plan in place and documented. As the old EMS saying goes, "If it was not documented, it did not happen."

Once you choose a route of documentation and planning, start to think about what your goals and objectives will be for the course. Simply stated, the goals of training are not tasks, but rather a philosophy or idea of what we should be at the end of the training. Any training that does not directly support the goals of the emergency response organization is secondary to your department's mission. Another key aspect is to relate your goals and training to incident responses when preparing training materials so that the information and skills will remain with the learner after the class.

Determining Location

As you are preparing your lesson plan, review the location of where this training will take place. The location of training may alter your plans when considering the use of classroom space versus the training ground. Factors to consider when selecting the location and most suitable environment for your classroom include:

- Ability to see, hear/listen, and participate
- Preparation time and demobilization time
- Available resources
- Class size
- Nature of training event (lecture or skills)
- Hazards
- Weather or outside conditions/distractions

The training office must be aware of the pros and cons of conducting a class inside versus outside. Both options have advantages and disadvantages, but both are extremely valuable if prepared properly. Ensure that you reserve ample time for preparation and trial your class in the selected environment prior to conducting the first live session. It is also important to have a contingency plan in place in case unforeseen situations arise with your first plan. The most significant failures that can occur are a student being injured due to a substandard teaching environment or a student becoming disengaged because he or she cannot see, hear, or actively participate.

Preparing Equipment and Materials

One of the most pertinent things an instructor can do to ensure success of a presentation or class is to prepare the necessary equipment and materials. In order to prepare, these items need to be collected, tested, built in some cases, and staged prior to the class starting. It is typical to prepare and stage materials days, if not weeks, in advance.

As all instructors eventually learn, the more materials available to a student, the more apt they are to enjoy the class, have a chance to participate, and allow learning to take place. An instructor's lessons plan should notate the use of each item, the subtopic, length of lesson, learning domain used, time allotted for student participation, and purpose. In addition, if there are any special instructions for instructors, this should be included as well.

The level of attention an instructor places on the preparation of his or her class is a direct reflection of the success of the final result: Did the transfer of knowledge and skills take place? On a final note, always have a contingency plan in case something fails and make sure your instructors are aware of the options.

Preparing for the Adult Learner

The next challenge in building a successful education program is to get the students engaged in learning, which shifts the discussion to andragogy. Andragogy is the art and science of helping adults learn. Malcolm Knowles, who is considered the father of andragogy, developed five assumptions about adult learners. These assumptions should be considered with each class that you teach and implemented into the class. The five assumptions are:

1. They have an independent self-concept and are capable of directing their own learning.
2. They have developed several life experiences which will help learning.
3. Their learning needs are closely related to changing social roles.
4. They are problem-centered and interested in immediate application of knowledge.
5. Internal factors motivate them more so than external factors.

There are several variables that may cause you to alter the lesson plans slightly from class to class when teaching adults. Understanding how adults learn will enhance the effectiveness of your class based on how the individuals in each class learn. In order to do this, take a good look at the people in your fire department and perform an audience analysis. A few items to consider when preparing for a class include:

FIGURE 6-6 Image of advanced fire fighter techniques being taught during the semi-annual Georgia Smoke Divers Program.

Courtesy of Brian Ward/Georgia Smoke Divers Program.

- Life and work experiences
- Current knowledge level
- Current skill level
- Culture
- Language barriers
- Educational background
- Internal motivators
- Past positive/negative training experiences

Decide what they need to learn and where they need to review. Determining these items will assist you in keeping the fire fighters interested, involved, and excited about training, because you made it personal to them (**FIGURE 6-6**). However, there is still more that can be done to keep the adult learner engaged. Selecting the correct delivery method is vital to facilitating the learning process.

Adult Learning Theories

Group Learning

Group learning is the concept of breaking a class into small groups of 3 to 4 individuals and allowing them to select a real world problem or project to complete. As the students work on the problem or project, they are integrating the needed materials simultaneously. If you refer back to Knowles's assumptions, you will note that the group learning concept allows students to use their experiences, the class is now self-directed, and it is problem-centered. This concept generally works best with students who have at least some working knowledge of the subject. If nothing else, they at least have the knowledge to research the topics they are actively working on.

In order for this concept to work, the facilitator must be available to help guide the students and answer any questions. The facilitator does not teach the materials; instead, they mentor and coach, a very nontraditional, hands-off approach in the fire service. It is important to analyze your audience prior to the class and attempt to place students together in order to balance out the students' knowledge and skill sets. This will help ensure that group dynamics are even and it teaches teamwork. There is also the benefit of students learning from each other's experiences. This type of learning is also suited to multiple domains of learning. One of the downfalls to this concept is that some students do not engage well with groups. In addition, some students will struggle with completing the task, yet still learn the materials.

As an example, imagine a group of fire fighters coming together to discuss modern day building construction versus 1970s construction. As a veteran group of fire fighters, they should be able to use their experiences, knowledge, and skills to research the problem, which is closely related to their social roles as a fire fighter. There is even the internal motivator of this being a life or death situation if the materials are not learned. The only job of a facilitator at this point is to help format and provide resources.

■ Experienced-Based Learning

Experienced-based learning is where the students learn by performing. Instead of the teacher standing in the classroom with a projector and PowerPoint, the class is taught while the students are preforming the task. This concept allows the student to see the actual tool, process, or behavior while simultaneously receiving the knowledge needed to perform the task. The students must also be allowed to use their psychomotor skills and manipulate the item or behavior. Any instructor can tell a student that something is hot, but actually feeling the heat reinforces this information. This is why realism, with the appropriate amount of stress added, in training drills is extremely valuable for adult learning.

Consider, for example, SCBA training. Have all of the students take a knee in the bay and watch you teach about the components and limitations of an SCBA. As you present the information, you should physically touch and manipulate that component. The next step is to give every student an SCBA and have them complete the process of identifying and manipulating each of components. This allows the student to use multiple senses, which increases the knowledge retention rate.

For the facilitator, it is extremely important to plan appropriately for a location with limited distractions and to have adequate resources. The facilitator must also be extremely competent with the information being presented. They should not only be able to identify the equipment or behavior, but to understand what makes the parts work or why certain actions are performed. As this type of learning is more hands-on, safety must be a primary concern, especially if introducing new topics to inexperienced fire fighters. Another benefit of this type of learning is the reduced amount of time it takes over the traditional delivery styles, which makes this the preferred learning style in most volunteer departments. However, keep in mind that not all topics can be delivered this way; for example, you would not conduct live fire training on the first day of recruit school.

■ Self-Directed Learning

Although this is not a teaching or facilitating style, it is considered by many theorists to be a main component for proactive and motivated learners. You could argue that motivated individuals are not the students you need to worry about. However, these students need just as much mentoring as any fire fighter. The primary reason is to keep them interested; the secondary reason is so that they are challenged.

As a training officer, these types of individuals may not need to attend every class, but they need to have accessible resources in order to explore various aspects of the fire service. The Internet is a valuable tool, but there are many lessons that cannot be obtained online, such as hands-on training. Even for the information that can be found on the Internet, it needs to be reviewed for accuracy. This is why it is important to invest in libraries, video collections, conferences, and publications. It provides a bank of information for anyone to use, and the self-directed learner is the one who needs it.

Challenging the self-directed learner is a valuable concept and one that can pay large dividends for all. Once you find a true self-directed learner, offer them a problem to solve and watch what happens. This individual will succeed eventually and produce a good product. This will help the learner grow and they will feel a sense of accomplishment and pride for taking on the additional responsibility.

Learning Process

Learning is defined as an observable change in behavior that occurs as a result of acquiring new information or skills and putting it to use through practice. The knowledge and skills change of behavior is generally the first and easiest to recognize and measure. However, the most difficult and most critical is the change in the attitude or affective domain. As a rule of thumb, this can be witnessed best when an action is being performed correctly without the pressure or instruction of a ranking individual or "boss."

Learning comes in various forms, from the introductory level or the most advanced and experienced level. For adults, experience plays a major role in allowing the individual to connect or associate personal life experiences with those on the fireground. For example, many individuals have used chain saws, power tools, and hand tools; however, most have not used these tools in the environment of a fire fighter. In essence, this allows the adult to adapt the knowledge and skills they already possess to a new environment versus learning an entirely new skill. This is why our training needs to be as realistic as possible, while always maintaining safety. In the chapter, *The Application Step*, Dr. Richard Gasaway explains how a fire officer thinks and how to construct training based on this thinking.

However important experience is, it cannot be the sole means of learning. We can sometimes ask for assignments at the busy stations or take part in specialty companies; however, the vast majority of experience comes with time. In essence, we cannot rush experience, but training and education can be received now, while we are gaining experience. For this kind of learning, practice is essential. Review the learning cone for an explanation of how we learn (**FIGURE 6-7**).

Creating Motivation

It is a known fact that little or no learning occurs unless people have a desire or need to learn. Motivation is the most important aspect an instructor can

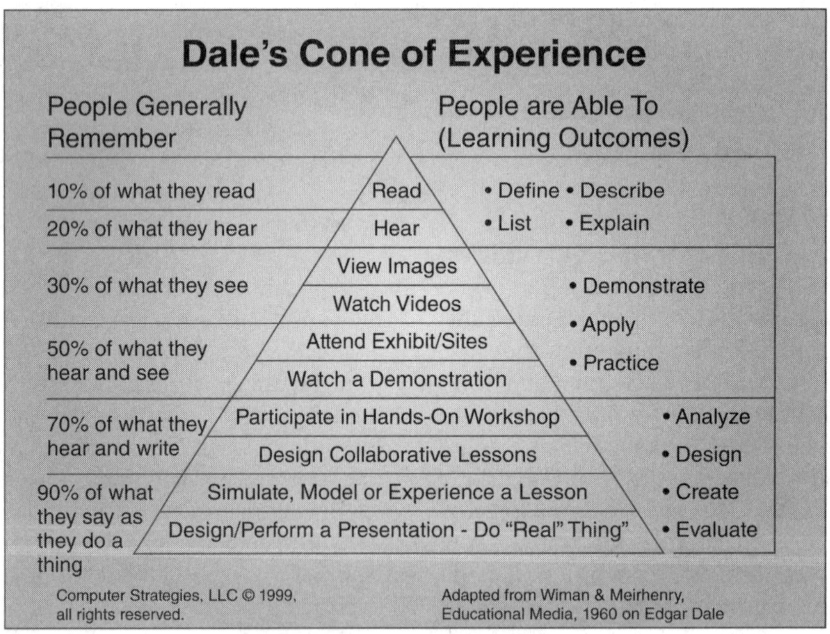

FIGURE 6-7 Illustration of Dale's Cone of Experience.

Modified from: Edgar Dale's Cone of Experience, Audio-Visual Methods in Teaching, 3rd ed. (New York Holt, Rinehart, and Winston, 1969). Reprinted with permission from Raymond S. Pastore Bloomsburg University, Bloomsburg, PA 17815-1301.

bring to the training session; if the instructor is not motivated to share knowledge, the students will not be motivated to learn. In many cases, having a meal before, after, or during training is a motivator, as does working with people the students like. Others simply enjoy learning new tools and techniques, so you may be challenged to discover new ways to deliver traditional material.

As motivation is key, you need to challenge yourself to develop training and instill the desire to improve in your students. Demonstrate to them the level of competency and effectiveness they should reach, challenge them, show them the difference they can make, the citizens they could save, and what it takes to save one of their own. Refer back to the affective domain of learning. While this domain is the most difficult to grade or measure, if you as the training officer can touch the fire fighters' hearts, values, and beliefs, they will be motivated to learn.

As a training officer, being motivated yourself goes a long ways toward changing a behavior or helping facilitate learning. Motivation is contagious and the students can read your body language. Motivation starts with you! One key to making every training event successful is to make it personal; if it is personal, you will care, and because you care about sharing information, the students will start to care.

■ Learning Characteristics

Several characteristics should be considered during the preparation stage for adults. Think of ways in which these adult learning characteristics apply to your crew, fire fighters, citizens, and department.

Problem-Centered

Adult learning should be centered on problem solving. Adults by nature tend to accept the responsibility and challenge of solving problems when they are equipped with the necessary information and skills. Generally, adults possess an internal drive to feel the sense of accomplishment upon solving a problem. In the fire service, everything we face is a problem for someone; it is our job to decipher the message, figure out what is happening, and make the necessary decisions to mitigate the situation. We should train this way as well.

Experienced-Based

The more time we can spend as instructors bringing the experiences of the students into the equation, the faster learning takes place. We are no longer learning new knowledge or skills, but relating new knowledge and skills to the old. Everyone brings their own personal and professional experiences with them; it is up to the instructor to discover and tap into these resources. An excellent example of this on the fireground is to conduct postincident critiques and review what went right as well as what could be improved. These experiences can be both positive and negative. If a negative situation is used, be respectful and explain how we could have possibly changed the outcome. The National Firefighter Near-Miss Reporting System is based on this concept. We encounter a situation that turned out differently than was wished for, so we in turn use this situation as a learning experience to reinforce safety and correct behaviors. National Institute of Occupational Safety and Health (NIOSH) reports and case studies from around the country can be used to accomplish this. Remember to stay positive and focus on the learning points of how it could occur in your department.

Participation

Occasionally, the best way to learn is the hard way. As we have discussed, all individuals learn best through actively participating in the training. When we are placed in the situation and forced to make our own decisions and physically perform the work, we tend to create muscle memory and retain the knowledge and/or skill. The instructor's role here is to coach and guide the student without giving them the answers. Imagine if instructors only told students that a fire was hot and stressful. How would the students feel on their first real fire?

As instructors and training officers, we have to learn to facilitate lessons instead of simply lecturing. This is more difficult to accomplish when introducing a new topic, but it can be fairly simple when building upon a foundation of requisite knowledge and skills. The key as a facilitator is to introduce the information, keep the lesson on track, and answer any questions. This method allows the student to think through the information, offer suggestions and input, and ask questions based on their experiences and training.

Provide Feedback

Feedback during and after instruction is crucial to the progression of learning for every fire fighter.

The key is to practice the skill until perfection is met, and then keep practicing to maintain the skill. The longer an instructor allows the student to perform the skill in the wrong manner, the less likely the student will acquire the skill in a reasonable amount of time. However, after a student has been shown and guided through demonstrating the proper technique, there comes the time where the student will need to make their own decisions and work through the problem. It is unfair to allow a student to perform a skill incorrectly and allow them to carry it to the emergency scene. Negligence on the instructor's part can cause increased workloads, lack of efficiency, and safety concerns for the students.

Know Your Audience

Background, professional and personal experiences, education, training, and culture are all different for each student you instruct; the key is for the training officer to determine the proper speed that allows all students to succeed. One of the greatest challenges in fire service instruction is to identify each learner's pace, facilitate the instruction to meet each learner, and remain cognizant of this throughout your training. There are several techniques that can assist you in keeping the students' attention, regardless of the speed that they learn.

From time to time, you will have exceptional students who either have a prior background in the topic or catch on very quickly. Use your resources and allow these types of students to assist you to helping other students. This will allow the fast learner to enhance their skills and stay active, at the same time using a one-on-one technique for the student who may have difficulty with the task.

Connect the student's background to your presentation. Use the student's prior knowledge and skill set, even if it was in a different occupation, to explain new information. This will help the student relate (law of association) the new information to old information. In this way, you will be building upon a foundation instead of creating a new slate.

Another method is to tailor your level of information to coincide with the experience level of the class. The methods and techniques used in recruit school cannot be the same as those used for an experienced officers' class. The instructor needs to gauge the level of facilitation and challenge the experienced officers through questioning techniques learned in basic instructor school. As discussed above, adults are very much problem-centered, so create problems for them to solve based on their experience levels. This will keep your students engaged and the learning process active.

Start With an End in Mind

A training officer or chief officer within the department generally determines training schedules and topics. In addition, training can be based on NFPA, ISO, state/province, or other criteria. Regardless of the reasoning behind your topic selection, ensure that you, as the training officer, understand the end goals. What is the purpose of this training and what will the students be capable of performing after successful completion? These questions and others will help you determine the necessary steps in building your course structure. Start every class with the end in mind. State the objectives and goals of each class upfront. This allows the students to focus on what they are expected to achieve. This is also a motivating factor, as students want to know what they will be able to accomplish once they receive the training.

Whether it is increased skills, improved confidence, higher level of understanding, increased safety, or a promotional requirement, all of these can be associated to Maslow's Hierarchy of Needs (**FIGURE 6-8**). These are the items that motivate us as humans. As you review the diagram, you will see that our basic

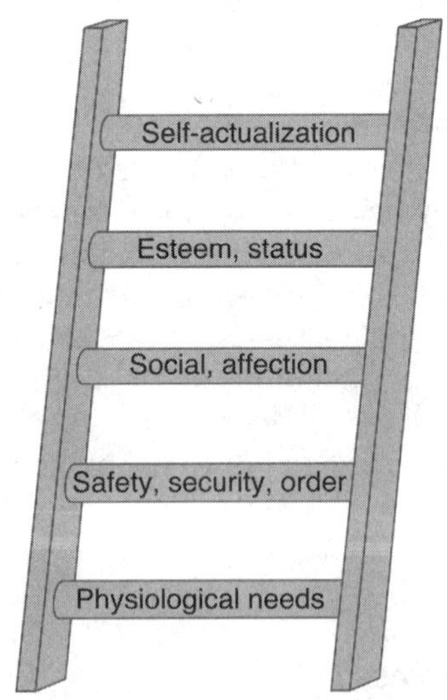

FIGURE 6-8 Illustration of Maslow's Hierarchy of Needs.

survival needs transition to self-fulfillment as you travel up the diagram.

Final Preparation

Lastly and perhaps most importantly, are you ready to teach? In the above sections we have discussed all the things you need to make sure are ready, but what does it take to make you ready?

You should have the objective and the lesson plan, but have you thoroughly reviewed it? Even for a course you may have taught several times, conducting a thorough review of the material is important. While conducting such a review, you may also want to determine what, if any, parts of the course may need to be updated. Effective lesson plans are dynamic and change based on the needs of the department, best practices, and new standards. They should also be reviewed based on the individuals in the class.

Are you mentally ready to teach the course, and how is your attitude? If you are having a bad time, are not ready to teach the course, or do not like the course, that will show in your delivery. You may need to consider some alternative delivery methods, or to give the program off to someone else to teach.

On the day of the delivery, have you prepared yourself? Did you get a good night's sleep, or are you trying to teach a class while fatigued? What does that do to your delivery? What about your appearance? Admittedly, rarely will students be concerned over appearance while conducting live fire training, but in a traditional classroom situation most people will likely not respect someone who looks like they just woke up after sleeping in their clothes.

Your appearance also has a strong mental impact on your students. If it is very hot outdoors and all your students are wearing a full set of turnouts, it is not appropriate for the course instructor to be wearing shorts and a T-shirt. If it is dangerous enough that they need the equipment for their protection, you do too. There is no quicker way for an instructor to lose credibility than if they are not willing to do what they require of their students, not willing to wear what instructor requires their students to wear, and act in a manner in which they require their students to act.

Not only must you have all the materials ready, you must be ready to be successful.

Key Points

- Through analyzing emergency response tasks, the essential elements can be distilled. Identifying what constitutes mastery is essential prior to designing the learning episode. Once this is defined, the planning of the preparation step can begin.
- By identifying the gap between current performance and mastery, the specifics of the training program are identified. This gap analysis is the basis of the preparation step. The training is identified through development of a lesson plan that includes goals and objectives.
- Goals are the statements that describe the intended outcome.
- Objectives, which include ABCD and SMART objectives, identify the requisite steps.
- Developing and creating a solid plan based on a gap analysis is important. A systematic and professional training program, built on solid preparation, leads to success on the emergency scene. Drills and practice are essential; however, they are only a part of the process. Preparation builds the foundation for all training success.
- Study how adults learn and how you can alter the lesson plan to fit the needs of that particular class.
- The preparation step is only the first step in successful training, but an extremely important one. If you do your ground work well at this step, training will flow more smoothly in the following steps of instruction.

References

Anderson, L.W., et al. (Eds.) *A Taxonomy for Learning, Teaching, and Assessing: A Revision of Bloom's Taxonomy of Educational Objectives*. Boston, MA: Allyn & Bacon, Pearson Education Group; 2001.

Bloom, B.S. *Taxonomy of Educational Objectives*. Boston, MA: Allyn and Bacon, Pearson Education; 1984.

Dale, E. *Dale's Cone of Experience*. Weman & Marbenry Educational Media; 1960.

Dave, R.H. *Developing and Writing Behavioral Objectives*. (R.J. Armstrong, Ed.). Tucson, AZ: Educational Innovators Press; 1975.

Everyone Goes Home. http://www.everyonegoeshome.com. Accessed July 22, 2014.

Knowles, M.S., et al. *Andragogy in Action: Applying Modern Principles of Adult Education*. San Francisco, CA: Jossey Bass; 1984.

Krathwohl, D.R., B.S. Bloom, and B.B. Masia. *Taxonomy of Educational Objectives: Handbook II: Affective Domain*. New York, NY: David McKay Co.; 1964.

Laird, D. *Approaches to Training and Development*. New York, NY: Perseus Book Group; 2003.

Maslow, A. *Motivation and Personality, Second Edition*. New York, NY: Harper & Row; 1970.

National Fire Protection Association (NFPA) 1001, *Standard for Fire Fighter Professional Qualifications*, 2013. http://www.nfpa.org/codes-and-standards/document-information-pages?mode=code&code=1001. Accessed June 10, 2014.

Stewart, T. Mystified by training? Here are some clues. *Fortune* 2001:143;184.

Thorndike, E. *The Fundamentals of Learning*. New York, NY: Teachers College Press; 1932.

CHAPTER

7 The Presentation Step

Alan Brody

Introduction

In the presentation step of your program, there are several methods you can use to get the information out to your students. The importance of a well-designed lesson plan cannot be overstated at this point. A lesson plan will be required to stay on track and to organize all necessary materials for the presentation. A thorough lesson plan will go a long way in the successful delivery of your lesson.

Methods of Instruction

■ Lecture

The lecture can be the most nerve-racking experience an instructor has. Standing before a class of your peers or strangers can be challenging. As an instructor, you have surely witnessed all types of lecturing styles. Some lectures may have been brutally boring, while others were captivating. As an instructor, it is good to find the lecture style you like best. Sometimes it can be very difficult to take a subject that is very mundane and make it exciting. However, it is your job to find a way to connect to the students and grab their attention.

The lecture method is well-suited to delivering the objectives of the subject matter and an overview of the topic. The key to a successful lecture is to make the subject interesting and present it in such as way so that it is easy for the students to understand. Another advantage to the lecture method is that it can be prepared shortly before class, as it does not require any training aids. This method is effective for both small groups or a large crowd, and it is very cost effective. There are some disadvantages to the lecture approach. Depending on the size of the room, the number of people, and whether sufficient audio/visual equipment is available, it can be difficult for the students to see and hear you if the classroom is not set up appropriately for a lecture. It is also limited in terms of the number of senses that it can reach, so it may not be the most effective learning method. The subject matter being taught may benefit from another approach if you wanted the students to smell, hear, or taste an experience as part of the lesson.

During the lecture process, students expect the instructor to have certain traits and skills. To be an effective instructor, you should:

- **Be a content specialist.** You need to study the subject inside and out, understand it, and be comfortable with your knowledge of it. During the first few minutes of your lecture, the students will determine if you are (1) a competent speaker, and (2) have a thorough grasp of the subject. For you to keep the class engaged, a thorough understanding of the subject is imperative.
- **Be respectful.** Adults have busy lives. Their time is typically limited, and this is true of all members of the fire service, regardless of whether they work for a career or volunteer department. It is

imperative that you, as the instructor, respect both your students and the time they are putting into the class. This especially is true when answering questions. You will be able to determine whether learning has occurred if you plan a well-organized, logically structured presentation that will involve the students with open-ended questions.

- **Be enthusiastic about the subject.** The class will respond in a positive manner if the instructor is able to capture and hold their attention. Again, this can only happen if you have a firm grasp of the subject. Using examples that are relevant to the fire service industry will capture the attention of your students. The best lectures have the students interested in the subject—even dry subjects can be interesting if presented in the right way and with enthusiasm.
- **Use humor.** The use of humor in the classroom should not be overlooked. Effective instructors will use a humorous ice-breaker at the beginning of the class or after the return from breaks. Humor used throughout the lecture will help create a sense of levity with the subject, and it will make the session more fun for the students. Students enjoy lectures when they are made fun and enjoyable. However, be careful not to offend anyone with your humor. The old adage applies here: "If you wouldn't say it to your mother, you should not use it in class."
- **Consider timing.** It is important to keep your lectures as short as possible. Students sitting through a 2-hour lecture may have problems with grasping a theory or a concept if they are required to sit for an extended length of time. Lectures that will last longer than approximately 30 minutes need to have scheduled breaks. A good rule of thumb is to take a 10-minute break on the hour. Using a variety of teaching aids will also help students stay engaged in longer lectures. Use time appropriately. Do not waste time with handing out materials or by going over material previously learned in another session.
- **Constantly evaluate your students' learning.** You should apply various types of assessments as you present your lecture. Examples of ways you can informally and formally assess learning include asking questions, giving quizzes, or reading students' body language or tone to ascertain whether they are confused or unsure. Ask questions on the material that you feel represent what is important for the students to learn. At the end of a learning session, you should evaluate the students' overall learning. These evaluations can be formal or informal. One great tool to determine if the student has learned or has additional suggestions is to have a written evaluation at the end of the class. Informal evaluation should be frequently used on an ongoing basis during the lecture. This can be accomplished by posing open-ended questions to the students.

The Illustration

The illustration method can and should be used in conjunction with other methods of instruction, as it is more of a supplemental approach. Essentially, with this method, the instructor will provide a visual stimulus for the lesson. Some type of visual, whether it is a prop or a visual projected on a screen, will be used in every class. A good speaker can "paint a picture with words"; however, even with the best description, sometimes it is still easier to understand something if you have a visual. Having the students visualize an item or a procedure is also a great way to engage the class. Illustrations are typically inexpensive and can be found around the station in most cases.

Good illustrations should:

- Have clear visual aids or props, pictures, or writing that can be easily seen by all participants
- Be appropriate to the subject being taught
- Be appropriate to the attending audience
- Have colors that look good together and do not distract from the presentation
  - There should be appropriate use of contrast when using colors, such as black words on a white background.
- Be simple and concise, particularly when drawing or printing on flipcharts, white boards, chalkboards, etc.
  - Excessive amounts of time should not be required to draw or print the information.
  - Preprinted flip charts will save time during class and are easily accessible.
  - Also, lightly sketching your flip charts with a pencil so that you can trace over them later

will help you save time during the presentation. Also, the information will already be on the paper, so you will not have to review notes while writing.

The Discussion

During any training session, individual or group discussions that will enhance the session and allow students to bring personal views or experiences that are relevant to the topic at hand should occur. Keep in mind that in any discussion there will be questions and stories. Depending on the subject, you may have to curtail the stories to keep the session on track or on time. As an instructor you should anticipate areas of your training session that would allow or require a discussion and prepare, if possible. The discussion technique is the most widely used method among instructors. Facilitating a discussion gives students a chance to share their own knowledge and experiences and learn from each other. In order for a discussion to be successful, however, it is best if the following conditions are in place:

- The students, especially veteran or seasoned personnel, have some knowledge of the topic.
- The instructor is familiar and comfortable with the materials and is willing to let students explore concepts, brainstorm, and bounce ideas off each other.
- The instructor has a strong presence in the class and the ability to keep the class focused and on topic.

Notes or bullets on a slide can help keep the discussion focused. During discussions you should remember to repeat questions brought forward by the students. Never assume everyone in the class has heard the question or answer. Another way you can use discussions to engage the class is to split the class into small groups and give each group flip charts or note pads with notes or questions to stimulate a discussion. Questions can be open-ended or closed-ended; however, open-ended questions will generate more discussions as they require a longer response than a simple yes or no. Furthermore, the instructor can glean more information on what the students have learned from the class with the use of open-ended questions. Based on the students' responses, the instructor can determine whether he or she has achieved the desired learning outcome. When using discussion as a form of assessment and a method of learning, it is also important to consider the dynamics of the class and make sure that you hear from all students. Sometimes you will have students who are naturally more vocal and overpowering paired with students who are quieter and tend to recede into the background during discussions. It is best to make sure you give the quieter ones time and space to share, even if you have to pry the answers out of them. The whole class will benefit from hearing different points of view.

At times, you may need to work a little harder to engage the class in a discussion. People for the most part do not like to speak up in class. Quieter individuals may be shy or may not know the material. Sometimes you can get participation from these folks by going around the class asking for answers or suggestions. While it is helpful to respect the student's privacy, you should ask questions to make sure that the information is getting across. As students become more familiar with their instructor, the more likely it is that they will open up.

The Conference

By bringing in guest lecturers or known personalities, the conference is a great way to lure staff to attend a special educational session. The gathering of your staff at one location at any time for a training session could constitute a "conference" type of event. A conference does not have to require travel to various parts of a state or out-of-state. These conferences could even take place at your department's training facilities.

One organization that offers free classroom sessions is the Metro Atlanta Fire Fighters Conference (MAFFC) (**FIGURE 7-1**). This organization was developed by a group of highly motivated fire fighters throughout the Metro Atlanta area. Through assistance and donations they offer free classroom sessions with speakers, such as David Rhodes, Ray McCormack,

FIGURE 7-1 The Metro Atlanta Fire Fighters Conference logo.
Courtesy of Metro Atlanta Fire Fighters Conference.

Skip Coleman, and others. (However, please note that Hands-on-Training [HOT] is approximately $25 per class.)

When we have the ability to "gather the troops," it is best to use this time to convey important information that everyone needs to hear. A conference-type setting can make this possible. If you are scheduling a conference, set the best time that will get you the attendance you are seeking.

A cost-effective method is to host an internal conference with various departmental speakers. This approach can also be a motivating experience for those speaking, as it provides them with an opportunity to share their expertise and experience. It also breaks up the monotony of these training sessions, which can involve reviewing the same recertification training PowerPoints seen on a regular basis. A typical schedule for an internal conference may include: a light breakfast; 4–6 one-hour training sessions, each with a different speaker and topic, spread over the course of the day; and a lunch. The value these conferences bring to your department in terms of improving morale and cooperation among members versus the money that would be spent on an outside conference can be invaluable. You may also want to consider inviting neighboring departments, as this can bring multiple departments together and provide opportunities for sharing ideas among communities. It can also help your department move out of its insular "box" and potentially expose them to new ideas and ways of doing things.

When hosting a conference, you should prepare the following:

- **Agenda.** Prepare a detailed agenda that includes the schedule for each session and breaks, as well as a description of each session.
- **Handouts.** For each session, prepare handouts that are relevant to the lesson but do not cause the participants to become distracted. Always review materials to make sure they are up-to-date. Students generally like to receive handouts or other materials to keep after a conference.
- **Room preparation.** Ensure you have enough room for the participants to sit comfortably. Tables are a great option if available. Ancillary equipment, such as audiovisual (A/V) equipment, light systems, and projection equipment, should be checked before the presentation. You should also make sure that the number of attendees will not exceed the occupancy requirements of the facility. Additionally, you should remember that the more people you have in a room, the warmer the room may get, so you should prepare for this. Generally, a slightly cooler room will keep people awake.

The Demonstration

The demonstration method of instruction is the backbone of fire service instruction and provides the first step toward reaching the psychomotor domain of each student. In a perfect world, students would attain some theory on the subject being learned, have opportunities to practice the skill, and then have the opportunity to demonstrate their proficiency of said skill. It is ideal if the student has an opportunity to show the instructor what he or she has learned, because it facilitates learning and it allows the instructor to assess the student's understanding of the skill.

There are different ways to conduct a demonstration. For example, one way to conduct a demonstration may be:

1. The instructor shows and tells the student.
2. The instructor shows while the student tells.
3. The student shows and tells the instructor.

Whenever demonstrations are in session, safety is your number one priority. When demonstrating a skill, it is best to have a training safety plan (TSP). A TSP will help you identify possible safety concerns during training activities and ways to prevent injury.

It is important to determine what the desired outcome(s) should be when designing a presentation so you have a way to measure the participants' progress. As you perform the demonstration, it is important to describe the desired outcome(s) to the participants, so they know the ground rules for the skill. You may want to attach ratings to the students' performance of the skill, especially if the skill has high life-safety implications. Skills that require several steps to complete can be tracked through the use of a signature-required document. Your sign-off documentation should include:

- The objective to be learned
- The skill to be completed
- Applicable department performance criteria
- Evaluation criteria

- A checklist detailing the specific, measurable steps or procedures of the skill to be learned
- Name/date/signature lines for the student and the instructor

Good documentation, such as a signature-required checklist, ensures that the training officer and the department used due diligence in the training of the department members. During injury investigations, training documentation is one of the first things the investigating agency or health and safety representative will require for review.

Choosing and Using Classroom Aids

As an instructor, you have a variety of training aids at your disposal. When projecting visuals from a computer or device, it is best to have a back-up system. As a goal, every training officer should strive to have at least three different types of audio and visual aids incorporated into a class.

■ Dry Erase Boards/Chalkboards

Dry erase boards or chalkboards are great for presenting last-minute notes. They can be used to list items for further discussion once students return from a break or at the end of the day. One challenge with using chalkboards or dry erase boards is that they will accumulate residue from the marker or chalk used, which can make the board harder to read as the instructor continuously wipes and writes over old notes. One advantage that dry erase boards have over chalkboards is that you can easily use multiple colors that can be seen from a distance, whereas chalkboards can be difficult to read from the back of the room. Both types of boards are easy and inexpensive to maintain. However, remember to remove any permanent markers situated near dry erase boards, as the use of these will ruin the board. Also, use caution with using certain colors, such as yellow, especially if there is a glare from the sun, as this could make it hard for students to see the writing.

■ Flip Charts or Easel Pads

Flip charts and easel pads are excellent to use when you break students into groups to share and list ideas or brainstorm. Instructors can also use flip charts to sketch out ideas for visuals for their presentation (remember the pencil sketching tip from earlier). When using a flip chart, it is essential to use flip chart markers, because other markers will not work as well. Flip charts and easel pads provide an excellent break for students, particularly in a heavily laden visual/lecture class with lengthy PowerPoint presentations.

■ Charts and Posters

Charts and posters are a great passive learning tool. The educational value of a poster that has been placed in a well-traveled area of a fire station should not be underestimated. It is an excellent resource that fire department members can easily access at their leisure. During a training session, charts and posters can be used to reinforce a subject and also serve as excellent handouts. Posters from vendors that supply pumps or equipment can be used as supplemental information materials in specialized training sessions, and they typically can be acquired free of charge from equipment vendors. Posters from educational stores can also be used to reinforce technical or complex presentations.

■ Overhead Transparencies or Projection Tables

There are several ways to project images on a screen. If you are using an older projector, you should ensure you have a spare bulb. Regardless of which style machine in use, the machine should be set-up prior to class and focused. Transparencies can be preprinted. Projection tables will project any material placed under the camera. When using a projector, you need to be aware of projector placement in relation to student seating. You should make sure that the head of the projector does not block the screen from students sitting behind or in close proximity to it. Additionally, the screen should be positioned above eye level.

■ A/V Equipment

A/V equipment is used regularly in fire fighter training. When viewing potential training materials, it is good practice to have your Incident Safety Officer hat on. You should exercise caution when downloading a video or audio file from the Internet for training, and check the content for accuracy and credibility. We should not be showing procedures and skills that are unsafe or do not adhere to accepted practices. It is imperative that audio and visual additions to your program be previewed with a keen eye for discrepancies or errors.

Anyone can create a high-quality video production with the use of relatively inexpensive cameras or phones and computer software. Numerous fire departments have incorporated in-house videos into their training plan. Software such as Movie Maker by Windows, Roxio MyDVD VideoLab, and Adobe Premiere Pro, is available for purchase to edit your own videos and can be used on most computers without the requirement of tremendous upgrades. Mobile phone apps such as Splice and iMovie will also allow you to record a video, take photos, add music, and edit, all from your device. Both of these options can produce content for standard viewing directly to a computer or through podcasting and/or webcasting. Some cities and counties have their own communications department. Utilizing their services presents the perfect opportunity to build a relationship across departments.

You will also need to be aware of potential issues when purchasing professionally produced DVD/Blu-ray productions. These programs can be very expensive to obtain and can become outdated quickly. Be sure to preview videos before purchasing to ensure that the video is up-to-date and worth the investment. Additional details on developing videos can be found in the chapter *Alternative Learning*.

Podcasting and Webcasting

Technology has provided several alternate means for information to be shared across the globe. One of the newest methods to share the latest in adult education theory and training techniques is the use of podcasting and webcasting. These methods allow an organization, such as the International Society of Fire Service Instructors (ISFSI), to disseminate knowledge and information on advanced topics, such as fire dynamics research, using subject matter experts. The member can get the information straight from the source, even before it is officially published.

One of the benefits of incorporating such technology into the training program is that, once recorded, members can watch it later or save it for further reference. It is much easier to master a topic and ensure you know every detail when you can return to it as often as you like. Plus, the instructor presenting in a webinar or podcast is always "perfect." The presentation is well rehearsed and executed—the instructor never has a bad day. Training officers can also edit the presentation to ensure the instructional delivery is at its best.

Key Note/PowerPoint Presentations

Key Note and PowerPoint presentations are the most widely used teaching tools. Sometimes, however, these tools are overly used and become a crutch for instructors. The instructor should only use these tools as an aid to the class and not to read from each PowerPoint slide word for word. Each key point from the PowerPoint should be mentioned, and then the instructor should provide additional explanatory knowledge. Plenty of information is available online to assist the instructor with the production of these teaching aids. The first step in creating a Key Note or PowerPoint presentation is to create a storyboard, or outline, of what you would like to teach. From this point on, you can add additional information; insert photos, videos, or audio clips; and manipulate color and text to make your presentation more appealing.

Remember that there is nothing more unprofessional than not knowing how to operate your equipment. Granted, there is truth in Murphy's law in that anything can go wrong at any point in time. Even the best prepared instructor can have a bad day. However, we should take it upon ourselves to learn how to use the equipment prior to class. To go in with the attitude that you are going to "wing it" and then expect students to sit and wait while you fumble around with the equipment gives the impression that you do not respect their time.

In addition to needing to be prepared ahead of time, there are a few traps instructors may run into with Key Note and PowerPoint presentations. You should be careful to avoid the following:

- **A presentation that is too long in length.** Copying a chapter from a textbook verbatim will quickly bore a class. These presentations should have key triggers that will allow you to expand on the information from your knowledge. Remember the 6 × 6 rule: no more than six words per line and no more than six lines per slide.
- **The use of too many graphics or animation.** This can lead to major distractions to the class that will not allow them to focus on the information being presented. Excessive graphics or

animations include fading slide transitions, spinning slides, loud explosions, and the like. Slides that have fast or different transitions can also be somewhat distracting to some students, although others may enjoy them. You should know what works best for your audience. Be sure to embed all videos in the PowerPoint presentation, if possible, and remember that video files have to be placed inside of the PowerPoint file folder in order for them to work.

- **The use of all capital letters.** It can be hard to read presentations when the words are in all caps. Caution should be used not to have too many capital letter words.
- **Poor use of color.** Color can make or break your presentation. Black letters on a purple background may look great on your computer; however, when you project it on the wall, it can be impossible to read. Always have contrasting colors between the background and the lettering. Inserting a background photo can cause similar problems.
- **Too many or too few words per slide.** There are differing thoughts on this subject. Generally, a 24-point font should be the smallest font used, but that may not always be practical. If a smaller font needs to be used, try to limit the information to a few lines. If there is a great deal of information that you are trying to get across, several slides with a larger font would work best, or a handout with details may work even better.
- **Facing the presentation instead of the class.** When using these types of presentations, it is best to speak to the audience and read off a computer screen or teleprompter that faces you. These programs use multiple screens that will allow you to use a notes page for additional information. Remember, no student wants to talk to the back of your head. Be sure to make eye contact with your students.

Instructor Transitions and Distractions

As instructors, we are expected to be able to perform certain tasks, such as transitioning topics, staying on schedule, knowing the materials, and being aware of ourselves and our mannerisms. Being able to transition thoughts, stories, visual aids, and topics while standing in front of a closely watching class requires practice and a firm grasp of the subject. As you segue into different sections of the presentation in an orderly manner, make it clear what activity or new topic is coming. You should also consider the sequence of topics and activities. For example, before you can talk about smoke explosions, backdrafts, and flashover, you need to explain the products of combustion. You could provide an overview of the products of combustion first and then give a small pause to allow the students to process the information and analyze it. After pausing for 5–10 seconds, ask the students an open-ended question. Discuss the answer with the students in a positive manner. Now, to transition to the next activity, you say to the class: "Let me show this video to help explain products of combustion and the pyrolysis process." As you transition from one activity to the next, explain to the students what will happen next. Care should be taken to make the transition as smooth as possible.

Also key to effective training is the use of appropriate timing. A lesson that is not paced well can be distracting for students. Timing breaks and the length of activities can be difficult to judge. However, a good rule of thumb is to provide one 10-minute break every hour. The more opportunities you have to teach a particular session, the better you will get at timing breaks and stopping points. Another potential issue with timing is knowing how much time should be allowed for questions and answers. We want the students to ask questions and to become engaged, but you also do not want the question-and-answer session to take over the class. Certain classes, such as a strategy and tactics class or a session involving the presentation of completely new information or research, typically require the instructor to field more questions than normal. You should practice the timing of each training session before your presentation by rehearsing it. You can do this by presenting it in front of another instructor or colleague or you can make a video of yourself.

The ability to be a great presenter can be summed up with one vital piece of advice: *practice, practice, practice*. You need to practice at being an instructor to hone the craft. Some people come by certain skills, such as speaking, naturally. Being an instructor, though, is not just about what you say and how you say it. Everyone has certain areas that they need to work on, whether it is speaking, transitions, writing, procrastinating, or timing. It is best to run through a lesson so you can anticipate any potential problems, conflicting information, or repetitive material. Practicing beforehand will help you ensure that you know the information thoroughly. A good instructor needs to be aware of the needs of the students, the classroom setting, potential safety concerns, and the aids required to run the class smoothly and be safe.

Lastly, your students will judge your abilities as an instructor, and they will base these judgments on your interaction with the class, the amount of distractions you allow, and your ability to manage the behaviors of the students in the class. Several distractions can cause the class to derail and stop the learning process. There is some preplanning that training officers can do to curtail distractions, such as using foresight and eliminating institutional and environmental distractions. Another tactic you may want to consider is to create a very professional and organized classroom, as this will set a specific tone. You should also establish your expectations up front and hold your ground on these expectations for the duration of the training session.

For better or worse, students will take note of any distracting mannerisms that you may have. Your students will judge your performance regardless how nice of a person you seem to be. If you struggle with certain tics or mannerisms, you should practice your presentation ahead of time to avoid causing any unnecessary distractions that will stop the learning process. Furthermore, the students in your class may also behave in ways that are distracting to other students, or even to you. In general, there are three categories of distractions, which are:

1. **Human behaviors**
   - **a.** Human behaviors that instructors may exhibit that can be distracting for students include fidgeting; playing with a pen, pointer, eraser, or other teaching tool while speaking; or the use of filler words, such as *ah*, *um*, or *like*, on a continual basis. Instructors should video themselves every so often to review what habits they have developed over time.
   - **b.** Students may also behave in ways that are distracting for others. We have identified three types of students whose behaviors can make for a very distracting learning environment:
     - **i.** **The talker.** This person will not be quiet during your session. He or she chats with others while you are trying to speak or maybe even speaks out of turn during class. There are several ways to curtail this behavior without causing disruption to the rest of the class. You can walk over to where the talker is sitting and continue presenting from his or her location in the classroom. This passive approach is sometimes enough to make the person quiet. You could also engage the talker by asking him or her a question. A less-subtle intervention you could use is to stop talking and stare at the person; this will usually shut the talker down.
     - **ii.** **The texter.** This is the student who has his or her phone out while the class is in session. It is important to lay the ground rules of a class at the outset. Be sure to use the word RESPECT in the announcement. The silencing of a phone should be an automatic request at the beginning of any class session, whether it is in a classroom or training ground. If the texting continues, you may want to talk with the student during a break.
     - **iii.** **The "know it all."** These people can be difficult for an instructor to deal with, especially if they are empowered enough to steer the class in another direction. Try to use their expertise when appropriate, but do not allow them to alienate the rest of the students. Sometimes it is best to ask the person to let someone else answer the question or respectfully challenge them. Be careful to not offend a student with this approach, however. The whole class may turn on you.

2. **Environmental**
   - **a.** The environment of a classroom or a training session can become distracting for students.
     - **i.** **Storms and inclement weather.** Sometimes the weather outside can wreak havoc on our training plans. Outside activities should never be attempted during lightning storms. If a lightning storm occurs, you should not resume training activities until 30 minutes after the last strike. Even when the students are

situated in a classroom, bad weather can be distracting for students, because they can become preoccupied with what is going on outside or concerned about their home or family. Furthermore, the possibility of an emergency incident requiring the response of members increases during times of bad weather, and this could impact your session as well.

   ii. **Extreme temperatures.** When experiencing extreme temperatures, such as those exceeding 100°F or below 32°F, it is important to plan for these conditions ahead of time and be sure you have the appropriate equipment. You will also want to use your common sense when planning training sessions and remember that we will train differently in different temperatures. A good rule of thumb is to add 5°F if in gear. This will help determine your work-to-rest ratios. Other alternatives include using a wet-bulb detector, which takes heat, humidity, and wind speed into account, or using weather flags based on the current heat levels and what activities it is safe for trainees to perform. Also, make sure to prepare students to bring sufficient water and food to training sessions. You may want to review proper nutrition with your class, in terms of calories required for upcoming activities and hydration.

 b. Outside noises occurring near or around the training session (i.e., construction, traffic, people talking in the hallways), can become distracting for students, especially if the noises are so loud that students cannot hear you.

3. **Institutional**
 a. Classroom ambience and equipment can become distracting to students. Florescent lights may cause headaches or the air conditioning might stop working. Make sure you check the classroom well in advance of the training session to ensure that it is properly equipped and ventilated.
 b. You may begin a class with the intent of using A/V equipment and then discover that the room lacks A/V connections or speakers, or that the projector you have does not work well. As discussed previously, computers, projectors, and cameras will go down at the worst of times. Be prepared to do the session without the electronic aids. You may have to work off of your lesson plans or handouts.

When preparing a class, try to place yourself in the shoes of the students. Pay attention to what environmental elements may cause a distraction during your class. If possible, try to have these removed or fixed so that the student has the best chance to succeed in your class. For example, if you see a light blinking or fading out, make sure this is fixed ahead of time. You may not be able to anticipate every problem, but if you properly plan ahead of time, you are more likely to have a successful training session.

Key Points

- A well-thought-out and produced lesson plan will go a long way in the successful presentation of your lesson.
- Some of the advantages to the lecture approach are that it can be prepared shortly before the training session, it does not require any training aids, it can be a very effective method for delivering instruction to small groups or a large class, and it is very cost-effective. Some of the disadvantages are it may be difficult for students to see or hear the instructor if the class size is very large, especially if audio/visual equipment is not used, and it may not be the most effective learning method as it utilizes a limited number of senses.
- Incorporating discussions into the learning experience is the most widely used method among instructors. Facilitating a discussion provides the opportunity for all students to share their knowledge and expertise and allows them to learn from each other.
- A training safety plan (TSP) will help you identify potential safety concerns of training activities and help you develop solutions to prevent injury.
- The ability to relate certain outside materials to the training session ultimately requires a deep knowledge of the material and vast experience.

- There are two types of distractions: human and environmental.
- As a training officer, refrain from using crutch words such as *hmm*, *huh*, *um*, *OK*, etc. These pauses and filler words can be very distracting for the student.

References

International Society of Fire Service Instructors (ISFSI). http://isfsi.org. Accessed July 23, 2014.

Metro Atlanta Fire Fighters Conference (MAFFC). http://maffc.org. Accessed July 23, 2014.

CHAPTER

8 The Application Step

Richard Gasaway and Brian Ward

Introduction

In simple terms, the application step is the portion of the training process in which the student practices and demonstrates the skill the instructor has taught. This step can also help the instructor confirm whether the student understands the skill and is achieving the objectives of the training. As a training officer, you will need to consider the different domains of learning—cognitive, psychomotor, and affective—in order to achieve the best possible results. During the application of one's training, psychomotor skills commonly become the focus of the exercise. However, an appropriate mixture of all the domains allows for the greatest amount of knowledge to transfer from the instructor to the student. The methods that we use as instructors during this step can make the difference between a good class and a great class.

The application step, arguably, is the most important step in the process. As fire fighters, it is imperative that we gain the appropriate knowledge, but what is even more important is how we *apply* that knowledge. As the instructor, we can tell students what it will be like to stretch a 1¾-inch hose line 150 feet down a long, dark hallway with heat pushing down on us, and the students may remember some of it. We can even try to describe the feeling of it using our senses—how the ears tingle and how the burning materials crackle and snap—to give students a sense of this type of environment. None of these words will truly convey or help the students understand the experience as much as placing them in a hot, pitch-black, burning building, bunkered down with a tool and a hose line. The stress and adrenaline that comes with this experience is something they will never forget. This is your application step.

Now, as effective as the application step can be, it may be the most difficult step to plan for and maintain control of. Safety has to be the top priority of every application step, whether the task involves tying knots at the station or participating in a heavy rescue class. Fire fighters getting hurt or even killed in the line of duty (also known as line-of-duty deaths [LODDs]) during a training event are unacceptable; however, it is reality. Approximately 10 percent of LODDs occur during a training event. It is our responsibility as training officers to protect those who are learning the job and to provide an environment in which they can succeed. As you plan the application step of the training, you will need to consider the type of task at hand, as this will dictate the type of application experience that we use. When planning this step, you will have different types of application experiences at your disposal, including practical situations, demonstrations, hands-on activities, role play, and scenario-based situations. We will also discuss recognition-primed decision making as well as the use of tactical decision games (TDG).

In addition, as you are planning for the application step, there are several potential obstacles you must consider. These obstacles, which we will discuss in more detail later in the chapter, include the limitations or potential issues concerning the training environment, time restraints and schedule challenges, the number of assistants or resources available, and weather conditions. Each of these areas should be addressed in your overall training plan. Remember our discussion in the first chapter concerning the use of foreseeability in legal matters and safety. Using foreseeability to mitigate these obstacles is critical during the application step in the training process.

The Action Plan

When developing the application step, the training officer should create an action plan for each training session. The action plan serves several purposes. The first purpose is that it keeps the class on track and organized by providing a blueprint of activities. Student objectives should also be developed and inserted into the action plan as part of the blueprint. The objectives set the expectations for what the students should have learned at the conclusion of the training. It is also helpful for the students and the instructors to be aware of what the objectives are. By knowing the objectives, they can understand what the end goals are for the class. Typically, for the application step, the student objectives are constructed using Bloom's taxonomy action verbs (e.g., demonstrate, examine, conduct, perform, etc.). For more information regarding objectives, please see the chapter *The Presentation Step*.

The second reason for creating an action plan is to document the training. Having documentation on the action plan will come in handy when planning future drills. The written action plan can also serve as documentation if an audit was to be performed on the training content and hours. In addition, if an accident was to occur during a training session, investigating authorities (i.e., National Institute for Occupational Safety and Health [NIOSH], law enforcement, or subject matter experts) will ask for a detailed plan and check for adherence to the plan. To determine whether the training division was liable for the accident, most courts will examine the situation to see if there was a plan and if it was followed.

The action plan should include the resources needed and the amount of time required for the training, the student-to-instructor ratio, student objectives, instructor assignments, safety concerns, appropriate briefings, and any walkthroughs to be performed. After the action plan is developed, all involved instructors should review and agree on the plan. In order to accomplish this, it is a good practice to have instructors meet and review the action plan prior to the class start date. Having instructors meet beforehand to discuss the action plan will also give them the opportunity to alter the training based on the instructor's capabilities. This will help them deliver the best training experience possible.

Application Step Placement

Typically, there are numerous application steps within a lesson plan, and they can be placed anywhere within the lesson. The key is to determine the appropriate timing. The combination of the cognitive and psychomotor lessons should be tailored towards the general capabilities of the students when possible, as not every class is the same. More importantly, the placement should be based on the students' knowledge level and skill sets. For example, would you teach a recruit in the same format that you teach an experienced fire officer? Fire officers and recruits may use similar drills, but the focus and skill level in which they are operating will be completely different. Therefore, the focus of your training will also be different. For example, in a drill, recruits will focus on how to execute fireground tasks, such as fire attack, a skill that is taught very late in a fire fighter program. Officers, on the other hand, will focus on tactics, which would be taught at the very beginning of an officer program.

In some cases, it is best to focus on the cognition aspect of learning, or the acquisition of knowledge, at the very beginning of the lesson. In other words, you may want to begin the class with a lecture or presentation that exposes the students to the new material or technical subject. With this approach, you introduce the students to the skill before asking them to perform it. At this stage, students can learn the reason behind or justification of a specific task. This format is generally best when applied to students who do not have a working knowledge of the topic at hand. An example of this approach would be if the instructor taught the students about fire behavior before placing them inside of a burning building. Our goal as instructors is to have the students pick up key elements during the demonstration or lecture, and then apply this knowledge when it comes time to perform the actual skill. This concept is known as the law of association, which is when the students associate new information with old information when it comes time to apply new knowledge.

Another option for placement of the application step is to demonstrate the activity at the beginning of the class and then follow up with a lecture or a discussion to explain the connections. By having the demonstration at the beginning of the class, students can form a picture in their head of how the finished product should look, and the instructor can explain the steps necessary to reach the final product. The last step of this approach is to repeat the practical with 100 percent involvement from the students. An example of this format would be to begin a class by showing the students how to tie a figure eight on a bight. Then, the instructor would slow down the process and provide a detailed, step-by-step explanation of the process. After the lecture, each student would tie the knot.

A slightly more complicated application step would be where the instructor shows a scenario with key points in mind and allows the students to actively participate in the drill. The decisions that the students make would be the driving force behind the lesson that you provide. The decisions, right or wrong, should be addressed through the lesson and compared to the key points the instructor had in mind when designing the drill. This type of activity will also provide relevance to the students on why the cognitive lesson is just as important to them as the practical lesson. This method generally works better when the students have some working knowledge or understanding of the topic.

A third option is to plan the training so that demonstration/lecture and application are delivered on the drill ground simultaneously. However, in order for this format to work, the instructors must plan specifically for this drill. Several distractions could derail the class if the instructors do not plan to keep the students' attention. One of the benefits of using this approach is that you can have the students discuss the skill being taught while they are seeing it performed in front of them. Since the students are using several senses at once to retain the new information, they can personally connect the movements with the technical information. If you examine the cone of experience, the more realistic we make the drill (the more real time we provide), the more apt the students are to remember the skills that they were taught. One important consideration with this approach is to remember the safety of the students. As instructors, we want the training to be realistic, and so do the students. However, if we allow the drill to get out of control, a student or instructor could get injured.

Incorporating Realism Into the Practical Application Step

Once the action plan and placement of the actual drills have been determined, the instructor should determine the appropriate level of student interaction for the application step. The objectives that the instructor develops will determine which practical situation the students will take part in. Student interaction can be broken into several different categories, including demonstration, hands-on activities, scenario-based activities, TDGs, role play, and simulation. Safety is always a concern with these activities. You may be wondering why these activities are included as a part of the training when they pose an element of risk. Research on realistic training and drills has shown that fire fighters learn best when they have opportunities to practice the skills they are being taught in environments that mirror those where the training would be applied. Furthermore, having fire fighters train in realistic situations requires them to practice the decision-making skills that they will need to use when operating on the fireground.

There may be no greater skill required to ensure the safety of first responders than the ability to make competent decisions. Many training programs do an inadequate job of preparing responders to make decisions in high-stress, high-consequence environments. The ability to make timely and effective decisions on an emergency scene requires a combination of knowledge, training, and experience. As training officers, it is imperative to ensure students safely receive knowledge, training, and experience in balanced proportions. A shortfall of one attribute and an abundance of another make the responder incomplete. It is also very important that responders retain what they are taught after the training is over. It is beyond the scope of this chapter to discuss the psychology of learning, but suffice it to say, practice and repetition are key ingredients.

Naturalistic Decision-Making Theory and Its Influence

The inclusion of realistic drills as part of fire fighter training is in large part due to the research conducted by Dr. Gary Klein. Originally, Klein was hired by the U.S. military to conduct research on how people learn and what approaches increase the knowledge retention and the decision-making skills of the soldiers. After his stint with working for the U.S. military, Klein continued his research in Cleveland, Ohio, where he followed expert fireground commanders and analyzed how they made their decisions on the fireground. It was with this study that a new concept emerged called naturalistic decision making (NDM). This concept has helped us better understand how fire fighters process an emergency scene and make decisions on the fireground, because it involves studying fire fighters in their natural work environment. By studying how expert fireground commanders make decisions on the fireground, Klein was able to develop a model that we consider when developing practical situations, or drills. This model is called recognition-primed decision making (RPD). It is a thought process that is now used in the fire service—and in other professions—to train people who work in high-risk, high-consequence environments.

The Recognition-Primed Decision Making Model

The recognition-primed decision making process is an eight-step approach, which involves the following sequence:

Step 1: Conduct a rapid size-up. Sizing up the incident involves thinking holistically about the clues and cues you gather at an emergency. When you size up an incident, you not only consider the current conditions of the scene (i.e., the weather, wind direction, the structural design of the involved building, the presence and behavior of smoke and fire, etc.), you also use prior knowledge that you obtained over the course of your lifetime. Although a great deal of information is considered during this step of the process, size-up is amazingly brief and may be completed in as little as 10–30 seconds. The reason this step is performed so rapidly is because the consequences for a slow response are too great and the conditions rapidly change, which can make observations obsolete in a matter of seconds.

Step 2: Focus on the most relevant information. To accomplish the task of size-up with expediency and accuracy, the decision maker must understand, in advance, which pieces of information are the most important and seek to identify and evaluate those clues and cues. You will need to help responders develop the skill of determining the most relevant pieces of information in advance. For example, at the scene of a house fire, the most important clues and cues are the fire and smoke conditions, the building's construction and the risk assessment of savable lives and property, and a resource needs assessment (i.e., personnel and water). Each piece of information will need to be considered carefully. Smoke will need to be analyzed in terms of color, volume, velocity, and density. The number of personnel immediately available will need to be determined. In this step, the decision maker makes a short list in his or her mind of the most important pieces of information needed.

Step 3: Compare the situation to what is typical. Every responder on the scene of an emergency (hopefully) brings some level of training and experience along with him or her. Once the clues are gathered, the next step is to compare what is going on at the incident with what you have seen or been trained on before. In this step, the decision maker will look for things that match the current situation to his or her previous experience and training. The more incidents the decision maker has experienced, either actual or simulated, the greater his or her ability to match typical patterns will be. The matching of current information to previous experience provides confidence to the decision maker and helps him or her move forward with a plan. Conversely, the inability to match or align the current situation to previous experiences can also serve as a significant warning sign that something may be wrong.

Step 4: Run mental models of options. In this step, you, as the decision maker, create a mental picture in your mind in which you envision the future of the incident based on your decisions. In this step, the decision maker considers the likely consequences of his or her proposed plan and imagines how the action plan will turn out. If the decision maker determines a course of action, runs the mental model, and the outcome is good, he or she will most likely proceed with the plan. However, if the mental model has a bad outcome, he or she should develop a new plan and run a new mental model. This process should continue until a viable plan is identified. While care should be taken to ensure a safe outcome to an incident, the decision maker may want to avoid modeling the perfect solution, as this is time consuming and can potentially set up unrealistic expectations for providers. Instead, decision makers should seek a plan that meets basic objectives and will keep responders safe.

Step 5: Form expectations about the outcomes. After running a mental model in his or her mind, the decision maker formulates expectations about the outcomes of his or her plan. It is during this process that the decision maker determines whether the action plan will meet the objective. In the context of emergency incidents, the decision maker should formulate two principal expectations: (1) What will the outcome look like? and (2) How long will it take for the outcome to occur? The ability to form an expectation for an outcome is rooted in the training and experience of the responder. The more outcomes stored in his or her memory, the better the responder will be at accurately predicting what will occur. Determining how long it will take to execute the plan or complete a task will depend on many factors, such as the size of the structure, the occupancy type, the fire and smoke conditions, the floor the fire is on, and so on. The decision maker's prior knowledge and experience come into play here.

Step 6: Detect unusual situations that require unique consideration. The decisions made at this step may vary depending on the decision maker's level of experience. The more training and experience a responder accumulates, the less likely it is that they will witness a truly novel situation. However, when unusual situations are detected, it is important to both recognize the situation as such and to avoid treating it as a trivial concern in the broader scope of the overall situation. Likewise, newer responders who have less experience need to be reminded that they should keep their feelings of aggression and the need to act in check, as this can lead to decisions and the implementation of action plans that are not supported by their level of training and experience. This can lead to dangerous outcomes because the responder may be challenged to run effective mental models and to form realistic expectations about incident outcomes.

Step 7: Make the decision. At this step, the decision maker decides whether he or she should proceed with the plan. While experienced company and command officers will readily admit that some decisions are made easily and with great confidence, they will also say that other decisions can be very difficult to make and are made with great hesitation. While confidence is an important attribute of a decision maker, this step should not be taken lightly. In the public safety domain, people could live and die by the decisions made at incident scenes. As the training officer, you should teach and coach decision-making skills and provide constructive feedback about the decision-making process and outcomes.

Step 8: Start the process over again. In a dynamic emergency situation, the decision-making process rarely ends when the decision is made. Once implemented, the decision maker must continually assess the progress of the incident in the context of the effectiveness of the decisions that are driving the action plan. Officers will admit that it can be very difficult for a decision maker to critique, in real time, the effectiveness of a decision. The decision maker may benefit from soliciting feedback from others on the effectiveness of the plan. Others may see things the decision maker missed, or they may see the same things but interpret their meaning differently. Differing assessments of the situation may cause the decision maker to realize that something was overlooked or overemphasized. The decision maker may choose to consider a new plan, in which case, the recognition-primed decision making process begins again.

As you work through drills with the students, you should teach them to use the RPD process as they perform the drill. The more they rehearse this skill, the more likely they will retain it in their memory. This process has proven useful because it requires responders to use their memories and recall prior knowledge and skills, which helps them make effective decisions. In fact, the more knowledge, training, and experience a person acquires, the more likely they are to be able to make high-quality decisions with little effort. Just as a professional athlete makes his or her performance seem effortless, so can a fire fighter make it seem with his or her decision-making skills when he or she is properly prepared. As training officers, it is our job to help fire service members learn the process for making a decision in a dynamically changing environment.

Using Postincident Evaluations After Practicals

In addition to teaching the RPD model, you should also train your students to conduct a postincident evaluation of their decision-making performance. There is always something to learn even when things turn out well. Opportunities to assess decision-making performance should not be viewed as a critique of one's performance; rather, they should be viewed as a tool for learning and self-improvement.

It can be unnerving for a decision maker to invite others to evaluate his or her performance. No one enjoys having flaws pointed out, let alone having those flaws pointed out by those who were involved in the incident. However, when the decision maker has high self-esteem and a drive for continual self-improvement, much can be shared and learned in this process. The decision maker and others have much to gain from this thought process.

Developing Expertise Through Realistic Training

It was noted at the beginning of this chapter that in order to make effective emergency incident decisions, the responder must possess expertise. This expertise is acquired through knowledge, training, and experience. You may think that some types of emergencies happen on such an infrequent basis that it can be difficult to obtain adequate experience. However, you can simulate these scenarios in your training. Responders who receive realistic training store those experiences in their brains as if they are real. By putting the student in situations where they experience emotions as if the incident was really happening, they are more likely to recall this experience later on when it actually counts.

Through realistic training, you can, essentially, preload these scenarios into your students' memory before they experience them on the actual fireground. It is during these drills that you teach students to hone the skills of the RPD process to help them improve the quality of their decisions.

Setting Up and Executing Realistic Drills

One of the goals for training officers is to create realistic scenarios in a way that ensures students build valuable experiences while ensuring their safety. For some, this may seem like an obviously simple task. They may suggest, "Use a burn building to light fires and send fire fighters in to put them out," or, "Acquire some cars and practice cutting them apart."

Although the development of scenario-based training may appear simple, the setup actually requires time and effort to develop and build, especially if you want to do it correctly. Unfortunately, there are instructors who will take shortcuts when developing or executing these drills to make this process easier for themselves. Some instructors justify their actions by saying to the students, "We're going to show you a simpler, condensed version of what you will actually see or do on the real fireground in this training, but when you get out into the real world you'll do it a different way." This is a set-up for failure. Here we present three examples of training scenarios in which an instructor has set up an unrealistic drill and instructed students to perform a task in an unorthodox way that is not practiced on the actual fireground, and then we present an alternative way training should be done.

Scenario #1: Fire Attack Decision. In this drill, the instructor has built a fire load consisting of class-A materials (e.g., wood pallets and hay) in either an acquired structure or a burn building. The instructor lights the fires and sends the students inside. Once inside, the instructor tells the students to blacken the fire down, but that they are not to extinguish it. Is this training realistic? The answer is no. If this was an actual fire, we would not partially extinguish it. The reason the instructor took this approach is because it is easier to relight the fires for a future training if they are not completely extinguished. This shortcut provides convenience to the instructor, but it shortchanges the students on the actual training.

Scenario #2: Hose Line Decision. Borrowing the fire scenario from the previous example, the instructor tells the student to advance a 1¾-inch attack line into a burn building or acquired structure to attack a fire. The exercise is repeated many times until the students gain proficiency with this skill.

Although this drill seems as if it is legitimate, lurking in the shadows is a flaw that can lead to a poor hose line decision at a real emergency. In training sessions, students may be taught to advance a hose line using

Continued

a small diameter hose for two reasons: a smaller diameter hose line is easy to maneuver, and the training fires are never allowed to become large enough to require a larger diameter hose line, such as a 2½-inch hose line. For these reasons, the instructor has students advance a small diameter hose line repeatedly, instead of training them to advance the variety of hose lines they will actually encounter at a real incident. While this training has value in the sense that it can help students develop a valuable skill set—advancing a hose line—it limits the students' exposure to the types of fires they will actually encounter and the types of hose loads they will have to carry. For students to be prepared for real-life incidents, it is imperative that instructors expose them to the actual elements that are present in big fires as well as small fires. Students are typically adept at identifying small fire clues and cues. What they need to learn is what big fire clues and cues look like. You do not want students to have their first exposure to a large fire be on the day of an actual incident. They may finish training with the thought that pulling a small diameter hose line will be sufficient for any fire. Teach them better.

Scenario #3: Search and Rescue. The instructor is training the students on a search and rescue mission in a low or zero-visibility environment. As is done in many fire fighter training programs, the instructor "blacks out" the students' face pieces and sends them into a structure to conduct the search and to find the fire. Although there is nothing wrong with teaching students how to find their way through a low or zero-visibility environment, training students this way can set them up for failure. Hydrocarbon-based products make up the majority of the burning contents in buildings. When they burn, they are smoky and create a tremendous amount of heat. During the decomposition process (the stage at which the contents begin to burn), they produce large volumes of thick, black smoke, which produces lots of fuel. This smoke/fuel is as volatile as any hydrocarbon-based fuel, such as gasoline. When that fuel is raised to the proper temperature with the right mixture of oxygen, it will ignite. Oftentimes, it ignites violently. A fire fighter located in a zero-visibility environment is in a position of great danger. If a crew finds themselves in this predicament, they should exit the environment immediately. Most instructors who stress the need to keep fire fighters safe would agree. If the environment becomes a zero-visibility situation, leave! And leave fast! But, strangely, not all fire fighters do this. Even after losing complete visibility, some fire fighters continue to press onward, feeling their way through the smoke/fuel. A crew in this situation is in grave danger, yet they continue onward. You may be asking yourself, "Why would they do this?" The answer may be in how they were trained. If a fire fighter's training includes blacking out a face piece and entering a structure to conduct a search or attack and he or she was taught to keep going despite the lack of visibility, the fire fighter learns to press onward in zero-visibility conditions. This strategy is incorrect, as the fire fighter should be exiting in zero visibility conditions. In this case, the fire fighter was trained to fail.

That being said, we do not want to discontinue training fire fighters on how to maneuver in zero-visibility conditions. If this training on these scenarios stopped, crews caught in such conditions would never learn how to get out. Instead, fire fighters should practice learning how to exit in zero-visibility conditions. To train fire fighters on how to exit a building in low or zero-visibility conditions, begin by having the firefighting crews enter the structure with good visibility. At some point in the evolution, have the crew stop, black out their face pieces, and tell them the conditions have changed drastically. They are now in a zero-visibility situation and they should immediately exit the building. Have the fire fighters practice how to extricate themselves quickly from the room containing burning contents. When executing this drill, do not let the subtle nature of this drill get lost. Remember, you are training the mind and muscles for automatic decision making under stress. It is about the task and direction of travel. When students are improperly trained for this task, the focus is on performing search and/or attack and moving deeper into the structure where they will be exposed to massive fuel. When students are properly trained for this task, the focus is on rapid extrication from the environment and moving quickly toward the exit.

Using Repetition to Generate Automatic Decision Making

As students practice a skill, they develop muscle memory. Muscle memory is when a skill is repetitively performed to the point where it is unconsciously known, as if the muscles remember the movements that were taught. Remember, muscle memory is accomplished by performing the task over and over again, not from receiving verbal instructions. This is a very important distinction.

It is also important that instructors understand that while teaching practical, hands-on skills, they are also programming certain decisions or approaches into the student's mind. As the instructor teaches and has a student repeatedly practice a skill, the student becomes more likely to automatically perform the task the same way in a real scenario.

How a student is trained will affect the decisions they make at an actual incident. Think of it this way: every skill taught is a preprogrammed decision. You can teach your students to make better quality decisions in high-stress, high-consequence environments when you train them properly and you provided them with a variety of solutions to a wide range of problems. This means that you cannot take short cuts when setting up the training scenario; you train them on a variety of scenarios in which they will need to make decisions on what would be the best tools and strategies to use.

Examples of Application Step Formats

Tactical Decision Games

TDGs were originally created for the military as a way to train soldiers without using live ammunition. The purpose of TDGs was to create an environment of stress by providing a scenario with questions that have a sense of urgency placed on them. During these games, a time limit is imposed, which increases the sense of urgency. This is where the decision-making skills of the participants were tested, as it is human nature to forget to consider key factors or misinterpret the current situation when being rushed to make decisions. TDGs provide a way to test the decision-making capability of an individual with minimum cost, and they can be conducted while sitting at the kitchen table.

Online sources such as YouTube videos are available as well and can be great teaching tools. Before presenting the video, the instructor should preview it and establish the benchmark parameters, where the teaching points start and end, and indicate key highlights within the video. The instructor may also want to develop some role-playing exercises in relation to the content. During class, the instructor would then play the video with the plan to stop it at a predetermined point and present several situational and decision-based questions for students to answer. The instructor would then time the students as they develop their answers. Once the time limit is up, the instructor would then show the outcome of the video and follow up with a discussion on the outcome and the students' answers to the situational and decision-based questions. In this discussion, the instructor should have students evaluate why their approach(es) would or would not have worked. The use of critique is essential in this style of training.

Demonstration

Demonstrations are conducted when the objective of the training is to have the student understand a process, but performance is not required. For example, an instructor demonstrates how water flows through a centrifugal pump through a series of pictures or a cut-away pump. The students do not have to perform the actual task; however, they should be able to understand the process in the event that they need to troubleshoot their pump at an incident scene. One way this can be accomplished is to show how the equipment works while explaining the pump's function, its parts, and how it works. By simultaneously showing and explaining how the equipment works, students can immediately relate the information they are hearing to what they are seeing. This approach has proven effective as students have demonstrated a preference for learning this way. They also tend to have better rates of recalling information using this format, as opposed to the traditional method in which students are lectured within the confines in the classroom and then shown a demonstration after the lecture. As with any practical session, a simple three-step process should be used and adjusted as necessary per the individual student:

1. Instructor shows and tells the student.
2. Instructor shows as the student tells.
3. Student shows and tells the instructor.

Instead of the slow and methodical demonstration in which you talk the student through each step of the process, you may also want to consider showing the task in real time first. With this approach, the student can develop a mental image of what the operation or task should look like once it is perfected. After the initial real-time demonstration, the instructor should slow the process down so students have the opportunity to grasp how each step is performed within the specific task. The student should be able to slowly pick up the process and build speed and confidence with each repetition. Remember that not all students learn the same way or at the same pace. It may be necessary to repeat the demonstration several times in order for students to grasp the concept. This is especially true during new or complex operations.

Hands-On Drills or Training

The hands-on, commonly referred to as a hands-on training (HOT) drill, is a type of demonstration. Using the hands-on method allows students to actually practice the skill. As with any demonstration, the same process mentioned above should be used; however, with the hands-on approach, students should be allowed to work through the process. This approach allows the students to test themselves, and the instructor can also determine whether the knowledge was sufficiently transferred. Essentially, the students must demonstrate that they can perform the skill with an adequate level of competence.

Safety is a key component during HOT drills. As the percentage of retention for what has been learned goes up as the training becomes more realistic, so do the hazards involved. HOT drills require instructors to pay a certain amount of attention to detail and preplanning in order to mitigate and prepare for involved hazards. Another key consideration is that most HOT drills occur outside, so instructors must consider the weather conditions when planning. Additionally, the instructors will need to ensure that enough instructors and adequate resources are available.

Role Play

The use of role play necessitates the student to act in real time in response to a theoretical situation and generally involves interacting with another person, such as the instructor or another student. Role plays can take on several forms, and they are easily incorporated into other types of drills such as HOT, scenario-based, or demonstrations. The key advantage to using role play is that it allows the instructor to test the thought processes, presence, and interpersonal skills of the student as he or she demonstrates his or her response to a particular situation and/or other participant. As is the case in real life, a clear-cut answer to every issue that may arise does not always exist. With this in mind, the evaluator should be sufficiently trained to a perform in the particular situation being tested and should have an in-depth understanding of the situation at hand. The evaluator should be sure to review any department standard operating procedures (SOPs) or key benchmarks that each student should exhibit before conducting the role-play drill. Generally, the instructor should provide a script that includes these key benchmarks and any other necessary information for alternate actions based on the student's response.

This type of involvement can be used within scenario-based simulations, such as a group of peers sitting behind the student who is being evaluated who perform tasks as they are assigned (similar to on a fireground) by the student being evaluated. The instructor would provide the group of peers with scripts that they would follow that would provide vital information such as the heat and the smoke and fire conditions found within an involved structure.

Role play can be used to train personnel for a variety of situations, not just for emergency response. For example, role play could be used for personnel management training in the context of human resource issues. Although most employees in the government system get to go home at 5 o'clock each evening, fire fighters can be contained within a fire station for 24 hours per day. It is not uncommon for fire fighters to feel penned up and confined, and this can create some issues. To contend with these types of problems, some departments have implemented human resources training that includes role plays on how to handle situations such as station arguments, citizen complaints, and other altercations. By working out these situations through the use of role play, the student builds a mental model of how to handle these types of situations before they occur. Additionally, following up role play drills with a proper critique of how an individual responded will also enhance learning outcomes. The critique is also a good time to review SOPs with the students.

Scenario-Based Simulations

Scenario-based training requires a slightly higher level of involvement on the student's part. The student must be able to examine the situation, decide on the appropriate action, and execute that decision while understanding the effects of his or her decision on the outcome of the scenario. Scenario-based simulations are generally timed and objective-based. There are several ways to extinguish a fire, so key benchmarks should be met by the student.

There are a few different formats that can be used for scenario-based simulations. A traditional method is to illustrate the incident on a dry erase board and then read the scenario to the student. This is the most popular and safest method. As the students' skill level and situational awareness progresses during the training program, the instructor may want to build role plays using picture- and video-based scenarios into the training. Since these methods are more realistic than the dry erase board, the student's retention rate of the material taught will also begin to increase.

Most officer schools do, and should, provide some form of scenario-based training for their up-and-coming officers. Otherwise, we are setting up our officer candidates for failure if we do not properly prepare them and assist them in building mental models of how incidents should be handled. Scenario-based simulations can be some of the most dangerous drills that we perform simply due to the realism that this type of drill involves. However, it is the training officer's responsibility to foresee and mitigate any and all hazards associated with the training. There also has to be a well-defined plan that encompasses assignments for each individual participating in the drill, including the objectives to be met, safety concerns, and what to do in the event of an emergency.

In addition, as a safety measure, it is best to incorporate these types of drills toward the end of classes, unless this is the focus of the class itself, in which case several other factors need to be considered such as prerequisite training, certifications, and authorization from their chief. The primary objective with the use of this training is to introduce the information in the safest possible environment. This will allow the student to practice and determine which method works or does not work for him or her.

There are several ways the scenario-based training method can be used. Some examples are provided below:

1. **Hot seat scenarios.** The drill begins with one student playing the role of a particular subject, in this case an officer. The instructor projects several images pertaining to the incident (e.g., pictures of a heavily involved multiple occupancy structure) for the student to study. In addition, a panel of officers and/or instructors acting as different apparatus and performing assignments may be included as part of the scenario. By placing the student in the hot seat where he or she needs to respond in real time to the simulated incident, the instructor recreates the stress, the sense of urgency, and information overload that comes with performing at such an incident. It exposes students to the complexity involved with such incidents and also the task of tracking assignments for all involved personnel. It also allows the student to practice communicating with crews that he or she cannot see. Generally, scenario-based training simulations are timed events, which include several benchmarks denoted for each individual participating in the drill. Instructors do not need to develop these scenarios from scratch; there are several computer-simulated programs and tools (such as CommandSim, StageIT Emergency Response Fire Simulator, and Fully Involved Fire) available that provide hot seat scenarios that instructors can use in their training. If budget is a concern, however, instructors can create a simple slideshow using Microsoft PowerPoint to provide this type of training.
2. **Live fire scenarios.** This type of drill involves a live fire, so the participants' safety is an inherent concern. To develop a live fire scenario, begin by writing down the objectives of the training by stating exactly what you want accomplished by the end of the scenario. These objectives will drive the direction that your drill takes.

Below is an example of a list of objectives for a live fire scenario in which an incident commander coordinates a class-B kitchen fire (The audience and degree are assumed below based on the national standards and department SOPs, which should be explained during the lecture portion of the class.):

1. Find and completely extinguish a class-B kitchen fire.
2. Perform a primary search and rescue operation extricating all victims (one dummy victim per drill).

3. Perform a coordinated roof ventilation operation with the fire attack crew and command.
4. Have backup fire fighters assist in deploying and flaking out the attack lines.
5. Establish a rapid intervention crew (RIC) with the necessary tools for operation.
6. Designate an officer to perform initial size-up and verbalize the size-up to responding crews.

Next, based on your objectives, determine how many instructors you will need to execute this drill. If the risk is high or you exceed the required ratio of five students per each instructor, add additional instructors. Make sure that the instructors you choose to include are qualified to assist in live fire drills. Then, you should document and address the assignments and safety factors based on the location, instructors, resources available, and type of burn building. Be sure to include instructor and student briefings. Lastly, write your scenario. Several examples of scenarios based on the above objectives are listed:

Each scenario presents a different problem that has to be solved by the individuals involved. This process teaches students to practice situational awareness, fireground decision making, and good communication skills. Live fire scenarios can be used for recruits, line officers, and command level officers. They can also involve multiple agencies, such as emergency medical services (EMS), law enforcement, and neighboring departments. As is the case with any drill, the key to making scenario-based training successful is to end each drill with a critique explaining what went well and what could be improved upon.

Potential Obstacles Associated With the Application Step

As you will find in any training program, there are numerous obstacles that have to be considered during the application step. These obstacles may interfere with the method that the instructor chooses, including the limitations or potential issues concerning the training environment, time restrictions and scheduling issues, the number of assistants or resources available, and potentially challenging weather conditions. You will need to consider these obstacles when developing the application step for your lesson.

Limitations or Potential Issues Concerning the Training Environment

The location of the training can pose various challenges. For example, if the drill needs to be performed outside, the element of weather conditions comes into play, as well as additional distractions such as busy

Scenario #1. Engine 7 (E7), Engine 8 (E8), Truck 10 (TK10), Squad 10 (SQ 10), Medic Unit 1 (M1), and Car 1 respond to a reported structure fire. Resident stated there was an oven fire and that the family is evacuating now.

*Upon arrival, the officer is informed that a child is still inside the structure (located in B/C corner).

*The fire is located in the kitchen with rollover.

The incident commander (IC) should make assignments as necessary, coordinating each operation. The IC should inform crews when the apparatus has arrived.

Scenario #2. E7, E8, TK10, SQ 10, M1, and Car 1 respond to a reported structure fire. A child resident stated there is a lot of smoke in house and does not know where his grandfather is.

*An adult victim is to be placed on second floor.

*The fire is located in kitchen with no rollover; heavy smoke is present.

Scenario #3. E7, E8, TK10, SQ 10, M1, and Car 1 respond to a reported structure fire that is called in by the alarm company.

*An adult victim is to be placed on first floor with his arms hanging out of C-side window.

*The fire is located in kitchen with no rollover; light smoke is present.

traffic, pedestrians, or the sounds of construction. Conducting drills inside a classroom may limit the amount of room available for the drill, but the instructor may feel that he or she has more control over the learning environment and the students. When planning the application step, try to reduce as many of the distractions as possible and keep students busy by reducing downtime. Your goal is to help students keep their minds focused on the task at hand.

■ Time Restrictions and Scheduling Concerns

If the amount of time that the instructor has to conduct a drill is an issue, it may limit the interaction that the instructor has with the students. With career fire fighters, more structured in-service trainings occur; however, when planning training for volunteers, time is a major factor. The time allotted for the application step, regardless of whether you are training careers or volunteers, must enable the students to achieve the required level of proficiency. If planned appropriately, using demonstration with hands-on applications, the students will learn just as much as in a cognitive session, but in a more-involved way. When scheduling the drill for the application step, consider your audience:

1. **Station personnel.** Training should be based on the likelihood that an emergency call will occur sometime during the day and interrupt the lesson. However, weekends are usually acceptable.
2. **Staff career personnel.** Training is usually conducted during the weekday schedule that personnel typically follow. However, you can schedule the training to occur on a weekend if necessary.
3. **Volunteer personnel.** Scheduling training on a typical work day (i.e., 9 a.m.–5 p.m., Monday–Friday) is generally not an option because volunteers usually have a regular, full-time job. Training has to be conducted at night or on the weekends. Volunteer programs taught in-house generally take longer to complete; however, the training schedule should try to accommodate the schedules of all volunteers, as appropriate. Training officers may also have to consider volunteers who work second or third shifts and determine how these volunteers can be trained given their schedules.

■ The Number of Assistants Available

The number of assistants available may dictate exactly which drills you chose to accomplish. The recommended student-to-instructor ratio for high-risk activities, such as live fire drills, is 5 to 1 per National Fire Protection Association (NFPA) 1403, *Standard on Live Fire Training Evolutions*. If the drills are low risk, a lower ratio can be used, determined by the instructor-in-charge. The key to making the application step a success is to ensure that you have enough instructors to oversee all the students, offer advice, and provide direction to correct any deficiencies. Safety is a key priority when considering the number of assistants, and with enough assistants, team teaching is an option. Using the team teaching technique requires that each instructor use the same lesson plan, terminology, and know the skills of the other instructors so that they can work within each others' strengths and weaknesses.

■ Planning for Weather

There are several issues to consider when planning an application step outside with regard to weather conditions. Consideration will need to be made regarding the time of year and the geographical location. Some departments have implemented SOPs that dictate the level of activities students can participate in depending on the weather. When planning for the weather, instructors may want to include the use of weather charts and flags in conjunction with wet-bulb readings. (The wet-bulb temperature incorporates ambient temperature, humidity, air flow, and other factors.) Another method is to add 5°F to the temperature for activities that require turn-out gear. The instructor can then compare this temperature against a weather chart to determine work-to-rest ratios. The instructor should also make backup plans in the event that the weather is not cooperating on the day of the scheduled activity. If it is raining, the simplest solution may be to move the drill inside or move to an alternate location with shelter. If it is snowing, the instructor should take care to remove any slip hazards or change lesson plans with another day's topic. Performing activities in lightning and heavy rainstorms should be avoided at all cost.

■ Adequate Resources

Ensuring an adequate amount of resources for certain drills can be difficult without proper planning. For example, how many departments have spare stocked engines available for training without taking them out of service? Even with the proper planning, if the department cannot afford to purchase the equipment or enter into a use agreement with neighboring departments, it may not be possible to conduct the drill the instructor wishes to implement, or the instructor may need to construct props from scratch. If the props are constructed, make sure that they are safe to use and are operable. In addition, do not take any shortcuts when developing them. If the students are to be tested at the conclusion of the course, the same equipment used during the application step should be used for the testing session. It is quite unfair for a student to be trained on Hurst tools during the application step and then be asked to perform the same techniques using TNT extrication equipment during the evaluation stage.

Training for Adaptability

A fire of any kind is very dynamic and can change at a moment's notice. The same is true for emergencies that occur on the fireground, whether it is a collapse, flashover, smoke explosion, rescue attempt, or any number of other situations. The key is to be able to decipher these events and the information available. The decisions you then make are based on your ability to adapt to a given situation. In order to accomplish this, instructors must learn to train their fire fighters for adaptability.

As fire fighters, we must use our past experiences and training to maneuver through and mitigate each emergency response. One particular program that is available to fire fighters is the Georgia Smoke Divers program. This program instills the following:

1. Attention to detail
2. Learn your limitations
3. Learn your equipment's limitations
4. Learn to adapt based on the situation

The first three items set you up with a firm foundation on which to base your decisions. The fourth item is carrying out your decisions and applying hands-on training with heat, fire, smoke, and chaos affecting your decisions.

As a candidate, there may not be a tremendous amount of new skills discovered; however, a major factor is being able to answer the question, "Can I do it when I'm distracted, mentally and physically exhausted, and when it counts?" The training scenarios and the realism of the Georgia Smoke Divers program extend beyond most other training programs available. As with any training, a critique afterwards to receive thoughts and opinions on how to perform more efficiently next time is important.

Key Points

- Using an appropriate mixture of all the learning domains (cognitive, psychomotor, and affective) in the application step will allow for the greatest amount of knowledge to transfer from the instructor to the student.
- It is our responsibility as training officers to protect those that are learning the job and to provide an environment where they can succeed.
- The action plan keeps the class on track and organized. It also provides a blueprint of activities and serves as documentation of the training delivered.
- Combination of the cognitive and psychomotor lessons should be tailored towards the general capabilities of the students when possible, as not every class is the same.
- The ability to make timely and effective decisions at an emergency scene requires a combination of knowledge, training, and experience.
- The more knowledge, training, and experience a person acquires, the more likely it is that they will be able to make high-quality decisions with little effort.
- The mental model is where the decision maker forms a mental picture of how the action plan will play out.
- The percentage of retention for what has been learned goes up as the training becomes more

realistic, so does the hazards involved. Safety is the number one priority.
- The decisions you make are based on the information you receive from the situation and your ability to adapt to a given situation.

References

Georgia Smoke Divers. www.GeorgiaSmokeDivers.com. Accessed June 26, 2014.

Klein, G. *Sources of Power: How People Make Decisions.* Cambridge, MA: MIT Press; 1998.

National Fire Protection Association (NFPA) 1403, *Standard on Live Fire Training Evolutions*. 2012. http://www.nfpa.org/codes-and-standards/document-information-pages?mode=code&code=1403. Accessed May 5, 2014.

Zsambok, C.E. and Klein, G. (Eds.) *Naturalistic Decision Making*. Mahway, NJ: Lawrence Erlbaum Associates; 1997.

CHAPTER

9 The Evaluation Step

Alan Brody

The Evaluation Step

The evaluation step can be broken down into two main focus areas:

1. Evaluation of the instructor
2. Evaluation of the student

Both aspects should be evaluated equally. Within the instructor evaluation domain, items such as the learning session, which include the content, the course, the instructor capabilities, and the physical setting, all provide important information for an instructor to evaluate. This chapter will also discuss how to evaluate the students' progress and will provide testing procedures.

Training officers have to continuously improve in order to stay abreast of the most current practices and technology. Ongoing evaluation will assist in achieving this goal. Evaluations can be measureable or qualitative in nature, but they need to be fair and should not leave room for instructor interpretation. There are many items to be considered when preparing the instructor's evaluation. First is the obvious question: What objectives do we test? There are also the not-so-obvious factors that need to be considered, such as safety, environment, set up, pass/fail points, scoring system, and test administration. As we progress through the chapter, we will discuss these concerns as well as other important factors to help you, as the training officer, prepare for the evaluation step.

Evaluating the Instructor

Evaluating the instructor is an important process from which we can glean much needed information. By assessing the instructor's delivery, we can determine whether he or she is succeeding at the key components of teaching, such as how well students retain the information or how well the learning process is going. The evaluation process can take on several different forms, such as formal or informal and formative or summative. For example, the training officer could conduct an interview with a student where they review the course materials at the end of a course (summative) in an office setting (formal). Or, the training officer could sit in a class and provide feedback during a break (formative and informal). The training officer could incorporate any combination of these types of assessments into the evaluation process. Regardless of which method you use to assess instructors, be sure to provide feedback in a timely manner and preferably before annual evaluation time. Conducting ongoing assessment and continuously providing feedback to instructors will allow them to work on weaknesses and improve their teaching skills.

Instructors should be capable of properly preparing the classroom and their lesson plans, props, activities, and other teaching-related tasks for each class. Instructors need to remember how important it is to review the content and materials before entering the classroom to ensure that they are knowledgeable about and comfortable teaching a particular subject. This level of proficiency can be accomplished by simply reviewing the lesson plans a week prior to the class actually starting.

Areas of Focus During the Instructor Evaluation Process

Adherence to Safety Guidelines and Creating a Safe Training Session

Instructors are responsible for conducting training that is safe for all members, and ensuring that the training is being conducted safely is the number one priority of any training officer. When evaluating the instructor's ability to create a safe learning environment, questions that the training officer may want to ask are: Is the classroom, training ground, training props, and equipment in a safe condition? Is there a training safety officer at the training site, and is this individual trained sufficiently?

Successful Learning Outcomes

It is important for training officers to determine whether the student is learning something from the training. You know a training program is successful when the students complete a course and have demonstrated competency in meeting certain certification requirements and have achieved the objectives set forth by the instructor. This will be discussed in more detail later in the chapter, but learning outcomes can be evaluated in many different ways, ranging from a simple question-and-answer session to a more formal written test.

Classroom or Training Session Set Up

The time, effort, and attention to detail instructors spend on classroom set up is another important consideration in the evaluation process. Attention must be paid as to whether the equipment that will be used for the training is set up appropriately and safely for class. You should inspect the props to be used in the training session for operational readiness and safety. Check whether the instructor has provided ample room for student and instructor interaction. You should also evaluate how the seats in the classroom are arranged for the current lesson. If the instructor is going to be conducting a lecture with a PowerPoint presentation from the front of the classroom, are the seats arranged so the students can see the instructor and presentation? Other items to consider when evaluating classroom set up include, but are not limited to:

- Whether the instructor ensured that the classroom has adequate lighting.
- Whether the instructor ensured that the classroom or learning environment is at a comfortable temperature (environments that are too cold or too hot can cause distractions and affect the focus of students).
- Whether the instructor ensured that there is comfortable seating for all students.
- Whether the instructor arranged the room so everyone can see the speaker and/or any visual aids.
- Whether the instructor provided adequate breaks (i.e., a 10-minute break for every hour of learning).
- Whether the instructor stayed on schedule.
- Whether the instructor ensured all training props would be available for the training session well in advance of the training.
- Whether the instructor implemented measures to ensure minimal interruptions.
- Whether the instructor ensured all training material is on hand well in advance of the training session.

Anticipation of Group Needs

The students' needs can be planned for by explaining the procedure they were expected to accomplish, enforcing the safety requirements and concerns of the course, and providing for rehabilitation of the students who experience high work stressors. Also, ask the students for suggestions on what could make the class better. All sessions should begin with an introduction in which the instructor explains the possible safety/injury concerns and the need to understand the skills being discussed. All instructors should use due diligence when explaining any procedures that may be new or foreign to students.

The Instructor's Ability to Motivate the Class

The training officer should evaluate whether the instructor provides feedback and uses strategies to motivate the group. The group needs continual motivation from the instructor as well as other students. The instructor should provide constructive criticism in order to help build morale among the group and the confidence of each student. As the students gain confidence, the group gains it as well. Once several students are motivated, other students may catch on, as motivation is contagious.

Creating and/or Implementing Activities That Are Based on Course Objectives

The training officer should evaluate the instructor on whether he or she has created and/or implemented activities that are directly related to the objectives of the drill. It is important that instructors create meaningful, well-executed activities that are based on the objectives of the training session, as this will help the instructor achieve the desired outcomes for the lesson. The training officer should also confirm whether the instructor discussed the objectives with the students. It is important that the instructor explains the objectives of each drill to the students, because the students need to understand what skills should be achieved by the end of the drill and why they are completing the activity. At the very minimum, students should be able to demonstrate the required level of competency by the end of a drill. However, we can also evaluate whether the instructor exceeded expectations with the training. As the training officer observes the class, questions that he or she could ask himself are: "Did the participants exceed the objectives of the session by completing the tasks in a short period of time or by demonstrating exceptional attention to detail? Did the students provide additional insight to the objective outcome?"

Ability to Activate Prior Knowledge and Learning

Another thing that the training officer can look for is whether the instructor attempted to activate students' prior knowledge and learning during the training session. It has been demonstrated that activating prior knowledge and learning leads to better retention of information because the students are able to relate the current lesson to a previous experience. Students may experience more depth to their learning if they can relate a particular skill or drill to previously learned knowledge, particularly if it helps them understand why they are learning a particular drill or approach. One way the instructor can activate prior knowledge is to facilitate discussions among students. Discussions provide opportunities for students to ask questions and share experiences, which in turn may help the students relate the current topic to their requisite knowledge.

Use of Critiques to Enhance Student Learning and Outcomes

The training officer should evaluate the instructor on his or her use of critiques to promote learning and improve outcomes. Any criticism should provide feedback while being respectful toward the students. It is important that the critiques are honest and constructive, not only so students understand their strengths, but also what areas they need to work on.

■ Conducting the Instructor Evaluation

All instructor evaluations should be conducted in a private manner that allows the students to complete them truthfully and honestly. Generally, the instructor should leave the classroom while students are completing instructor evaluations. The instructor evaluations should be collected and the information compiled into a document, which is given to the instructor for review. The training officer should then follow up and meet with the instructor to discuss the positive and negative outcomes of the evaluation.

For the instructor, it is important to take some of the feedback from students with a grain of salt. The instructor should remember that it is nearly impossible to please every student, but the instructor can, however, learn from the comments and use them in a constructive way to help him or her make the class better next time. It is also important for the instructor to recognize that a comment takes on relevance if it indicates a pattern. If several students are mentioning the same items over and over again, it may be worth looking into how that particular problem can be fixed or the skill can be improved upon.

Evaluating Students' Learning

Just as instructors are evaluated for their performance as trainers, students are evaluated to determine if learning took place. Exams or assessments will help the instructor identify areas of weakness in his or her teaching or areas that may not have been covered in the course. These exams, also known as criterion-referenced assessments, show a correlation between the criteria or standard tested against and the student's mastery of a topic. Generally, a score of 70 percent is considered passing by most departments; however, some departments and state/provincial testing agencies are moving away from using a scoring system and are using a pass/fail system instead (discussed in more detail later in the chapter).

In addition, some departments have used norm-referenced assessments for promotional testing in order to rank individuals based on their test scores.

Developing Assessments

Regardless of which type of assessment is used, the test must be validated to ensure that the assessment is accurate and well-founded. There are a few items that should be considered when developing an assessment. The first question to consider is, "What do the objectives say?" The assessment questions should reflect the objectives established for the course. For example, if the objective says the student should be able to explain a task, then the assessment should ask the student to explain the task.

It is important that your test questions also match the objectives delivered at the beginning of the course, as these represent the direction and framework of the course. Since the students base their studies on those objectives, it is only fair to test them on what was taught and learned in the course. Students need to have clear and concise test instructions. The scoring methods and the criteria for spelling, grammar, and terminology should also be clear in your instructions. Tests using department-specific or fire service jargon must be reviewed for consistency and validity.

The instructor also needs to consider the amount of timing allowed to complete the test or skill. Typically, the authority having jurisdiction (AHJ) will determine the time or number of attempts allowed for the exam, except in cases where a state/provincial or national accreditation is being used. All results from testing need to meet local confidentiality legislation or regulations. Caution must be used when discussing results with students. It is not acceptable to post the names and scores of each student in a place where everyone can see. When posting scores, other methods should be used to conceal the students' identities, such as using the last four digits of a social security number.

Next, the instructor should match the assessment items, or test questions, to standards, such as NFPA standards or State EMS Scope standards, or they should be matched to best practices outlined in the manufacturer's guidelines or the department's standard operating procedures or standard operating guidelines (SOPs/SOGs). Doing so ensures the assessment items or questions are valid. If a question is challenged, you will be able to reference a standard or guideline to confirm the correct answer. Matching assessment items to standards or the department's guidelines on best practices also limits the possibility of conflicting information coming from various sources. The instructor should make sure that the students know what bank of information the assessment will come from. Once test questions are matched to standards, a group of peers should review each question/statement, the distractors (false answers), and the correct answer. If all peers agree that each assessment item can be matched to a standard or best practice, then the test item is validated. When matching items to a standard, be sure to use the specific job performance requirement (JPR). This process allows for quick referencing of materials. This process should also be completed for any performance or skills-based assessments. Agencies such as the National Professional Qualification (NPQ) require this process to take place for any agency issuing NPQ certificates upon completion of certification courses, such as Fire Fighter I, Instructor II, or Incident Safety Officer.

Frequency of Evaluations

One of the first items to consider is the frequency of your evaluations (**TABLE 9-1**). How often evaluations

TABLE 9-1 Frequency of Tasks Performed on a Quarterly, Annual, and Biannual Basis

Frequency of Tasks	Example
Quarterly	Self-contained breathing apparatus (SCBA)
	Defibrillator
	Standard company evaluations
Annually	Tasks unique to jurisdiction
	Hazardous materials
Biannually	CPR

	Date	Candidate Signature	Evaluator Signature
5.3.13 (B) Breaches a Masonry Wall			
5.3.13 (B) Breaches a Metal Wall			
5.3.14 (B) Salvage Cover Spread			
5.3.14 (B) Salvage Cover Fold			
5.3.14 (B) Salvage Cover Roll			
5.3.14 (B) Construct a Water Chute			
5.3.14 (B) Construct a Water Catch All			
5.3.14 (B) Stop Water from Sprinkler Head			
5.3.14 (B) Close and Open Main Control Valve (OS&Y)			
5.3.14 (B) Close and Open Main Control Valve (PIV)			
5.3.15 (B) Assemble and Connect Equipment for Drafting			
5.3.15 (B) Transfer Water Between Portable Tanks			
5.3.16 (B) Extinguish an Exterior Class A Fire			
5.3.16 (B) Select and Use Fire Extinguisher			
5.3.17 (B) Set up Generator and Lights			
5.3.18 (B) Shut Off Electric Services			
5.3.18 (B) Shut Off Natural Gas			
5.3.18 (B) Shut Off LPG			
5.3.19 (B) Extinguish Ground Cover Fire Using Water			
5.5.2 (B) Accordion Hose Load			
5.5.2 (B) Horseshoe Hose Load			
5.5.2 (B) Flat Hose Load			
5.5.2 (B) Donut Hose Roll			
5.5.2 (B) Twin Donut Hose Roll			
5.5.2 (B) Hose Drain & Carry			

FIGURE 9-1 Georgia Firefighter Standards and Training Council, *Georgia Basic Firefighter Candidate Task Book.*
Data from: Georgia Firefighter Standards and Training Council Core Competency Taskbook. Georgia Firefighter Standards Training Council.

must be performed is determined by state/provincial or local codes, local laws, or the AHJ (i.e., the fire department, city council, or board of directors). For example, some agencies may require that a competency-based task book be completed each year for state/provincial certification, in addition to the end-of-course evaluations required for certification (**FIGURE 9-1**).

Selecting a Type of Test

The next step in the evaluation process is to determine what type of test to conduct. There are various pros and cons to each type. Test usage, scoring, time, comprehension, and intended learning levels all play a part in selecting the most appropriate assessment.

Types of Tests

The first question to ask is, "What type of test(s) are you going to give?" You will also need to consider the objective of the assessment as you select a type of test. Are you checking for a desired knowledge level prior to class, the students' understanding at a particular point in the class, or mastery of a topic at the end of a course? To thoroughly evaluate the students, you will need to use a variety of assessment types as discussed above, such as formative, summative, formal, and informal. The different types are described below and will provide you with a sense of when each type may be given.

Prescriptive/Pretest

The prescriptive or pretest is used to determine whether a student has the prerequisite knowledge levels before the class begins. Some courses require an advanced knowledge in certain subjects prior to entry. The pretest can be either norm-referenced or criterion-referenced assessments depending on the objective of the test. For example, an objective for a criterion-referenced pretest may be: "All students must achieve a score that is equal to or greater than the 80th percentile to be accepted into the class." In contrast, an example of an objective for a norm-referenced test would be: "The top 20 scores are accepted into the class."

Progress Test/Quiz

The progress test or quiz is used to check for student understanding at various points throughout the course of a program. Some programs, such as paramedic training schools, require students to maintain a certain grade point average in order to advance to the next level of instruction. Progress tests or quizzes can also be norm-referenced or criterion-referenced assessments.

Comprehensive Test/Final

The comprehensive test is an all-encompassing exam, which takes place at the end of a course and measures the students' overall knowledge of the subject covered in the course. Taking a comprehensive or final exam is similar to the experience of taking the National Registry Boards for EMT status or taking the Fire Fighter I test. The comprehensive tests are criterion-referenced assessments that are based on standards and best practices.

Test Formats

After selecting the type of assessment, the training officer needs to select the test format to check the students' level of learning. There are a wide variety of formats that the instructor may choose from, including written, oral, and/or performance. Recall that each level measures something different, and as you increase the difficulty of the test, the more difficult it is to properly grade.

Written Tests

In general, a written test is used when the objective is to test the students' recall, recognition, or comprehension of the material (e.g., remembering information from the course textbook). The written test can take on many forms, such as multiple choice, fill-in-the-blank, short answer, and essay. Although the written test helps the instructor measure students' ability to recall information, it can also measure whether higher levels of learning were achieved by using forms that require the student to not only recall facts, but also apply them. The essay form is one way to measure this level of knowledge. This is called the evaluation learning level (see the level of difficulty list below). For the first three types of written tests—multiple choice, true or false, and matching—the answer is provided to the student in some form, making it easier for the student to provide the correct answer. These types are also extremely easy to grade. The level of difficulty increases as you move down the list, however. The forms that follow—fill-in-the-blank, short answer, and essay—do not supply options to the student. These types of assessments are much more difficult for the instructor to grade as they have to interpret what the student is saying. For these types of assessments, the instructor should develop a rubric that notes the benchmarks or key points that should be stated in the students' answers. The different formats of written tests arranged in ascending level of difficulty are:

- Multiple Choice
- True or False
- Matching
- Fill-in-the-Blank
- Short Answer
- Essay

When developing questions for written tests, it is important that each question is stated in simple, clear, and concise language. The instructor should also use approved terminology and refrain from using jargon. The wording of the questions and answers needs to be grammatically correct, and the style of the wording should be consistent as well.

Multiple Choice The multiple choice test includes questions that are followed by four or five possible answers that the student may choose from. This test assesses the student's ability to recall and recognize information and is considered to be a lower level of learning assessment.

In the list of possible answer choices, the option that contains the answer must be the only possible correct answer. The other answers should be distractors, in that they would be close to the real answer, but not exactly correct. This is done to avoid ambiguity. Additionally, the test question should not include any "clues" that would give the answer away, and the position of the correct answer in each multiple choice list should also vary. The use of answers such as "All of the above" and "None of the above" should not be overused. Lastly, each question should be autonomous—in other words, no question should give any clues or provide the correct answer to any of the other questions in the test. An example of a multiple choice follows.

Multiple Choice Read each question below and circle the correct answer.

1. SCBAs are required to be inspected
   A. Once a day
   B. Once a month
   C. Bi-annually
   D. At the beginning of each shift

True or False For the true or false format, the instructor provides a list of statements and the student has to determine whether each statement is true or false. This type of test can be very difficult to construct as the question must be entirely true or false. These tests also provide the student with a 50/50 chance of guessing the correct answer. This form is best used in combination with other forms, such as multiple choice or short answer, on a larger exam. In some cases, the student is asked to make a false question true by providing explanations as to what makes it a false question. An example of a true or false question follows.

True or False Please read each statement carefully. If the statement is true, write *True* beneath the statement. If the statement is false, write *False* and write a sentence explaining why the answer is false.

True or False: The captain has one bar on insignia.
Answer: False—Based on department SOPs, the captain rank has two bars on insignia.

Matching The matching section of a test consists of a column of questions and a column of potential answers that are listed in random order. The student has to match the question to each answer. Matching sections are also good for checking the students understanding of key terms. In this case, one column would include the terms and the definitions would appear in the other column. As is the case with the two previous forms, the matching sections test the students' ability to recall material. An example of a matching section is displayed below.

Matching A list of key terms is listed in column A below. Locate the definition of the key term in column B. Indicate the answer by writing the letter accompanying the definition next to the key term in column A.

Column A:	Column B:
____Adapter	**A.** Plumbing accessories for connecting hoses and pipes of incompatible diameter, thread, or gender.
____Ladder	**B.** Pressurized water source for fire apparatus (typically part of a municipal system).
____Water hammer	**C.** A structure of wood, metal, or aluminum, commonly consisting of two side pieces between which a series of bars or rungs are set at suitable distances, forming a means of climbing up or down.
____Hydrant	**D.** A pressure surge or wave resulting when a fluid (usually a liquid but sometimes also a gas) in motion is forced to stop or change direction suddenly (also known as momentum change).

In a matching test, all the answers should be conceivable alternatives for each question.

Fill-in-the-Blank/Short Answer Fill-in-the-blank and short answer sections require the student to provide a short response. With the fill-in-the-blank section, the student must supply a word, or possibly several words, to complete an incomplete sentence. A short answer section consists of questions that the student must answer in several sentences. There are several ways to write a short answer test, and most have a fill-in-the-blank component; however, some could require a sentence or two to complete. Fill-in-the-blank sections generally only require students to recall information, whereas the short answer sections typically require the student to demonstrate some comprehension of the material as they are asked to explain or justify the answer. Examples of fill-in-the-blank and short answer questions are below.

Fill-in-the-Blank Please read each sentence below and supply the missing term.

1. The ______ apparatus is required to have a combination of several sizes of ladders.
   Answer: truck
2. The spanner wrench can be used to tighten _____ when there are severe leaks.
   Answer: hose couplings

Short Answer For each question below, please write a response that is 3–5 sentences long.

1. Explain the process of flashover.

   Answer: Everything in a room reaches its ignition temperature simultaneously. High heat and rapid environmental changes precede the flashover. Research has shown that most flashovers occur generally within the 7- to 10-minute mark. However, there are existing variables that can change the timing.

Essay The essay form is typically the most stressful part of the exam for the student to complete, and for good reason, as it requires the student to demonstrate a higher level of learning. For an essay question, the student is required to supply a short detailed statement, which may include several paragraphs. The essay is a useful assessment, as it can demonstrate the depth of learning the student has attained during the course or lesson. When developing the essay question, instructors should have specific key points or benchmarks in mind that they want students to include in their answers. The essay question moves away from the objective-based questions that we have discussed. A certain amount of subjectivity is part of the essay assessment, which can make it more challenging for the student to answer and also for the instructor to evaluate. An example of an essay question and a student's answer is below. Following this sample is a short list of key points, or a rubric, that the instructor will use to evaluate the student's answer.

Essay Please read the question below carefully and write a thorough, detailed response. Be sure to provide examples in your answer.

Question: Discuss the practice of coordinating fire attack and ventilation.

Answer: Fire attack and ventilation should be coordinated in order to properly assist the attack crews and save any possible lives that may still be in the structure. The fire attack crew should locate the fire before performing ventilation operations, according to the situation on scene and information received from interior crews. If ventilation is performed prematurely, the fire may be drawn toward the attack crew, develop beyond the capabilities of the attack lines, or harm savable occupants. If the two tasks are coordinated between the ventilation and fire attack groups, the fire risk profile will be reduced, temperatures will drop, and visibility will increase.

Key points that should be considered when evaluating this essay question:

- Coordinated effort must be indicated between the fire attack and ventilation crews.
- Improper ventilation may draw the fire toward or behind attack crews, blocking the means of egress.
- The two functions coordinated together will lead to reduced temperatures, increased visibility, and increased chances for occupant survival.

The Oral Exam

Although oral exams do not involve writing, they are similar to essays in the sense that they can be difficult to grade and require the instructor to develop a rubric or answer key with key points/benchmarks indicated. Many promotional assessment centers use oral exams. With promotional testing, it is best to film the student's answer. It is very easy for instructors to become distracted during oral exams and miss a key word or phrase that would affect their evaluation of the student's response. Generally, oral exams should not be used as the only form of assessment; some other form of testing should also be used to determine a student's complete understanding. Sometimes, an oral exam is necessary to accommodate students with a learning disability. Using an oral exam to accommodate a student's needs should be reviewed by your legal counsel and human resources department. Training officers and instructors are not allowed to diagnose disabilities. The designation of a disability generally requires a doctor's letter.

Performance Testing

In the fire service, the use of performance testing is necessary to determine whether the student learner has comprehended and mastered a skill. Because there is the potential for fire service members to sustain injuries, or even die, while performing job-related duties, it is absolutely critical for fire departments to test the skills performance of all employees. This testing is done to ensure safe operation of a fire department, and it is also required under some states'/provinces' legislation and is part of a state/provincial license or test. As is the case with any skill testing, the student must be able to practice the skill before being tested. The performance test checklist should provide the following:

- Step-by-step instructions for the learner, including the criteria expected.

- Evaluation criteria; the AHJ may have specific times or ratings assigned to a skill.
- A summary of the skill and how it fits into the "bigger" picture.
- Itemized skills and step-by-step instructions, presented in bullet form.
- A place for the student to sign and date.
- A place for the instructor to sign and date.
- Date that skill was completed.
- Number of attempts to master the skill (a maximum of two is usually allowed).

The performance testing must be conducted in a safe environment under as ideal conditions as possible. The student needs to be provided with clear written instructions on how to perform the skill, and the testing environment should be adequately prepared so that the skill can be demonstrated. A training safety plan (TSP) should be put in place prior to conducting performance testing.

Some performance tests require more than one fire fighter to complete a task. During this type of session, fire fighters rotate through each position. After the testing, the student should be advised on how well he or she performed. The performance test checklist or associated document should also be completed at that time. Checklists are crucial to identify the skills tested during the skill testing session. It is generally accepted that in order to properly test all domains, a performance test and either an oral or written test should be given, at the very minimum. These types of tests assess the cognitive and psychomotor domains.

Other Evaluation Points to Consider

When deciding what should be evaluated or tested, there are several items the training officer/instructor should consider:

- Does a standard, regulation, or legislation require members to demonstrate a certain level of competency for the skill/knowledge?
- Is the knowledge/skill a life safety skill?
- Does the knowledge/skill provide a due diligence for the fire fighter, department, or municipality?
- Is the knowledge/skill a requirement under terms of employment?
- Is the knowledge/skill a requirement for a promotion?

When determining if the knowledge or skill falls into any of these categories, it is best to look at best practices in the fire service.

Providing Accommodations During the Evaluation Process

When conducting evaluations, you have to consider the needs of the students. Language, educational, and literacy levels can cause the evaluation process to be unfair to some students. It is important to set the ground rules early in your training session to let the students know what will be expected of them during the evaluation/testing process. This will give the instructors of students who require special accommodations the necessary time to make the appropriate arrangements. However, be careful to not diagnose students with disabilities. If a student claims to have disabilities, simply ask them to provide documentation so you can address their needs as appropriate. Typically, you will need to coordinate with the human resources department to ensure that the correct measures are taken.

Determining Success Levels

To determine success levels, a clearly identified and achievable objective must be prepared prior to the evaluation. With a clear and achievable objective, the instructor will be able to determine the minimum level of desired success. Levels of success may be acceptable in certain circumstances as elected to by the AHJ. Technical standards may have awareness, operations, technician, and specialist levels, and all have success levels that build on increasing knowledge and skills.

There are several tools the instructor may use to determine the success of a student. Depending on the subject or skill being taught, the evaluation tool could be oral, written, or performance-based. The evaluation criteria should remain the same depending on the method of evaluation used.

The identified content at the appropriate level of desired learning will need to be assigned a weight to be applied to the question or skill. To determine if the desired outcome is met, several questions or skills may be required.

When determining a success level, the knowledge or skill should be assigned a level of competence that is required of the student. Competency levels can be identified as:

- Highly skilled, moderately skilled, skilled, or needs more training
- Pass/Fail
- Mastery

Developing a Scoring System for the Assessment

Testing or performance demonstrations must be included in the evaluation process of a program. All programs should involve some type of testing so the instructor can determine if the objectives have been met and to provide the student with feedback on their progress. As you design the tests, you will need to consider not only the type of test but also the scoring system you will use to determine pass/fail levels.

When designing a test that will be scored, you will need to determine the weight of each question. The top score a student can achieve is 100 percent, with passing score usually in the 70–80 percent range. Questions must be weighted (usually by points) to achieve the top score. You may choose to weight each question equally, or you may base the weight of the question on the importance of the material. In the case of the latter, you will assign a weight to the question depending on the importance of the material covered, taking care to ensure that the total points for all questions in the exam equals 100. For example, if you give an exam that contains 25 questions, some questions may be worth 1 point, while other questions may be worth as much as 10 points (as indicated on the test itself). However, when you tally all the points up for the 25 questions, the top score should be 100 points.

Preparing the Evaluation Environment

Written Exam Testing Environment

For the written evaluation, the environment should be as nonthreatening as possible. Adequate lighting, comfortable chairs, and suitable materials (i.e., pencils, paper, etc.) are all necessary requirements for a successful test session. Washroom facilities should be in close proximity to the classroom, and students should be given specific instructions for using them. Other items to consider are:

- Distractions
- Suitable, ergonomic desks
- Ergonomic chairs
- Computer, printer, and scanner (if necessary for computer-based testing)
- Office supplies (as needed)
- File cabinet(s) (the instructor should have at least one lockable file cabinet to store tests securely)

Practical Exam Testing Environment

For the practical environment, safety should take priority. All equipment should be tested for proper operation and safety. The tools a student is tested with should be the same tools that they were trained with. A TSP should be implemented during all training activities, including the evaluation process. Instructions to the student should be read exactly as they are stated on the testing form, and the students performing activities should be required to wear the appropriate level of personal protective equipment (PPE) for the skill. When the learner is required to perform a potentially dangerous skill during a testing session, the student must be informed of the potential hazards and risks associated with that task. Furthermore, measures that should be put into place to reduce or mitigate any hazards include:

- The training area should be cleaned and maintained to avoid accidents.
- Ice build-up should be removed and sand should be spread around slick areas in cold weather situations.
- The examination area should be cleaned up once the drill is completed and it is safe to do so.
- Ensure that all live fire drills are conducted in accordance with NFPA 1403, *Standard on Live Fire Training Evolutions*.
- Adequate rehabilitation is provided for all personnel.
- A safety officer is present at the examination site for all drills.
- Adequate space is provided for all drills, apparatus, and personnel.

Test Instructions

Instructions must be clear, concise, and fair. They need to be brief and include the expected time required to complete the evaluation. It is acceptable for the instructor or proctor to read the instructions to the students. If there is a specific way to complete a score sheet or complete a specific skill, the instructions should be read or discussed with the students. If an electronic or computer-based scoring process is used, time should be allowed for additional questions or time to have the students fill in required background information on the forms.

Evaluation Administration

The administration of the evaluation needs to be non-biased and as fair as possible. During test preparation and administration, it is important to have evaluation security in mind. For evaluations to be fair to all the participants and for the instructor to be able to determine the effectiveness of the instruction presented, students should all go into the evaluation with the same expectations. Students who gain inside information on an evaluation will not give an accurate indication of the learned skill or theory. It is common practice in some states/provinces to not allow certification or state/provincial fire fighter testing to be given by any instructors or personnel who participated in training the students. At the very least, the state/provincial certification tests are given by someone from outside the department, and the tests must be checked out from the governing agency that keeps all tests under lock and key.

The use of proctors should be considered when administrating a written exam, as they are unable to provide students with the correct answers. Proctors should be a third-party and not directly responsible for the training of the students.

Administering the Performance Evaluation

After reading the directions, the evaluator should allow the student to attempt the skill and document what the student did. It is important to not tell the student what to do or give hints as to what step is next. However, if there is a safety concern, stop the evaluation immediately. Administration of the performance evaluation requires a TSP to be prepared with the identified safety concerns noted for the participants. The appropriate resources need to be available with equipment that is required to perform the evaluation safely.

It is important that the skill required be achievable for a competent student. It may be necessary for the instructor to personally practice and evaluate the performance to determine the viability of the skill in the evaluation setting. Performance evaluations can be somewhat subjective; therefore, it may be necessary to have more than one evaluator available at the testing site to mitigate the subjectivity variable in the evaluation process. It is a standard practice in most agencies that during a "retest" situation a different evaluator be assigned in order to remove any subjectivity.

Checklists for Performance Evaluations

The design of a checklist for performance demonstration is essential to maintain consistent criteria for the students performing a required skill. The checklist also gives the AHJ a document stating that a required skill was achieved at a specific point in time.

There are several checklists produced and available online. Any checklist used should meet safety and department-specific criteria.

The performance checklist may include the following items:

- The objective
- A statement explaining why it is important to be able to perform the skill
- An outline detailing the desired student's performance criteria, including the outcome
- Evolution description
- Evaluation criteria
- Maximum time allowed to perform skill (if applicable)
- Standard or SOG/SOP reference
- A section on the form for the evaluator and student to add comments
- Successful/unsuccessful indication
- Name and signature lines for both the student and evaluator
- Date skill was performed

Of course, additional items may be added to the checklist, but it is best to keep checklist sheets to one page, if possible.

After the Evaluation

The evaluation process is crucial to any training program. Evaluation of training provided should be considered during the design phase of any session. The ability to properly evaluate a session requires the instructor to consider a desired outcome with achievable results that can be measured in some way. For students to be adequately prepared for a final evaluation, they should be evaluated throughout the learning process, with instructors providing encouragement to the learner at all points in the process.

After the students receive the results of the evaluation, they should be given an opportunity to review the evaluation and appeal, if required. The instructor or training agency should use the results of the evaluation to guide the design or implementation of new department procedures or additional training for staff. The data gleaned from evaluations can provide the training division with a type of gap analysis to identify the department's potential needs. To be professional and ensure validity in the evaluation process, the evaluation process should be transparent to all participants.

Key Points

- The evaluation step can be broken down into two main areas of focus: the student and the instructor.
- The evaluation process can take on several different forms, such as formal or informal and formative or summative. For example, the training officer could conduct an interview with the student where they review the course materials at the end of a course (summative) in an office setting (formal), the training officer could sit in a class and provide feedback during a break (formative and informal), or the training officer could use a combination of these types of assessments.
- The instructor should match the test questions to standards (i.e., NFPA standards, State EMS Scope standards) to ensure that all questions are valid. If a question is challenged, the instructor will be able to reference a standard or guideline to confirm the correct answer.
- Many promotional assessment centers use oral interviews to evaluate students' performance. When conducting promotional tests, it is best to film the students' answers.
- The performance testing must be completed in a safe environment under as ideal conditions as possible.
- Instructors should be extremely careful to not diagnose students with disabilities. The diagnosis of a disability requires a doctor's recommendation.
- The tools a student is tested with should be the same tools on which they were trained.

References

Georgia Firefighter Standards and Training Council, *Georgia Basic Firefighter Candidate Task Book*. https://www.gfstconline.org/index.php/files/download/e6ebff253bff12b. Accessed April 21, 2014.

National Fire Protection Association (NFPA) 1403, *Standard on Live Fire Training Evolutions*. 2012. http://www.nfpa.org/codes-and-standards/document-information-pages?mode=code&code=1403. Accessed April 21, 2014.

CHAPTER

10 Alternative Methods of Learning

Skip Heflin and Brian Ward

Introduction

There are several methods for learning, including computer-based training (CBT), colleges, outside agencies, conferences, and other schools. As individuals, we all learn differently, and as instructors, we should remember that not all students learn the same way. As many of these options should be made available to your fire fighters as possible. Although some may be obvious, other avenues must be investigated and the value for your department determined before incorporating them into the learning options.

Computer-Based Training

The fire service has traditionally taught all types of training classes in one of two ways: with an instructor lecturing in a classroom or with an instructor directing activities in the drill field/fire station bay. Modern technology has opened new doors that some departments have begun to utilize. Whether it is driven by innovation or the ever-tightening budget, departments are jumping onto the technology bandwagon in droves and incorporating computer-based training into their training programs. It is time that you consider what eLearning can do for you and your department.

Training via computer programs is known by many names—computer-based learning, computer-based training (CBT), and eLearning—and all generally refer to using technology to augment the learning experience by differing degrees. For example, using a computer to play a video is considered by some to be computer-based learning. What you will discover in this chapter is that there are a variety of options available, and we are limited only by our imaginations.

Let us begin by discussing what CBT is not:

- **CBT is not a catch-all answer to all of your training needs.** Some subjects and skills simply do not lend themselves to training through an online environment. You could teach a lot of information about ropes in an online course, but sooner or later the student must put their hands on an actual rope to learn and master a rope skill.
- **CBT is not for all students.** The use of CBT results in a decrease in instructor–student interaction. Some students require more direct observation and interaction with the instructor than others. CBT students must be self-motivated to varying degrees, depending on the type of learning that is offered.
- **CBT is not inexpensive or free, in terms of money or time.** The software that is required to conduct CBT can be very expensive, although some software is available for free from open source sites. If you are not familiar with open source software, it is prudent to do some research. Open source software is software that has been released to the general public, typically through the Internet, and made available to anyone. The community can see the programming code that makes the software work. The community can also, if they so choose, make changes to the open source software by creating new versions

or adding features that eventually become part of the available software. There are many open source software programs available that are free to use. One major benefit of open source software is that, because the community can help develop the software, it can develop at a rapid pace and the community itself becomes the help desk.

- **Using a CBT program does not come easily to everyone.** CBT is one of those areas where the younger generation may have a slight advantage over the older population, because they grew up using computers. They tend to have a more intuitive grasp of how computers work. Also, teaching via CBT is different from teaching in-person; there is a learning curve in figuring out how to make the software present the message so that it is absorbed as intended.
- **CBT is not a lone fire fighter clicking screen after screen by his or herself with no interaction.** One thing to consider is that the agencies that require continuing education for licensing or certification purposes have differing opinions about how CBT can be used. For example, the National Registry of Emergency Medical Technicians (EMT) will not count an hour of students simply clicking slide after slide and reading to themselves as an hour of training. They consider such time spent to be "distributive education."
- **Some members in your organization may not readily accept CBT.** Implementing CBT into your training program may be met with resistance. Prior negative experiences or a lack of experience with CBT may discourage people from adopting the technology. Department members who lack basic computer skills may have difficulty using this type of training, and therefore may be inclined to oppose it. Having a concrete plan to introduce members to the computer-based environment can help mitigate most of these potential problems. You may want to arrange a test-drive session to allow members a chance to see how the software works. When conducting a test drive, it is best to start off small and gradually build complexity into the experience, as this will help keep participants from feeling overwhelmed and will allow them to develop their confidence as you increase the complexity and depth of the CBT program. Although using CBT can be challenging for some members, younger fire fighters in your organization will typically benefit from it. Many of our younger generation fire fighters have grown up using computers, and they often learn more effectively this way. This proficiency also gives younger fire fighters the opportunity to help those members who tend to struggle with technology. It gives them a chance to shine as they share their knowledge and experience with less knowledgeable members.

As you can see, there are some challenges and limitations with the use of CBT programs. However, as we noted in some of the examples above, CBT does have value and a place in training fire service members. Below we discuss the possibilities of CBT and what it can offer.

- **In terms of student–teacher ratios, CBT can be a very effective force multiplier for your training division.** Two to three training staff members can easily create content modules that could effectively teach your entire organization. It could also allow any staff member to be the instructor of a course without the common constraints of time or travel. The time it would normally take to reach the same number of students in the traditional format is exponentially less.
- **CBT provides content that is accessible 24/7.** The content that is produced for CBT can be made available around the clock. This convenience can be helpful to busy companies and to members who have been off-duty for an extended length of time. By making content available all day, every day, you may just find that participation increases because members can access the content at their convenience. This reduces, or even eliminates, scheduling conflicts that typically come up with planning arranged training sessions.
- **CBT creates a reference source for line fire fighters.** After content is created and used, it can be stored on a website to become a reference source for your members. The average fire fighter, due to peer pressure, may be hesitant to ask a question if he or she feels that everyone else knows the answer. Having a repository of training content provides another way for that fire fighter to get his or her answer, as he or she can now look up the answer on his or her own time and avoid any potential embarrassment. Company officers also find the availability of CBT content very useful when an incident occurs that indicates a need to retrain members, as they have access to a past CBT module that can provide this retraining. They may choose to use the module after the incident to refresh members' knowledge or skills.

- **CBT supplies a permanent archive of traditional training classes.** Training division members find the archive a convenient way to access past training sessions and build upon them. This makes the archive a great reference source for staff members as well as the line fire fighter. Some departments use their website, with the stored CBT content, as a shared repository of training material to be accessed by area associations or departments who could benefit from the use of the same training modules. However, this type of setup can be cost-prohibitive and is only used by organizations who have the money and willingness to pay for the website.
- **CBT offers a consistent way of training members that can be hard to achieve with traditional training methods.** CBT allows a particular class to be conducted in a very consistent manner across multiple organizations or regions. All participants receive the same content and training. It does not vary from instructor to instructor or course to course the way it would with a traditional classroom model. Any experienced instructor will readily admit that it can be difficult to teach a class consistently across a large department. During a multicompany drill, for example, the same class may be taught 3–5 times per week for a period of 6–8 weeks in order to run all members of a large department through the drill. Any instructor would become tired of teaching the same class material in such a short amount of time.

History of Computer-Based Training

Computer-based training is not new and has been used for some time. The tools that can be used to create content for CBT, however, have undergone dramatic changes in the past 10 years. In previous years, the production software that was used to produce content had a steep learning curve, and it was also cost-prohibitive for most departments. Advances in technology have driven production costs down and have made the software much easier to use. In the past, most organizations that wished to use CBT were required to hire a private company at a great cost. Now, any department with moderate amount of computer knowledge and determination can produce their own customized content in a very short period of time.

Some people have negative associations with the idea of conducting training using a computer. These attitudes are largely due to past experiences when the CBT content was considered to be nothing more than a PowerPoint presentation or a video posted on a website or saved to a computer disk. CBT as a training concept was criticized for lacking engagement, depth, true instruction, and evaluation. Most people who have completed an online college-level course in recent years, however, can offer a different picture of what computer-based education is like and will tell you that it can be a very valuable learning experience. Still, fire personnel sometimes confuse video-based training with CBT. Video-based training is typically a sample video that can be watched at any time, whereas CBT is a training program that contains a number of multimedia elements, including video, pictures, simulated activities, and so on. CBT could also contain activities that the student must complete in order to progress to the next stage in the module. Some form of evaluation is typically included as a part of the CBT program as well.

The fire service has been slow to accept CBT programs, although in the past few years, implementation and use has increased rapidly as many departments have been forced to cut their training budgets and production software has become easier to use. It is now commonplace to see CBT being utilized among federal agencies, state/province and local entities, and associations. CBT allows us the flexibility to either utilize a course as it was originally created or to create and add additional experiences for our students that will help them learn to do their jobs safely and effectively, while remaining cost-effective.

The first experience your members have with CBT may be the first time they have ever used a computer-based learning tool, and this experience could shape their future attitudes toward using CBT programs. It is much easier to establish a positive perception of CBT among members the first time they use it than it is to change a negative attitude after one has already been formed. Therefore, it is essential to get off on the right foot when trying to sell the idea of using CBT to fire service members.

Benefits of Using CBT

Utilizing CBT in your organization has the potential to increase student engagement in training sessions. You may find that those students who would normally not interact with others in the traditional classroom become very engaged in the online environment. Barriers such as social anxiety, which can cause these students to be silent in the classroom, become irrelevant in the CBT environment. CBT is more flexible than a traditional classroom environment and has the potential to provide a better learning experience for more students, in the sense that all students get to experience the best positions during the exercise because the virtual classroom time is much easier to manage.

In addition to CBT making your instructional staff available to members at any time and in more than one place, it also has the potential to increase the amount of content/information presented to your members. Savings, both in terms of money and time, can be realized when the training staff can teach more students because less time is needed to produce the program, so more time can be devoted to the business of actually teaching. The CBT format may also allow members to complete training while on duty. Members will appreciate this, as it saves them from having to spend time in the classroom while off duty.

With CBT programs, instructional staff will find they spend less time grading written quizzes or tests. If quizzes and tests are included in the CBT program, the students receive results immediately upon completion. Since the CBT program can grade assessments, instructors can spend their time on more important tasks related to their job. The same benefits can be gained from utilizing similar technology in written promotional testing. A learning management system (LMS, discussed later) can be used to manage many different test banks, as well as the tests themselves. The test banks produced by major publishers are now compatible with some learning management systems, and custom questions developed by the instructor can be easily added. LMS also has the capability of archiving copies of tests for promotional or re-certification needs. Because records are kept electronically, it eliminates the need for storage space that traditional paper files require and makes it an environmentally friendly solution for departments.

Since CBT can be conducted onsite and does not require companies to travel to other locations to perform multicompany drills or other activities, a reduction in the fuel budget can be realized. Live video classes and/or meetings can be utilized as well, depending on the staff's ability to utilize these features and the infrastructure of the organization.

Much research has been conducted on the younger generation of fire fighters and how we can best reach them. As noted earlier in the chapter, the younger generation stands to benefit greatly from this type of training because of their ability to use it easily, their interest in using it, and the fact that they learn well on it. Some departments have begun to utilize a website house training materials, standard operating procedures (SOPs), and interdepartmental communications, and have found that it becomes a central location and reference for members to use when preparing for a promotional test or following the activities of the department.

Requirements for Implementing the CBT Program

The requirements for utilizing CBT vary greatly, depending on the extent to which the organization wants to be involved. We have broken our discussion on CBT-related requirements into six categories:

- Internet access
- A hosting site
- Hardware
- Software
- Student motivation and time management skills
- Instructor requirements

As is the case with anything new, a learning curve should be anticipated when developing content for CBT software. However, learning how to produce educational content is similar to learning fire-ground skills: the more you apply these skills, the easier it will become.

Internet Access

Although access to the Internet is not required, it is preferable. If Internet access is not available, the CBT program can be produced and copied onto a compact disk for use on a typical computer; however, if using this option, keep in mind that there may be compatibility issues with the various computer operating systems. Departments that have Internet access across stations are able to utilize LMS. High-speed Internet access is preferred, as it helps the programs run more smoothly, and so is the ability to broadcast wirelessly at each location so that cables are not needed. One problem some organizations encounter is that while they do have Internet access at all stations, limitations on access to Internet sites are imposed by another department, such as Management Information Systems (MIS) or Internet Services Departments (ISD). Before launching a CBT program, you will need to look into these restrictions to ensure that the content will be accessible by the intended audience.

Hosting Site

Depending on the resources that you have, you may need a hosting site, which is a place to host your website and upload the CBT content. There are many different companies that offer this service, and the hosting site can be integrated into your organization's existing website. It is recommended that you shop around for the best deal. If you need to purchase a domain name (the name for your website, e.g., www.yourorg.gov), the company that provides your hosting site can also provide this service. These companies generally have several servers that back up your content, so it is secure. This protects you in the event of a server failure. A copy of your content will be safely stored and will not need to be reproduced.

Hardware

Video Recording Equipment You may want to create videos to upload to your online courses, podcasts, or webcasts (to be discussed in more detail below). Departments may film their own drills and activities and incorporate these videos into the CBT program. Anything from footage of actual incidents to demonstrating throwing ladders can be recorded and inserted into a CBT module for viewing by the students.

Fortunately, creating, shooting, and editing videos in a cost- and time-efficient manner has never been easier. An instructor or training officer can simply use their iPhone or other small video recording device to film a training video and produce it quickly with minimal effort. In addition, high quality cameras can be purchased for a fraction of the old cost. Most organizations have an information technology (IT) or communications department that may be able to provide cameras, editing software, and a space to edit the video at no additional charge to the department. It is preferable to use digital video cameras, rather than video cameras that use film or tape, as this makes it much easier to add the video to a CBT module. Many new cameras come with a built-in hard drive, and the videos can be easily transferred to a computer through the use of a cable.

In addition to the recording device, you may also want to purchase a green screen for filming. A green screen is literally a green-colored fabric screen that can be placed behind the subject being recorded. The advantage of the green screen is that it allows you to record the video in a quiet place, with the green screen serving as the backdrop. Then, when it comes time to edit the film, you can replace the green background with a different one. The new background could be anything from a hazardous materials spill to a raging fire. This way, it appears as though the original subject was recorded at the scene of an incident. As you can imagine, this feature has many obvious applications in the emergency services industry.

Computer Equipment While most organizations take for granted that they have access to computers, the reality is that not all emergency services organizations do. Furthermore, to implement CBT programs, the training division would not only need a computer that is dedicated to producing content for the CBT program, but they would also need computers at satellite locations where the members would access the CBT program. In addition, the type of computer one would use to create or develop the CBT program would be different from the ones that members use to access it. The computers used for accessing the CBT program would require a desktop PC with an operating system that is compatible with the content that is produced. The Windows operating system is compatible with most LMS and CBT modules that are widely used. The training division should consider the screen size for the monitors used for the training. While a 14-inch screen may be adequate for the person who is developing the CBT program, it would not work well for 4–6 fire fighters who are all trying to see the same thing on the small computer screen. One solution for this common problem is to connect the computer display to a television through the use of a video splitter switch, which projects the PC image onto the TV. Most fire stations have a large-screen TV in the day room or common area that could be used for CBT. Not only would the larger screen provide the group with a much better viewing device, using such areas for CBT may improve the experience as well. The computers used to deliver CBT content can be similar to those one would purchase for home use, with the addition of the video switch.

For the training staff, a laptop computer as opposed to a desktop computer may be more appropriate for developing CBT programs. The cost of laptop computers has decreased in recent years, and they are now competitive to desktop models. A laptop would enable training staff to utilize the computer in multiple locations while still having enough power to run all of the needed software. One component that should be augmented is the RAM (random access memory). A sufficient amount of RAM would be required in order to edit the videos. The memory requirements vary from system to system, so depending on the software purchased, you may find that an upgrade is required for computers that will be developing and storing CBT content.

Software

Software for Creating eLearning Content Certain types of software will be required to produce and use eLearning or CBT content. While many different programs that can be used to produce CBT content exist, we will only discuss two: Adobe eLearning Suite and Articulate Studio. The intent here is not to endorse one product over the other, nor are we trying to supply a user manual. Our goal is to give you an idea of what is available and provide a brief introduction on the capabilities of some common software programs.

Many different companies produce production software, and the price and ease of use of the software will vary from one company to the next. Some software can be so complicated to use that one must have previous software programming experience to use it effectively, while others are designed for use by those who possess only basic computer skills. Any reputable supplier should offer a trial period that allows you to use the software for free before purchasing it to see how you like it. You should take advantage of these trials as much as possible. As you explore different options, you will find that you are able to use some programs better than others, and this may influence your purchasing decision. The software programs we discuss in this chapter use Flash or HTML to produce the CBT content. These are computer languages that can be accessed with common Internet browsers, such as Google Chrome or Microsoft Internet Explorer.

Adobe sells a product called Adobe eLearning Suite, which can be used to produce CBT programs, and it is one of the most thorough and comprehensive suites available. However, it does require you to invest some time in learning the different programs available in order to be productive with it.

Another software suite is called Articulate Studio. Articulate offers a suite of programs that will produce eLearning or online courses. It has a more user-friendly interface for less experienced users. The instructor can create a PowerPoint presentation that includes pictures and videos and create a packaged training video for all personnel to view. Some of the features that Articulate includes are the capability of doing voiceovers, adding quizzes, and setting options for how participants navigate based on their assessment results.

Both of the software titles mentioned above include add-ons for PowerPoint that allow the creation of the base module in a program that most instructors are familiar with. Video editing programs are a part of both titles. Therefore, they essentially include everything one would need to produce high-quality CBT content.

Using Internet-based software to share the content will allow members greater flexibility in accessing the CBT course from a variety of mobile and stationary devices, including smart phones, station desktop computers, or personal laptops, without the worry of program compatibility. As long as the device or computer you are using can access the Internet, all you would need is browser to access the content.

Software for Editing Videos Several options for Macintosh or Windows products exist for editing videos that are either free or very inexpensive, including Windows Movie Maker (comes standard on most PCs) or iMovie or Splice for Mac (including iPhone and iPad) users. If using a digital camera, the software needed to edit the video is typically included with it.

Windows Movie Maker Windows Movie Maker (WMM) allows you to record a video on any digital recording device and upload the videos to your PC. As mentioned earlier, WMM is a free software application that comes standard on most PCs with Windows operating systems. After the videos are downloaded, you simply open the WMM application and a timeline will appear. The editor can drag the videos into the timeline and review the video. If there is a segment that the editor would like to remove from the video, it is a simple touch of the mouse. After the editor places all of the necessary parts in the timeline and removes the unwanted segments, WMM will walk the editor through the process of publishing the video. Several options exist including adding music, transitions, animations, titles, and credits.

Although the video you have created with WMM will not look as if it were produced by a major film studio, it will be a quality video produced with little more than a standard computer and a video recording device. Fortunately, most departments have both. This is a great training tool for showcasing new procedures or best practices. It enables you to provide a hands-on demonstration presented by one of your trained instructors.

iMovie and Splice iMovie and Splice are both Apple iPhone applications that you can download for use for a nominal fee. iMovie is available for purchase, and Splice is available for free, although there is an add-on option for an additional fee. All video recording and editing can be done on the iPhone (iMovie is also compatible with the iPad and comes standard on most Apple computers), and so it is not necessary to download the file to a computer. Furthermore, this system offers all of the same benefits as WMM. Both products offer tutorial videos that instruct the editor on the uploading and editing process. Without providing step-by-step instructions here, the process is basically the same as WMM. Once the video is complete, it can be saved or "exported" to your camera roll or library on your iPhone (or iPad) and uploaded to social sites, such as YouTube, or emailed or uploaded to a department website. This application provides another great option for capturing training at a moment's notice with just the cost of a smart phone. If you are planning on using your iPhone to develop and save videos on a regular basis, you should consider purchasing an iPhone that contains a larger amount of memory. If this is not an option, be sure to download the videos to a computer and delete them from your iPhone on a regular basis, so that the memory space remains open.

Camtasia Camtasia a software application used to create interactive videos and screen recordings. Its features enable you to edit videos and create screen casting. Some of the recording and editing options include adding music, recorded audio, and picture-in-picture options. This application is well-suited for creating webcasting videos.

Software for Creating Visuals

Microsoft Visio is a software application that you can use to create professional data-driven visuals, such as diagrams and flowcharts, which you can then upload to the department's intranet website. This software is compatible with Windows or Macintosh operating systems. For example, you could create a diagram of an engine using various shape and drawing tools to shows the parts of that piece of equipment. You can also link data, such as pictures, videos, summaries, or an Excel spreadsheet, to each piece of equipment. As the student clicks each part, he or she will see any notes, pictures, or video linked to it. Visio is a very useful tool for creating diagrams and flowcharts, or for illustrating processes that need explanatory materials.

Learning Management System Software Applications

A learning management system (LMS) is a software application that allows you to plan, deliver, and administer learning events, or training, online. The features of this application include online virtual classrooms, which are instructor-led courses that are concurrent with or independent of the main program. Students also use the LMS to access the various features that are utilized and the CBT content. The LMS can be hosted on a server owned by a hosting company or one owned by your organization. With LMS, the instructor can create accounts in the system for each user, so each user has a unique username and password. The LMS system allows you to build and store course content and class materials, including videos, activities, quizzes, and written tests.

There are many different companies that sell LMSs, but all the products offer similar features, such as calendars, text chat rooms, video meeting plug-ins, and more. You may already be familiar with a few popular LMS applications, such as Blackboard or Moodle. These systems are used by major universities, corporations, and emergency services organizations. While Blackboard and Moodle share many common features, one area where they differ in is price. Blackboard charges for use of their software, whereas Moodle is an open source project, and so the public can use it for free. Numerous resources are available, such as books or websites, that provide support and offer tips on how to use these LMS applications. Without an LMS, the CBT program would be incomplete, as the LMS application provides the necessary components to offer learning management.

Subscription-Based CBT Programs For the organization that has no desire to author or develop custom content for their department, CBT can still be utilized. There are many companies that will create content for a fee. There are also subscription-based programs available, which members may access for the duration of the subscription. These systems provide offerings similar to those of a typical LMS application, and may even offer record management solutions incorporated with the LMS. Any custom content would require a separate charge as well, and they charge a fee per user for access. The main disadvantage of subscription-based systems is that the content can be generic in nature and not specific to your organization. For organizations that do not have the desire or capability of producing CBT content, commercially available modules can be purchased on an ongoing basis, usually for a fee with the costs directly related to production time. These modules could then be posted to your organization's LMS to be accessed by members.

Student Motivation and Time Management Skills

In order for the CBT program to be successful, the students must be self-motivated. Consider homework as an example. Since the student is not in a physical classroom, the assignments for homework would be delivered through the LMS application. The student would be responsible for accessing and completing the quiz, homework, or other assignment before the deadline to meet the objectives of the course. Students who are self-motivated are more likely to succeed with this type of learning environment, while those who lack motivation may become frustrated or overwhelmed with tracking these assignments on their own. This can also become a source of frustration for the instructor, who wants his or her students to succeed. Students who possess strong time management skills are also more likely to succeed with CBT programs. Consistent requiring of skill demonstration on the part of the instructor should help to identify any weaknesses in time management skills of involved members early enough in the course to allow for coaching or corrections on the part of the participant.

Instructor Requirements

Just as certain learner characteristics help ensure success with the CBT program, the same is true of the instructor. To teach using an online environment, instructors should have a firm grasp of computer skills and be familiar with common commands used in modern operating systems. Instructors who have never taught a course online will find quickly that they must work to actively engage students. At the beginning of a course, instructors must put in a sincere effort to lure students into participating in course activities. As mentioned previously, this helps to identify students who fail to manage their time appropriately or who are not properly motivated. It also facilitates the learning process. Whether the activity is a quiz or a discussion with a question and answer session, the instructor must ensure the student is engaged.

Application of CBT

By now I hope you realize that this environment can be used for more than just training. Application of this technology is limited only by the imagination of those that apply it. In the last section of this chapter, we will discuss additional applications of the CBT environment by providing some examples.

- **Example #1: Monthly training modules to meet mandates and/or certifications.** In this example, a CBT training program is created to meet the various mandates required by the state/province and the guidelines set by the National Registry of EMTs for recertification. This model also could be applied to meet International Organization for Standardization (ISO) or National Incident Management System (NIMS) requirements.
  - Each month, modules are created to satisfy the various requirements set forth by the organizations listed above.
  - For each module, the training division staff uses pictures, PowerPoint presentations, videos, and interactive components to reinforce learning.
  - Quizzes are used to evaluate learning and are the terminal requirement of each section. These quizzes are uploaded to the LMS used by the organization through an Internet browser. The LMS delivers and administers the content for each module.
  - Either the company officer or his or her designee (also known as qualified instructor) uses the module to teach a class to the company members.
  - On their own time, participants use station computers to log into the LMS each month and complete the associated quiz.
- **Example #2: Initial recruit training.**
  - With initial recruit training, the LMS is used as a communication tool to provide learning resources, quizzes, and course administration for the recruit fire fighters during their initial training period. All of the course materials, includingthe quiz questions, are uploaded onto the LMS from the publisher's disk prior to the recruit class beginning. The instructor posts the quizzes, but conceals them from the participants until the date they are needed.
- **Example #3: Multicompany drill augmentation/Hazardous materials team monthly training modules.**
  - In this example, a blended learning approach is utilized. In other words, the class includes both the physical classroom and the virtual classroom components. The lesson plan is written to accommodate the blended approach, using suitable materials. The portion that will be utilized on the LMS is uploaded in advance of the class. The module will also include a quiz derived from the online materials to ensure that material has been learned/completed. The course is advertised and the dates companies shall report to the training facility for hands-on training (HOT) are scheduled.
  - Participants log onto the LMS using their account information and complete the training. An officer is responsible for assigning or teaching the material, thereby using the LMS as a delivery, course administration, and evaluation instrument.
  - The instructor ensures all members have completed the classroom portion, answers any questions, and then takes all companies out for the HOT component.
- **Example #4: Blended learning/hybrid class.**
  - Here, the blended learning approach is used again, but for a course that is normally taught in the classroom only. In this case, the LMS is used to augment or supplement the classroom.
  - The instructor adapts any materials that lend themselves to the virtual classroom and uploads these materials onto the LMS ahead of the class start date. All relevant quizzes and learning resources are included.
  - The class meets every few days in the classroom. On the other days, the virtual classroom is utilized, as members log in on their own time to complete assignments and quizzes. The student

uses the virtual classroom to augment the materials that are covered each class day.
- When the class is completed, the instructor archives the grades from the course and removes the students from the class to make the online course ready for the next session, whenever that is needed.

- **Example #5: Promotional testing.** This example demonstrates how an LMS could be used to conduct written testing and scenario-based promotional assessments for promotional testing.
  - Quiz banks are purchased and uploaded onto the LMS in advance. The LMS is set up to randomly draw the determined number of questions from each category (e.g., SOPs, Strategy and Tactics, Human Resources, etc.). All question orders, including answers, are mixed so that no two tests are the exact same.
  - Using CBT authoring/production software, the training staff creates several personnel video scenarios to assess candidates for testing. This module is included in the course. If the written test is passed, the candidate will be allowed by the LMS to see the scenario portion of the test.
  - This serves as the first phase of the promotional process. Candidates must pass this portion to move on to the next phase.
  - Candidates arrive on the advertised test date to the department's emergency operations center (EOC), where they have desktops utilized for testing.
  - As candidates take the test, the proctor sits in the back of the room and monitors the testing live as candidates complete it. Results are obtained within seconds of completion and the candidates that pass are scheduled for the next phase before they leave that day.
  - Automated, computerized grading also allows for consistency in grading. One instructor may grade one way based on his or her experiences while another could grade the quizzes a different way.
- **Example # 6: Task book tracking.** In this example, the LMS is used to collect, confirm, and communicate the completion of the state/province-required task book for fire fighter recertification.
  - The instructor creates task books using Adobe software. The format used is a fillable electronic PDF that can be filled out using Adobe Reader software, which is available on most PCs.
  - A master copy of the book is uploaded onto the LMS. The company officer can now download the book and save one for each of their members and complete it during the year as the tasks are completed.
  - At the end of the year, the completed task books are uploaded onto the LMS by the company officer.
  - The task books are confirmed by training staff as they are submitted. When a book is considered complete and approved by the training staff, a passing score is issued via the LMS.
  - The company officer can see the members assigned to them in the LMS. He can also check the grades in LMS to confirm that each task book has been received.
  - The training staff should download all of the books. The task books are signed by the training officer and archived for record keeping.

By presenting several examples of ways CBT can be applied, we hope this has encouraged you to think of ways that you can apply this technology to help your own agency. Never allow the fact that no other organization has used CBT in a certain way before hold you back from experimenting with new ways to utilize technology. Share with other organizations what you have discovered in your attempts, including what approaches did or did not work, so that others can learn from you. Through this dialogue, you will quickly learn new ideas and ways that you can increase the effectiveness of your own efforts.

The Threat of Deception

It is important to discuss cheating because this is a concern shared by many educators when it comes to online quizzes and tests. The threat of cheating can be mitigated by LMS, and there are several measures that can be taken to discourage such activities, including:

- The order of quiz questions and answers can be mixed so the order is never the same for each quiz.
- Test banks can be used. A test bank includes a variety of questions that could be used for each test. The LMS is set up to randomly pull a set number from each bank. As long as each bank contains an adequate number of base questions, the same questions may not be used on each quiz.
- Timers can be used for the quiz. It will tell the LMS to only open the quiz during predetermined hours of the day or only on certain dates and times. A time limit could also be set for each quiz. This discourages students from looking up answers

because there is simply not enough time to do that for each question.

- When a computer connects to the Internet, it uses an IP address, which is a unique identifier for the computer. The IP address can be set to be static, so it cannot be changed. Most LMS applications can be configured to allow only certain IP addresses to log into the system, such as those from a station computer located in a proctored setting. This allows you to control which computers can access any given test.
- Traditional proctoring can be used if necessary. This involves having a staff member or officer supervise the participants to ensure the test is taken in the specified manner. Proctoring is typically used for promotional testing.

Podcasting and Webcasting

The International Society of Fire Service Instructors (ISFSI) began using podcasting in an effort to reach a broader audience. Prior to its use, the ISFSI had relied heavily on printed materials that were sent via standard mail. Although the materials were excellent, relevant sources of information, they were only sent to members and there was no way to ensure that nonmembers could see them. In 2009, the ISFSI set on a course to reach more instructors than ever before with then-President Eddie Buchanan and Director Steve Pegram launching "The Voice," a weekly and monthly podcast. At this particular time, podcasting was new to the fire service, but had been used successfully by other organizations and groups to spread their message. At around the same time, FireHouse.com was also posting podcasts with a variety of fire service hosts, covering a wide range of topics.

The ISFSI used a mixing program application called Garage Band, which is used on Apple computers. The show was recorded and then posted on the web. "The Voice" podcasting program was a success. Soon the ISFSI was looking for other ways to use this new medium. While this type of podcasting was focused and fast moving, it was soon overtaken by a newer and better medium: webcasting. With webcasting, not only can you watch a presentation, you can see and hear the instructor making the presentation in real time. Sometimes you may even have opportunities to interact with the instructor, making it a more dynamic experience.

However, podcasting had a built-in audience due to its portability and ease of use. The Apple revolution was sweeping the country and it appeared everyone had an iPod or MP3 player of some kind. The podcast could be burned to a disc or downloaded from iTunes and put on an MP3 player. Fire fighters were listening to podcasts at a record pace and the technology was rapidly changing. Podcasting had become the easiest way for people and organizations to spread their message and it seemed like everyone was doing it, including local fire departments.

Around this time, FireEngineering.com joined the world of podcasting, but they enhanced the end product. The one thing that podcasting lacked, unlike webcasting, was interaction between the host and the listener. Most people were listening to podcasts on the go and not concerned about interacting with the instructor, but many others wanted a chance to ask questions or weigh in on the issues. FireEngineering.com rolled out *Blog Talk Radio* and introduced a line-up of "Who's who" of the fire service. The format was identical to that of a conventional talk radio show in the sense that it had hosts, guests, and callers. The shows were presented live according to a set schedule and recorded to be made available as a traditional podcast. The ISFSI took advantage of this opportunity and joined the line-up. Using an Internet radio service, Blog Talk Radio, Fire Engineering.com, and the ISFSI were able to reach a much larger audience than ever before. This new format caused a demand for more shows and on a more regular schedule. Since this method was developed, "The Voice" has done very well. Through this program, the ISFSI has provided news and information using ISFSI instructors and guest speakers with focus on particular topics.

Before podcasting involved receiving callers, this method of broadcasting was fairly simple and usually required only one person to make it happen. However, with the introduction of Blog Talk Radio, there was a sudden need for a second person to work as a call screener and to operate the computer interface. The ISFSI added a second segment to "The Voice" with the introduction of "The Training Office," a one-hour add-on special at the end that focused on developing and delivering training. As the shows progressed and the hosts became more comfortable with the system, most of them went back to a single host who operated the computer, screened the calls, and hosted the show. At this point, we are waiting to see what the next advancement in technology will be and how we can use it to continue spreading our message of training and education.

For the local fire department, podcasting and webcasting can offer several advantages over traditional training. These advantages include a reduction in travel time and fuel cost for companies, a wider audience platform, and a reduction in personnel costs. Furthermore, unlike a traditional classroom environment, these programs can be saved and replayed at a later time. If there is a specific subject that an instructor needs to cover with numerous individuals who work on different shifts in a short period of time, and it does not require a lot of discussion, podcasting is a very cost- and time-efficient option. If discussion or pictures are required to emphasize or help understand the topic, webcasting is a viable option.

The initial setup for developing a podcast for your local department involves purchasing an appropriate computer and the necessary software. One benefit to using Apple products, such as Garage Band, is that they supply fast-paced upgrades and options that are compatible with movie editing software like iMovie and Splice. For Windows, there are several software applications available that you may use, depending on the needs of your department. Another option, Blogtalkradio.com, is currently being used by ISFSI for their productions; it offers the following options free of charge:

- Host your own live talk radio show using any phone and a web browser
- Five simultaneous guests/listeners on your show
- Live chat with listeners
- Promotional tools, which include sharing on social networks and embedding on blogs/iTunes
- Shows are archived as podcasts.

For setup and development of a webcast, several options exist, depending on what your end use is. Many departments have used products such as Skype. Skype works great and can be free of charge; however, Google just released their version of webcasting software on Google+, which is called Hangout. Hangout can record what you do and will upload the recording to YouTube for free. The only drawback with Hangout is that it limits the number of connections that can participate at a single time to nine. Any software will be obviously influenced by your connection speed and the connection speed of the other parties. For live feeds, training officers can use Google+ or Skype. If you want to record a video on a personal computer for distribution, Camtasia (see previous discussion) works well.

iTunes U

iTunes U is an Apple application that provides access to complete online courses from leading universities and other schools on your iPad, iPhone, or iPod Touch. In addition, it also supplies the world's largest digital catalog of free education content. Anyone can access the courses. In addition to posting actual lectures, schools may also post course materials, such as books, syllabi, presentations, applications, and new ebooks for the iPad.

Key Points

- The CBT course content is available to students around the clock. This convenience can be helpful to busy companies and to members who have been off-duty for an extended length of time.
- Utilizing CBT in your organization has the potential to increase student engagement in training sessions.
- One component of computers that should be augmented when implementing CBT is random access memory, also known as RAM. A sufficient amount of RAM is required run the software used to edit and produce training videos. Check the needs of your operating system and

Alternative Training Methods

Brian Ward and the ISFSI made the following observations in a blog on Target Safety Training Resources:[1]

> *Looking outside the box is one of the most valuable tools that a training officer or any firefighter can use when implementing training. Learning from other individuals in different geographical locations can greatly influence how we mitigate certain situations. The key is to understand that we may never be able to use 100% of the expertise of say an individual that operates 1000 miles away; however, we can take parts and use them in our department. Concerning training,*

[1]Ward, B. *Alternative Training Methods Think Outside the Box*. Target Solutions. January 22, 2013. http://www.targetsolutions.com/blogs/fire-blog/alternative-training-methods-think-outside-the-boxall. Accessed September 7, 2014.

Continued

working in a fire department is similar to a scavenger hunt. There are pieces of the puzzle lying everywhere around us. It's our job as a training officer or instructor to find the pieces that fit and will make us a better department. Ultimately, keeping us safer and reducing firefighter injuries and deaths.

In 2008, the idea of bringing in Chief John Salka to speak to the department concerning Rapid Intervention, along with a full hands-on training day, turned into the first Gwinnett County Leadership and Safety Conference. The idea grew into several speakers including Chief John Norman, Chief Rick Lasky, Chief Kelvin Cochran, and others attending. Each speaker brought their own topic and area of expertise to the conference which made it a tremendous impact inside our department but throughout the Metro Atlanta Area.

Something on this magnitude had never been attempted before in the Atlanta Area and took a tremendous amount of help and determination in order to put together. The financial need for this endeavor was not allocated by the department during budget time and finding sponsors was a tough task but was successfully accomplished. One of the great benefits of bringing speakers/training into a central area is that it allows multiple departments to become involved and share the cost.

Some of the items that evolved from this training event to consider if your department wishes to put together such an event are:

- *Create a plan that includes who, when, and where. It's very important to answer all of the obvious questions prior to soliciting sponsors or buy in from outside departments.*
- *Look at all possibilities and combinations. Especially when considering time of year, speaker times, and event sites. You may choose to run two separate locations, one for classroom and one for hands-on-training.*
- *Understand that the timeframe to put a conference together usually takes much longer than anticipated.*
- *When looking for sponsorships, think way ahead. They have an annual budget similar to the government. They cannot allocate money at the drop of a hat, so it is important to get on their radar prior to their conference scheduling and budget meetings.*
- *Advertising is also important, especially if you are looking for participants to help share the cost. Look for mutual benefits with publishing companies. They have a very large distribution system, use it.*
- *Get help and lots of it—especially when it comes to keeping track of the financial documents and registration.*
- *Don't be afraid to follow through, big risk equals big reward. The greatest accomplishment of organizing a conference was receiving emails 6 months later saying how a firefighter or station changed their way of thinking or operating because of something that was said by a speaker.*

the software to ensure you have enough RAM to run the program.

- In addition to presentations and other visual, webcasting allows for an instructor you can see and interact with.
- Creating, shooting, and editing videos in a cost- and time-efficient manner has never been easier. An instructor or training officer can simply use their iPhone or any other small video recording device and produce a training video with minimum effort.
- Looking outside the box is one of the most valuable tools that a training officer or any fire fighter can use when implementing training.

References

Ward, B. *Alternative Training Methods Think Outside the Box*. Target Solutions. January 22, 2013. http://www.targetsolutions.com/blogs/fire-blog/alternative-training-methods-think-outside-the-boxall. Accessed September 7, 2014.

CHAPTER

11 Budgeting

Jason Hoevelmann

Budgeting

In the fire service, most members do not spend much time thinking about the budget until it is time for the department to announce raises or cuts in benefits. However, if you are at the staff level or are in charge of a division, the budget is your lifeline as to what you can and cannot do. To successfully navigate your annual budget, you must understand how it is developed and used. This chapter will give you an overview of general budgeting and accounting terms, as well as an introduction to several types of budgets and how to use each one to your advantage.

The budget is the guiding document, or plan, that outlines the estimated income, or revenue, and expenditures of the department. A straightforward and easy-to-understand budget scenario would be if the budget showed equal amounts of money going out and coming in. Unfortunately, budgets are not always that simple. In reality, a great deal of work will be involved to create a budget that will give your division adequate funding to achieve your mission and goals.

There are specific steps to developing and following the budget for your division or department. In this chapter we will discuss the types of budgets that are most commonly used in the fire service. However, you will be bound by the process used by your agency. As you will see later in this chapter, your budget is a legal document that is usually adopted by an ordinance, so you will need to be familiar with the legal obligations concerning a budget adoption.

Planning the Budget for the Training Division

Purpose of Planning

As with any task or item that we assume responsibility for, planning is of paramount importance for budgeting. Without planning, we have no way to justify or monitor the budget throughout the fiscal year. It takes some experience and foresight by the division head to ensure that our budgetary and division needs match. In most cases, it is prudent for you to budget conservatively in order to avoid overruns or shortfalls halfway through the fiscal year.

When planning a budget for a training division, attention must be paid to the training division's needs as well as the resources that are available to the division. You will need to look at the short- and long-term missions of your training division and your organization as a whole. Balancing the needs, resources, and missions of the training division can be challenging at times, especially when philosophical changes take place at the top of the regime and the new leadership has a different philosophy on budgeting than those of the past. However, if you can appropriately justify the division's needs and goals, the budget you develop should support the division's purpose.

As the needs of the fire service are constantly changing, you will find that the services provided by your training division will need to evolve, too. During the planning phase, you should look at the services the training division currently offers, along with the services that the division may need to provide in the future. For example, although your agency may not perform many technical rescues, changes to your industrial areas may require your division to provide training in this area within the next few years. Training in technical rescue may be required before these changes are actually implemented in your jurisdiction.

The Budget Planning Process

Planning the budget for the upcoming year takes time and a great deal of effort. The first step in the process is to review what the training division is currently doing. A training program must have a method for evaluating the members' performance that year. Without this information, you cannot know what areas you will need to target in future training. You will need to also weigh these needs against the time and resources you must commit to any mandated training or training on basic skills that need to be mastered. Examining the past year's evaluations of how fire fighters performed in prior training exercises will aid in the effort of identifying trends and areas of weakness in the department.

Having data on the members' performance available can guide you in determining the budgeting needs of your division, and thus the goals and objectives for the upcoming year. Knowing the areas in which the department may be underperforming helps you to identify classes or programs that may need extra focus and funds. For example, the division may require additional funds to hire an outside group or instructor to address a need for training on hazardous materials release containment. Coordinating the resources needed for this training can be painless if it is planned for and the division's budget is adjusted accordingly. However, if you fail to identify these needs during the planning phase for next fiscal year's budget and the problem is not noticed until halfway through the next fiscal year, it may be difficult to find funds to cover this expense. You may be forced to pull money from another program or class, which can cause that program or class to suffer.

After the needs of the training division have been identified and the historical data has been reviewed, you will need to look at the upcoming year and determine what funds will be required to accomplish the goals of the training division. This will require you to begin by estimating what your revenues will be for the fiscal year. This amount is usually established during the general budget planning session for the entire agency. Although it is not uncommon for the funds for a division or department to remain fairly stable, you may want to be conservative and use the same revenue figure of the prior year until your share has officially been provided.

During this stage, the division head will need to review and identify trends in the revenues and expenditures of the training division. Attention should be paid to how the division has spent money in the last 3–5 years. These figures can illustrate trends in spending and revenue and provide a fairly accurate barometer of how funds may fluctuate in the coming fiscal year. However, when planning your budget, use caution when it comes to depending on certain revenues, such as donations, as they do not always pan out. These funds can widely vary from one year to the next and may not be available every year.

The division's revenue will dictate what the training division can and cannot spend. Your division will incur a variety of expenses in the coming year. Depending on your department, these expenses may include, but are not limited to: classes, conferences, books, computer materials, lodging expenses, travel, outside speakers, and paid overtime for instructors. You should list these expenses in detail during the budget process.

Expenditure Detail

The tool that will be used to set the final budget is the expenditure detail. An expenditure detail is a list of the division's expenditures in line item form. This list provides a short description of each expense item and its fixed cost. The list can either be a simple spreadsheet or an elaborate chart and form. The purpose of the expenditure detail is to provide a map of anticipated expenses for the upcoming fiscal year.

It is important to have specific numbers and vendors in the details. Numbers that are too soft in one direction or another can adversely affect the division's ability to stay on budget. This tool is essential for effective and efficient budgeting. Budgeting is truly a matter of planning appropriately and sticking to the needs of the division or department.

Below is an example of an expenditure detail for a recruit training class (**FIGURE 11-1**). Each item, or expenditure, is listed, along with its associated cost. All dollar amounts are rounded up to an even number. (Rounding numbers may be used to make calculating totals a little easier.) Depending on when items are

2012 Recruit Training Budget				
Items	Number	Cost		Totals
Manuals	10	75		$750
Gear	10	$2,500		$25,000
Adjunct Instructors	100 hours	$33		$3,300
Water Bottles	10	$6		$60
Hay	20	$4		$80
Staff Instructor	400 hours	$20		$8,000
Misc. Expenses				$500
Office Supplies				$500
			Total	$38,190

FIGURE 11-1 A sample 2012 recruit training budget.
Courtesy of Brian Ward

purchased or paid for, there could be increases, and these may need to be taken into account.

The expenditure detail can be broken down into categories to separate the expenses for the total training budget into sections. Multiple expenditure details may be included within one overall expenditure detail for the training division to show how each category, or group of expenditures, comprises the entire budget (**FIGURE 11-2**).

An expenditure detail may also present the final budget for the training division. Each category would be presented as a line item, along with its associated costs. As explained above, each of the categories listed in the final budget below would have its own

2012 Training Budget				
Items	Number	Cost		Totals
Recruit School	3	$40,000		$120,000
Officer School	2	$1,500		$3,000
Drivers Program	3	$3,000		$9,000
Technical Rescue Disciplines	5	$4,000		$20,000
Special Projects	1	$6,000		$6,000
Conference	5	$1,500		$7,500
Operating Cost		$20,000		$20,000
Misc. Expenses		$1,500		$1,500
Office Supplies		$1,500		$1,500
Building Maintenance		$15,000		$15,000
Bay Renovation		$25,000		$25,000
Burn Building Inspection	1	$2,200		$2,200
Forcible Entry Door	1	$3,900		$3,900
Projector	1	$1,200		$1,200
			Total	$235,800

FIGURE 11-2 Sample 2012 Training Budget.
Courtesy of Brian Ward

expenditure detail. The Fire Fighter I and II class would require its own expenditure detail, just as a confined space and rope rescue class would be another category requiring its own expenditure detail. They then would be shown as seen in Figure 11-2 in the final budget line item.

When detailing all expenditures, it is very important to be as specific as possible. This is the document that will be used during the fiscal year and may be used to justify the budget that is being requested. It will also be used as part of the preparation process of creating the budget. Whether there will be enough money in the budget to cover the training needs of the department requires attention to these details.

Budget Preparation

■ The Process

Preparation of the budget is the stage in the budgeting process in which the cost and revenue projections, as well as any other pertinent information, are compiled and put into use. Many forms and worksheets are used in the preparation process. This is the point in the budgeting process where needs are outlined and a case is built for justifying the budget request for the training division.

As we indicated earlier in the chapter, there must be goals and objectives supporting each item in the budget request. During the preparation stage, a path must be created that clearly identifies the needs, goals, and objectives of the training division and shows how each will be accomplished through the requested budget. This is where specifics and accurate record keeping will play heavily in the final budget request.

When outlining the new fiscal year budget needs, training and classes that will be repeated or offered annually on an ongoing basis should be addressed first. These items typically do not change much and can be formulated into dollars and cents quite easily because of the records maintained on these classes. It should not take a great deal of time to determine the costs for these types of programs, and doing so will provide a quick snapshot of what money will be left for new programs or requests.

You will need to justify new programs and requests by supplying specific information pertaining to the needs of the division and the organization as a whole. The arguments for the new or supplemented programs should explain how the training division will benefit from these programs and, ultimately, how these programs will boost the performance and safety of the department. Use standards, best practices, line of duty reports, and other credible sources to support your claims.

When creating your format and presentation, keep it simple and use terminology that anyone could understand. Avoid using jargon and lingo when describing the items you are requesting and why they are important. It is important to remember that, in many cases, the governing body of a public service agency is staffed with nonemergency personnel who are not familiar with industry terms and theories. They must be educated about the items you are requesting. If you can get the members of the governing body to understand the importance of each request, then it is likely the community you serve will understand and support them, too. One key point to consider here is the cultural and educational background of the governing body. In your presentation, you will want to appeal to their perspective. For example, a governing body consisting of law, accounting, and business majors are more likely to understand arguments that will save the department money, such as "This training will save the department in the long run, because it will help the department cut down on the costs incurred from personnel injuries." Many times it is up to the training officer to explain how spending a dollar now will save two later.

Separating capital expenses from operating expenses is a good idea when creating the training division budget. This does not mean that a separate budget needs to be created for each—you just need to make sure each item is clearly identified and the categories are understandable. The reason why separating these types of expenses is a good idea is because capital expenses do not typically recur. Capital expenses for a training division could be the costs associated with building training facilities or purchasing new training equipment or props. It is possible your training division has earmarked revenues in their budget to save money over a specific period of time for the purpose of purchasing a capital item. Such items need to be shown separately from the operating expenses, such as salaries, classes, books, and travel.

Types of Budgets

■ The Line-Item Budget

It is important to know what type of budget the organization uses. In the fire service, the line-item budget is most popular. This is the type of budget used in the earlier diagrams (see Figures 11-1 and 11-2). A line-item budget breaks down the expenses for each department into specific categories, such as salaries, fuel, utilities, and uniforms. The estimated cost is allocated for each category. The final amount listed at the end of the expenditure detail for each category shows what will be available for the entire fiscal year. Once the budgeted amount for that category is spent, the money is gone.

This type of budgeting is preferred by most fire departments because it is easy to track and it shows specific uses of the funds. It is also simple to understand and use. A training officer can access and update this budget at any time. As expenses incur, they are subtracted from the funds allocated to the item's respective category. The training officer can then see how much money is left for that item or category. Line-item budgeting also allows for subcategories for each fund. As we saw in the Figure 11-2, rope rescue is part of the training budget. However, we could also have travel, lodging, gloves, helmets, and meals subsumed under the rope rescue line item. The line-item budget provides a versatile and specific method of tracking all expenses.

■ The Program Budget

Program budgets are not used very often in the fire service. This type of budget is program-based, and the budget is determined by the service each program provides, rather than by what the program might cost. With program budgets, the goals of the program are the primary mission. This is not to say that program budgets have no place in the fire service. For example, a fire prevention division may use a program budget to track the reduction of juvenile firesetting and/or burns. Some grant programs may require a program budget to be used to monitor the need and goal of the grant.

■ The Performance Budget

The performance budget is another model that may be used by the fire service organization. Some government agencies also use performance budgets; however, these are typically for specific departments or divisions within each government agency. In the fire service, a fire prevention division that performs inspections and does code enforcement may utilize a performance budget to track revenues and expenses for their division, which may be based on the number of permits and violation fees incurred by the jurisdiction. To accurately estimate and track the revenue from fees, you would pay close attention to historical data and current market trends. If construction costs are elevated and the number of new buildings is down in volume, the budget based on performance will be challenging. Typically, the expenses will be set to the amount of revenue that will be brought in based on the performance of the department.

■ The Zero-Based Budget

The final type of budget we will discuss is a zero-based budget. This budget is similar to a program budget, where the department will offer different levels of service with advantages and disadvantages of each. These service options are based on the amount of revenue coming in and the best utilization of that revenue to meet the needs of the citizen. This type of budget is time-consuming and very rarely used in the fire service. Fire inspections may be an example where zero-based budgeting could be utilized.

The zero-based budget assumes that no funds are granted just because they were allocated in prior years or for a previous program. The manager would develop the budget with zero dollars and add cost as necessary to fulfill the needs of the department on a per item basis, which can be extremely time-consuming. As you can see, the intent of this type of budget is to control cost and for managers to be able to scrutinize each cost. The best approach may be to blend or modify this approach and consider some items such as office supplies or utilities as a carryover from year to year.

No matter what type of budget format is used, it is important to know the advantages and disadvantages of each type and how each one works. The more familiar you are with budget formats and the preparation process utilized by your training division, the easier it will be to follow and adhere to throughout the year.

Budget Justification

In the budget preparation section, we touched on the need to justify the budget. Justification for a budget is something that takes forethought and time. Generally, the budget request must be accompanied by the reasons why the items are being requested.

The most effective way to justify a budget request is to have solid facts based on research. The research and sources must be credible and accurate. Sources you can use to justify the budget request include national standards, trade journal articles, trade publications, textbooks, as well as any other reliable written or online media. However, the most important source for information could be the organization's own data. Accurate and factual documentation of the number and types of incidents occurring in previous years will provide tremendous support toward justifying the need for proper resources.

As the training chief or training officer requesting the training budget for the upcoming fiscal year, the ability to identify weaknesses and the needs in the fire department is paramount. Having a training program that is evaluative in nature can prove to be a very powerful tool when it comes time to justify your budget requests. If deficiencies can be identified and documented during training evolutions over a set length of time, it will be easier to justify the need for new programs, classes, and equipment in the budget request. The same can be said about the entire training division. By incorporating evaluation methods of the members into the training process, you, as the training officer, can isolate needs the needs of the training division and organization as a whole and use these data when it comes time to plan the budget for the next fiscal year.

Formal Budget Adoption

Statutes that determine how a government body operates and spends its money bind most government bodies and organizations. These statutes play a role in the adoption of budgets and other ordinances. The adoption of the budget by the governing body will typically require a public hearing. Line items are rarely singled out during the adoption process; however, adopting a budget does take time. It is important to remember this when the budget planning begins to ensure that the training budget will be completed on time. A delay may hold up the entire budget process or result in approval of a budget that you did not request.

The adoption process will start as soon as the budget is passed along to a department or division head. This person will review the budget and confirm whether it is approved. This is called the internal review. Once the department or division head approves the budget, the budget gets pass along to the governing body for review. This is the time when some scrutiny may take place. The governing body may conclude that further justification is required for the budget submitted.

Understand that each organization establishes procedures and policies in accordance with the ordinances that dictate how the budget process is done. Know what is expected and ensure that you understand the process thoroughly so that you can gain approval for the budget that is needed. Follow the local guidelines of your local jurisdiction and be thorough.

Tracking the Budget

Once the budget has been adopted, it must be implemented and monitored. This is an ongoing part of the budget process to ensure that funds are being spent as proposed. Very careful monitoring will need to be done to ensure that overspending does not occur. You will also need to track the budget carefully to ensure that expected expenses are accurate and purchases are made at the appropriate times.

Shortages will occur if the budget is not monitored carefully. A budget that is almost exhausted halfway through the fiscal year is not indicative of good and careful budget control. Furthermore, careful monitoring also involves collecting data for the upcoming budget year. These data can help you produce more accurate projections for next year's expenses. The more accurate the projections and actual costs are, the easier it will be to justify budget requests in the coming fiscal year.

Using the budget format adopted by the organization, the funds can be closely monitored. Every purchase should be documented and the dollar amount subtracted from the available balance of the current budget. This helps you track how much money is available and whether the budget is staying in line with projections. Most of the tracking and monitoring depends greatly on documentation and careful scrutiny of purchases.

The figures below show two examples of spreadsheets that your department may use to track and monitor the budget. **FIGURE 11-3** shows a month-by-month expenditure report for the first 4 months of 2012. Expense items are listed in the left-hand column and each budget category is noted across the top. This report was constructed using a simple Excel spreadsheet. The numerical cells formatted with boldface contain formulas to calculate totals either for an entire column or row. For example, the maintenance costs for the month of January, appearing listed in the

2012	Compliance	Specialty Training	Maint.	Building Upgrades	MISC	Office	CONF	Monthly Totals
Januaray								
Operating			$1,666.00					
Phase 1 Renovation				$ 16,000.00				
Office Supplies						$265.00		
Grounds Maintanance			$1,400.00					
Recruit School	$ 8,800.00							
Total	**$ 8,800.00**	**$ -**	**$ 3,066.00**	**$ 16,000.00**	**$ -**	**$265.00**		**$ 28,131.00**
Febuarary								
Operating			$ 1,666.00					
Recruit School	$ 8,800.00							
Haz Mat		$ 1,200.00						
Ropes		$ 1,200.00						
Total	**$ 8,800.00**	**$ 3,500.00**	**$ 1,666.00**		**$ -**			**$ 13,966.00**
March								
Operating			$ 1,666.00					
Phase 2 Renovation				$ 9,000.00				
Forcible Entry Door	$ 3,900.00							
Total	**$ 3,900.00**		**$ 1,666.00**	**$ 9,000.00**	**$ -**			**$ 14,566.00**
April								
Operating			$ 1,666.00					
Grounds Maintanance			$ 1,400.00					
FDIC							$7,500.00	
Total	**$ -**	**$ -**	**$ 3,066.00**	**$ -**	**$ -**		**$7,500.00**	**$ 10,566.00**
Year End Total								**$ 67,229.00**

FIGURE 11-3 Sample monthly expenditure report for the first four months of 2012.

Courtesy of Brian Ward

fourth column labeled "Maint.", are totaled in the cell appearing in the fourth column, nine rows down. The overall expenses for the month of January, which are totaled for each category in the ninth row, is calculated in the "Monthly Totals" column, nine rows down. As the training officer updates the spreadsheet with new data, the totals automatically populate for the monthly and year-end totals.

This tool can be used year after year without ever needing to create a new spreadsheet. You can save a template onto the server with the non-reoccurring items and their respective dollar amounts removed from the column. When it comes time to create a new spreadsheet, you can download the file and save it with a new file name that includes the fiscal year. For departments with small budgets that cannot afford software programs, this spreadsheet provides a viable option.

FIGURE 11-4 shows the formula row highlighted in blue, which represents the monthly totals for each category. Scan to the right and you will see the "=SUM(B28:H28)." This cell contains the formula for the block series (highlighted in blue) that is being counted in row 28. Essentially, cell I28 is formulated

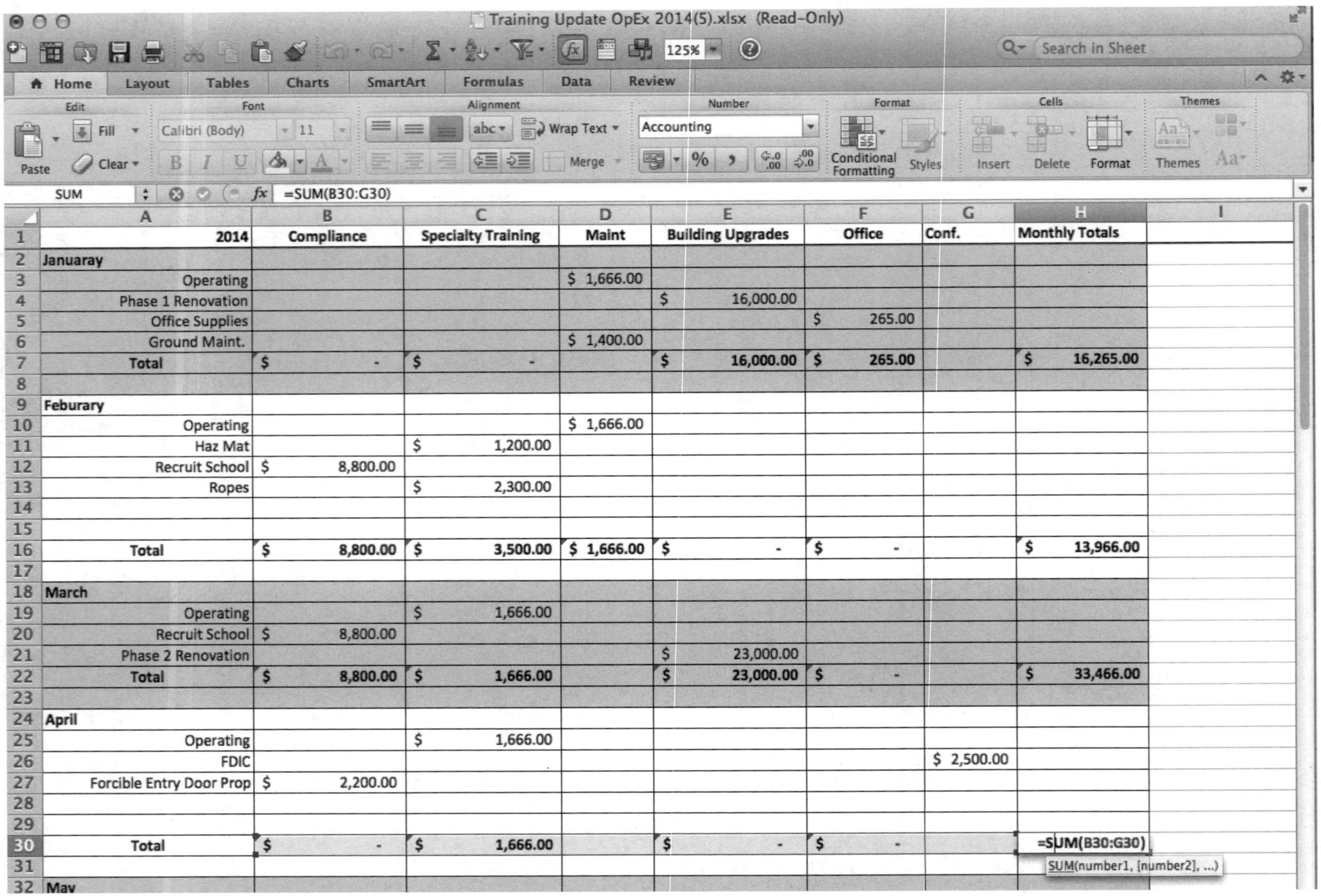

	A	B	C	D	E	F	G	H	I
1	2014	Compliance	Specialty Training	Maint	Building Upgrades	Office	Conf.	Monthly Totals	
2	Januaray								
3	Operating			$ 1,666.00					
4	Phase 1 Renovation				$ 16,000.00				
5	Office Supplies					$ 265.00			
6	Ground Maint.			$ 1,400.00					
7	Total	$ -	$ -		$ 16,000.00	$ 265.00		$ 16,265.00	
8									
9	Feburary								
10	Operating			$ 1,666.00					
11	Haz Mat		$ 1,200.00						
12	Recruit School	$ 8,800.00							
13	Ropes		$ 2,300.00						
14									
15									
16	Total	$ 8,800.00	$ 3,500.00	$ 1,666.00	$ -	$ -		$ 13,966.00	
17									
18	March								
19	Operating		$ 1,666.00						
20	Recruit School	$ 8,800.00							
21	Phase 2 Renovation				$ 23,000.00				
22	Total	$ 8,800.00	$ 1,666.00		$ 23,000.00	$ -		$ 33,466.00	
23									
24	April								
25	Operating		$ 1,666.00						
26	FDIC						$ 2,500.00		
27	Forcible Entry Door Prop	$ 2,200.00							
28									
29									
30	Total	$ -	$ 1,666.00		$ -	$ -		=SUM(B30:G30)	
31									
32	May								

FIGURE 11-4 A screenshot of a monthly expenditure report showing the formula you would use to calculate a series total in Excel.

Courtesy of Brian Ward

to calculate the monthly total for the totals highlighted in blue. The formula can be entered two different ways. You can manually enter the cell blocks [i.e., =SUM(B28:C28:D28:E28:F28:G28:H28)], or you can highlight the row by selecting the cells with your mouse, starting with the B column on the left down to the I column on the right, and then select the AutoSum button in the upper right hand corner of the task bar. Selecting AutoSum will perform the task of generating the summation formula for the cells selected for you. The same process can also be applied when calculating monthly totals (shown above in column I) and year-end totals.

Modifiying the Budget

As with anything that is planned at least 1 year in advance, changes will occur. Even with close monitoring and evaluating, the budget may need to be modified during the fiscal year. It is important to identify these needs before a crisis occurs. Modifying the budget is not uncommon. There are legitimate reasons for why a division may need to modify a budget in the middle of a fiscal year.

One reason you may need to modify a budget is because the projected revenue is lower than anticipated. A downturn in the economy could easily impact the revenues for the training division. A business may unexpectedly move out of the jurisdiction. Changes in revenue funding sources can be critical to the division's revenue, and they are also completely out of our control. In addition, a training prop or equipment could be damaged, resulting in an unexpected expense for the department. The augmentation or addition of services can also drastically change the budget. For example, a department's decision to add an ambulance service in the middle of a fiscal year would most certainly have an impact on the planned expenses for the training budget.

When modifications are needed, it is important to evaluate the changes and consider the needs of the department to justify why budget modification is required. The revised budget will need to be approved again and adopted, typically by ordinance, by the governing body. All the planning in the world cannot prevent the need for budget modifications.

Summary

The training division has many needs that need to be met in order to be successful. The budget is a dynamic legal document that must be closely monitored and evaluated. Diligence is needed to ensure that the training division budget stays on track and that the organization is utilizing funds in the most efficient manner possible. Planning, preparation, and justification are paramount when developing the budget to create a definite path for spending activities within the division.

Key Points

- In order to successfully navigate your annual budget, you must understand how budgets are developed and used from the perspectives of different governing boards.
- When developing our budgets, we need to look at the short-term and long-term missions of our division and of our organization as a whole. This can be a complicated task: a balance of both short- and long-term goals is needed.
- A review of data can guide us as we determine the budgeting needs of the training division. In doing so, we can identify classes or programs that may need extra support and funds. Additionally, reviewing the previous year's budget can help guide the next year's budget.
- Training officers should be familiar with division, program, and line-item budgets. Each of these may be called upon by the fire chief or governing board.
- Budget preparation is the stage in the budgeting process in which the cost and revenue projections, plus any other pertinent information, is compiled and put into use.
- When creating your format and presentation, keep it simple and use terminology that anyone would understand. Avoid using industry-specific jargon and lingo when describing the items you are requesting and why they are important.
- The budget is a dynamic legal document that must be closely monitored and evaluated.

References

International City/County Management Association. *Effective Supervisory Practices, Fourth Edition*, Edited by S. Wrighton. Washington, DC: ICMA; 2005.

International Association of Fire Chiefs, National Fire Protection Association, and M. Ward. *Fire Officer: Principles and Practice, Third Edition*. Burlington, MA: Jones & Bartlett Learning; 2015.

International City/County Management Association. *The Effective Local Government Manager, Third Edition*. Edited by C. Newell. Washington, DC: ICMA; 2004.

CHAPTER 12 Generations in Fire Service Training

David M. Wall

Introduction

Currently in the United States, we have the greatest diversity among generations in the workforce than we have had at any other time in recent history (**TABLE 12-1**). Each generation comes with its own point of view regarding work, values and ethics, change, and organizational hierarchy. These views are typically defined by events that occurred during each generation's childhood, from ages 5 to 18.

It is generally accepted that there are five generations alive today, with four generations on the job. Each generation possesses unique experiences that have shaped their values and beliefs about their career. For the purposes of this chapter, we focus on four generations: Generation Y (born after 1980), Generation X (born between the years 1965 and 1980), the Baby Boomer generation (born between the years 1945 to 1964), and the Veterans (born between the years 1925 and 1945).

Although the characteristics of the 20th century's generations have been developed and cultivated by researchers and everyday Americans, the years and labels used to define these generations are not always consistent among different authors and texts. This chapter will be a composite of several lists, remembering that these are generalizations with a correlating list of attributes, upon which other generalizations can be made.

Each generation has its strengths and weaknesses and offers different perspectives and traits to the fire service. We will discuss these traits and the ways in which each generation provides their own contributions or challenges to the fire service in the forthcoming sections.

A Firsthand Gen-Y Perspective

As a Generation Y employee, I can tell you firsthand that we learn differently and approach tasks in a different way than members of other generations do. When teaching Generation Y employees such as myself, it is important to remember that we tend to think we can acquire the knowledge needed to do manual tasks simply by reading how to do them. Many Generation Y employees feel confident performing any task—even the most complex ones—so long as we have been provided with instruction on how to do it. Providing step-by-step instruction can also be important to helping us succeed.

Because our generation has been surrounded by an abundance of technology, many of us have not spent a lot of time away from our computers and gaming systems and out in the proverbial real world, performing hands-on tasks and figuring out how to make mechanical objects, such as vehicles and equipment, work.

Continued

While we may be experienced with technology and tend to learn and use new software and technical devices with ease, we struggle when we have to perform more mechanical tasks. The baby boomer generation, on the other hand, may be wired the opposite way. For example, I remember a story my dad shared with me about a time when he and his brother took apart their lawn mower and used the engine to propel a homemade go-cart. After a couple days of fun, they put the lawn mower back together in time to mow the lawn. Many Generation Y workers, myself included, have never had experiences like the one my dad had. Consequently, Generation Y's mechanical knowledge is typically inferior to that of baby boomers.

Our life experiences, or in some cases lack of life experiences, can present some real challenges for instructors. For example, let's say you're giving a refresher course on pump operations and you have a room full of veteran fire fighters and one new Generation Y member who is fresh out of the academy. The discussion turns to pump cavitation. Chances are very good the veterans in the room will know what cavitation is, and they will know how to fix it. However, it is quite possible that the Gen Y recruit won't even know what cavitation is, let alone have any understanding of how or when it happens and how to fix it. Since the Gen Y recruit may have very little firsthand knowledge of how a pump works, it is likely that he or she will not know how to solve the problem.

This lack of knowledge about how a pump works may pose various challenges for the training officer, and it may even frustrate him or her. It is going to take extra effort and patience to teach the Gen Y recruit the mechanics of how a pump works. While the recruit may have the knowledge to pass the test, he or she may lack the deeper understanding of how the pump operates. For all members to succeed, it is essential for the instructor to keep these differences in mind and invest the time and effort to teach Gen Y employees the principles and practicals, and teach them in a such a way that it builds their confidence.

— Cameron Gasaway

Generation Y

There are approximately 70–80 million people in the United States who were born after 1992, which is approximately 30% of the population. This generation of digital natives, also known as Generation Y, Millennials, Echo Boomers, Generation Next, or Next-ers, consists of those who were born after 1980. This is the generation coming on the job today and for the next 15 years or so. Few people from this generation have ever used a rotary phone, and this generation has rarely used a landline. Computers have always been part of their life. With the use of social networking sites, such as Twitter and Facebook, they are more connected, more wired, and more aware of what is going on in other places in real time than are previous generations.

Because of its sheer numbers, Generation Y will dominate the workforce for years to come. The size of this group is the reason they are referred to as Echo Boomers, as they are the children of the Baby Boomer generation, and their population size is almost equally large.

TABLE 12-1 Table Presenting Population Size of Each Generation and the Composition of the Workforce for Years 2000, 2010, and 2020.

Generation	Born Between	Estimated Size	Percent of Workforce in 2000	Percent of Workforce in 2010	Percent of Workforce in 2020
Mature	1922-1945	75 Million	13%	3%	0%
Boomers	1946-1964	77 Million	48%	37%	20%
Gen X	1965-1976	50 Million	22%	22%	20%
Gen Y	1977-1994	76 Million	16%	38%	44%

Reproduced from: Behrstock-Sherratt, E and Coggshall, J.G. Realizing the promise of Generation Y. *Educational Leadership* 2010:67(8):28–34.

In general, Generation Y expects organizations to meet their needs on their terms. This generation is also known as the "me" generation, having parents who provided for them throughout their lives. These parents even earned names such as "helicopter parents" and "snowplow parents," because they cleared a path for the kids. Children were praised for minimal effort, and so as young adults, they tend to possess high self-esteem and self-confidence. However, due to the ways in which their parents protectively raised them, they also tend to be very reliant on their parents well into their adult years. This generation has grown up in good economic times and has spent more time in full-time education than previous generations. They question everything.

This generation is not afraid to challenge their bosses on the job, which can lead to some conflict in the workplace. Work for them is a means to an end. Work is not life; it is a place, and it does not represent their total identity. Gen Y'ers want flexible working hours, to be able to work from home whenever possible, and have plenty of time off. If the job does not meet their expectations, they simply move on to another one. This generation believes that they can have it all, and they are not embarrassed to ask for it. They are the future and the reason that Boomers need to pay their knowledge forward to a group that, in some respects, may perceive that they already know it all. Furthermore, this generation is more likely to become engaged with work when they can see the future utility, benefit, or relevance. One way to engage this generation is through the enhanced use of technology.

Some researchers find dramatic attitude differences in the younger generations. For example, it is unlikely that these future leaders will ever be able to think of workers having the same approach to life, work, or even to learning as their bosses did. It is generally accepted that this difference in attitude represents a cultural shift for private sector companies and their managers, with the public sector not far behind, simply because they are competing for the same employees to be hired. Generation Y'ers wants to make the right decision. They seem to favor getting as much information as possible and obtaining a general consensus from their peers before making any decisions.

Generation X

Of the current generations alive in the United States, Generation X, which comprises individuals born in the years from 1965 through 1980, has the smallest population size. They are also known as the latchkey kids because they tended to go home from school to an empty house, armed with a key to let themselves in, because both parents either worked full-time or their parents were divorced and working. The latchkey kids grew up independently and learned to fend for themselves. They take work seriously and are pragmatists, or people who are practical, and wish to get a job done. They tend to be resilient, independent, and adaptable, and they generally possess a live-to-work and work-to-live attitude. Generation X tends to be more skeptical, less loyal to their employers, and fiercely protective of their work/life balance.

This group is also technologically savvy. As with the Gen Y group, Generation X seeks a work/life balance. Their childhood experiences with divorce may have shaped how they view their family responsibilities. They have a "Just do it" mentality and a bias towards acting, even in times of insufficient information. While they initiated the movement for flexible work arrangements and telecommuting, it is likely that the Gen Y group will take this movement to the next level.

In a study focused on Generation X employees, women in leadership positions distinguished themselves from their Baby Boomer predecessors by making the connection between professional success and personal fulfillment. The Generation X women regarded their employment as a means to getting a fuller life and having closer connections to family and friends. In other words, the job represents a means to an end, not the end itself. Perhaps this is why Generation X members rate having job satisfaction higher than getting promoted. They focus on life outside the organization and are not willing to be turned into the workaholics that their parents were. It is important to remember that these studies were done on the non-fire service population. A study done on fire service employees may reveal different results that are driven by the fire service culture. Perhaps this could be a study for the future.

Baby Boomers

The Baby Boomer generation, born in the years 1945 to 1964, constitutes the largest numbers in today's workforce, but not for much longer, as a large percentage of this generation is looking to retire in the next decade. This group, possibly the last to experience stay-at-home moms, grew up in the economic prosperity of post-World War II. They witnessed their parents' strong work ethics and self-sacrificing behavior, and this, in turn, shaped their attitudes toward work. For this reason, once they joined the workforce, they hit the ground running. Competitive, extremely hardworking, and considered by many as workaholics, this generation is responsible for increasing the 40-hour work week. They generally value work and personal sacrifice and tend to be idealistic and driven, having made a significant impact on society in various ways throughout their lifetime.

The Gen X members of this group now hold management positions and maintain this strong devotion to work. They also have the expectation that their subordinates will do the same.

Veterans

The Veterans, also known as the Silent Generation or the Traditionalists, were born in the years from 1925 to 1945. Although they constitute only 3 percent of today's workforce, they were extremely influential on the Baby Boomer and Generation X employees they supervised and worked with. The Veterans grew up in hard economic times, did not take work for granted, and consequently became very disciplined and self-sacrificing employees. Their work ethic is defined by loyalty to their employers, discipline on the job, and knowledge. Some of this work ethic was passed along to Baby Boomer employees and older Generation X employees.

The Workplace Dynamic: Generations Working Together

When the Boomer managers are confronted with younger employees who do not fit their mold, this can lead to conflict. Although many Baby Boomers were hoping to retire early, they lost that opportunity in the recession of 2008–2009. Consequently, many continue to work, which means that these generations will need to learn how to work together. In addition, the economic downturn has also affected Generation X employees who were hoping to move up in the organization as the Boomers and members of the Silent Generation retired. Some researchers believe that these four generations of workers, who are currently competing for and holding positions in the workforce, may have to work with each other for as many as 10 additional years. These are profound changes in the workplace, providing an opportunity for the future training officers in a fire department to learn more about their students and themselves. These training officers will have to learn how to train individuals who may be from four different generations. To prepare for this new type of coaching, training officers will need to explore the characteristics of each generation to understand how best to motivate and train each one and ensure that they work together. This knowledge will affect the development of training content and how instruction is delivered.

Training the Generations

Training different generations is a challenge even for the most seasoned training officer. Today's recruit schools will contain various age groups, consisting mostly of Generation Next employees, but will also include Generation X individuals who are undergoing a career change. Others may have recently left the military after 10 to 20 years of service. Steven Covey, in his leadership programs and in his book, *The Seven Habits of Highly Effective People: Restoring the Character Ethic*, tells us first "seek to understand, only then to be understood." This is wise advice: we should listen more than we talk. This advice is important if we are to promote the Generation X fire fighters to the role of training officers, asking them to be responsible for training all of the generations at work in the fire service today.

How One Generation Can Help the Other Out

At a recent training officers' conference, a 20-year lieutenant remarked, "We used to hire kids like me who worked on their cars and drove tractors, had lots of outdoor activities, and did not think twice when asked to use hand tools. We just knew how. Now we are hiring kids who grew up in apartments or subdivisions, never drove a tractor or even a stick shift. They are not used to handling tools. We have to train them how to do all that. But they sure can handle technology and computers and can research anything I need, including downloading hazardous materials apps on the scene with their super smart smartphones!"

Perhaps the description made in this case study is an overgeneralization, but it does illustrate the change in demographics being witnessed in many areas of the country. What we have to realize is that by 2020, the new generation will equal the Baby Boomer generation in numbers, and be much younger to boot! Today's Generation Y is an unnoticed asset. They deserve and are in need of development by the Boomers and X'ers who are today's leaders and middle managers. Leveraging assets is how organizations accomplish their missions.

As you may have noticed in our brief descriptions of each generation, the characteristics of generations tend to overlap with its closest generation. For example, the first wave of Generation X employees is close in age to the last wave of Baby Boomers. Consequently, those who were born during this period will exhibit some characteristics from both groups. Another good example is the last wave of Generation X who witnessed the same advances in technology as the first wave of Generation Y. So while there is value in understanding generalizations, you do not want to overly rely on these labels. They can cause you to become biased, and you may typecast crew or class members as a result. To prevent ourselves from making this error, we should look for commonalities between generations and try to understand, not judge, their differences. Jack Parow, Fire Chief and President of the International Association of Fire Chiefs (IAFC), said it best when he asserted, "If we can harness the diversity between generations and use it to our advantage, we will have better and stronger organizations." This sentiment is well worth considering when working with a mixture of generations.

In their book on generations working together, *Generations At Work: Managing the Clash of Veterans, Boomers, Xers and Nexters in Your Workplace*, Zemke, Raines, and Filipczak discuss how organizations lack the ability to understand the differences of each generation. Failing to understand the characteristics of each generation makes it difficult for organizations to capitalize on them. In today's 21st-century globalized workplace, it is important to be aware of and savvy about generational differences because there are now more generations than ever before working together.

Case in point: when we say that Generation Y treats work as a means to an end, it is not because Generation Y lacks work ethic; Generation Y's work ethic is just different from that of a Traditionalist or Baby Boomer employee. Where the Boomers fear and resist change, the Gen Y members embrace it; they even require it. Generation Y employees need to know why something is the way it is, so it may be important for the training officer to be able to more fully explain the job to a new worker. The explanation of "We do things this way, because that is the way we have always done them," is not a sufficient explanation for Generation Y. Think of this as another step in situational leadership. Providing better directions at the outset of a lesson will provide you with a more empowered employee who will eventually need to think for himself or herself.

Why Officers Need to Have Situational Awareness in a Diverse Workplace

A recently hired fire fighter asked his company officer, somewhat naively, "When will I be promoted to Captain?" The Captain, who was from the Boomer generation, immediately thought to himself, *Why is this recruit asking me this question?* Unaware of the trait of forthrightness that Millennials tend to exhibit, he was surprised the new recruit did not hesitate to ask such a question and proceeded to lecture the new recruit on the importance of respect. The Millennial had thought it was completely appropriate to ask a superior on when promotions could be expected and was amazed by the captain's reaction.

Did the new recruit truly demonstrate a lack of respect? In what other ways could this situation be perceived? When faced with this situation, it is incumbent upon the Boomer managers to anticipate such questions and not overreact. Possessing the ability to anticipate such conversations means that you have situational awareness (SA) of workplace interactions. A situation such as this could be characterized as level 3 SA—in other words, the company officer needed to be able to predict what was going to occur next, not just notice or understand it.

The Technology Gap

One potential concern about today's wired-up students is that they expect to be able to continue their seamless texting and communicating with each other, even when they are on the job. This can be seen as disruptive in the fire service. A local pharmacy employs several workers in their 20s, along with a few older teens. The manager commented that she has to confiscate the smart phones at the beginning of the shift and place them in a basket by the door because of the interference in the workplace.

Differences in the ability to use or understand technology exist between generations. The Millennial generation, deemed the "digital natives" by Mark Prensky, grew up with digital technology, such as computers, cellphones, the Internet, and online gaming, and consequently has an intuitive understanding of technological gadgets and software. They learn and use new and existing technology with ease. In contrast, the "digital immigrants," which include the Veterans, Baby Boomers, and older members of Generation X, are akin to foreigners trying to navigate their way through an entirely different continent, and that continent is the ever-changing landscape of technology. Prensky claims that the reason for this difference could be attributed to the theory on the plasticity of the brain. The theory of plasticity proposes that the hours Gen Y employees spend on technology may have caused areas of their brains to grow and become more adept at certain cognitive skills. This is an area of research that has large implications for training officers now and in the future. Generation Y employees can effortlessly move from one task to another, from one Internet site to another, access smart phone applications, send texts, and update their Facebook status on the fly.

Communication

Generation Y is different from previous generations in the sense that they are optimistic and enthusiastic, future-oriented, and driven by opportunity. This is a generation that is ready and eager to make its mark on the world. They were brought up in an era of such rapid technological change that they embrace change more readily than previous generations. They will want rapid advancement and more responsibilities at a younger age. The way employers respond to these expectations and requests will determine whether the potential of this generation will be tapped in times of tight budgets and economic uncertainty or if they simply move on to another fire department or to a completely different field.

One study tells us that the Generation Y student is confident and motivated despite the turbulent times in the economy. This study also found that 57 percent of the Generation Y students are looking for opportunities for advancement, while only 8 percent are looking for better job security. This appears to be the opposite of the reason why many members of preceding generations, such as the Veterans and the Baby Boomers, went to work for the local fire department. For previous generations, a position in the fire service represented a job that provided security as well as opportunities for advancement. If that has changed, what does that tell the training officer about his or her students?

One way to deal with this may be to assign mentors to Gen Y employees early on in their career to foster the relationship and include a plan for succession. A mentorship program can be of great value to the fire department. It provides opportunities for multiple generations to work together and learn from each other. They can share with each other the different ways that they approach work, their views on ethics and values, their attitudes toward management, and their distinct and preferred methods of managing others or the ways in which they would like to be managed.

In the Los Angeles Fire Department (LAFD), for example, assigning company officers as mentors to new recruits has become standard. They are expected to teach and facilitate learning for their partner and help their partner seek and explore their own direction. Gary Ludwig, Deputy Chief of the Memphis Fire Department, asserts that assigning a mentor for each new recruit is an obligation of leadership. Having an older worker mentor younger workers allows them to share their stories and knowledge, while giving younger workers a chance to ask questions in a nonthreatening environment. A key responsibility of any officer is to teach skills and pass on their knowledge about the fire service to lower-ranked, newer fire fighters, as well as further his or her own skills and knowledge by learning the job above his or hers.

Mentorship is not a new concept. In fact, it has always been an important part of organizational training, and many fire service departments incorporate this practice into their training. The mentorship program in the fire service typically begins in recruit school or when voting in a new volunteer. Mentorship has many benefits. Newcomers to an organization who have mentors tend to become more knowledgeable about the organization than those without.

When training officers assign mentors to newer members, they should consider the individual's willingness to work with others. Unlike the Vietnam-era Boomer generation, Gen Y is willing to learn from and work with older mentors. Four things are needed to move organizations forward:

1. Open communications in a variety of methods.
2. Transparency and openness.
3. Willingness to honestly address any concerns.
4. Respect for the different values held by different groups.

Differences in Training Styles and Learning Preferences

When delivering training to a multigenerational group, the training officer is likely to hear different messages of learning style preferences from each generation. The Boomers will say to the training officer, "Just tell me what to do," while the Generation X employees will give a different directive: "Show me how to do it." Generation Y, on the other hand, wants to understand the reason behind the training and will ask the training officer: "Why do I need to learn this?" By paying attention to these differences in learning styles, training officers can develop a training session that meets the needs of the trainees.

When a group of 28-year instructors were asked what was most challenging about instructing new recruits and their views on Generation Y employees, they immediately addressed the challenges with keeping Gen Y members engaged in the classroom setting. They explained to their Gen X and Boomer bosses that the Millennials is the generation that grew up riding in car seats with Gameboys in their laps, technology at their fingertips every day of their lives, mega-channel cable TV, and the utilization of social media. Putting these members in classroom settings with nothing more than a PowerPoint presentation in front of them is the fastest way to lose their attention. Within 10 minutes, they are no longer engaged in the lesson, and no learning occurs.

So how do we engage this generation in training so that they retain enough information to progress to the next level? And how do we balance that with the learning styles and needs of the other generations? They are used to innuendo and nuance, so instruction that uses these techniques can be effective. If a PowerPoint presentation must be used, it may help to use flashy features in 3–4 second blocks to keep Gen Y members interested in the lesson. It does not even have to be showy, just interesting enough to keep them engaged. Engagement should also mean they are active and participating in their learning.

In a recent survey of Atlanta and Georgia fire service training officers comprising Generation X employees and older Millennials, the officers were asked, "How do you keep your classes engaged in the learning process?" The answers they shared were revealing:

- Do not give war stories from the past. Only share recent stories that fit the current topics being discussed. The story must be brief, occupying no more than a few moments, and it should pointedly illustrate the reason why a particular subject is being taught. It should also demonstrate practical application of the topic.
- Stay current with the business. A training officer who is not absolutely up-to-date with fire service techniques and with the technology will not be able to retain the audience's attention.
- Include the group and initiate active learning as soon as possible. Each instructional block is critical for the learning styles that are seen with today's workforce.
- Provide a method of getting a fire fighter student into an informal role and make each student responsible for his or her own learning.
- Study the learning styles and have your students assess their learning styles.
- Target the instruction to the learning styles that you have in the group. Although this is easier to do with smaller groups and with those you work with on a regular basis, you should make an effort to teach to the learning styles of your class. Even when multiple companies are brought together for a training session, the training officer can target the approach to training based on the learning styles of the class.
- Make training safe and effective.
- Remember, if it is not relevant, it will not be effective.

Research

The fire service is considered a highly reliable organization (HRO). Organizations that have people who challenge the status quo are known for their innovation and ability to adapt to change more quickly. As we have indicated in this chapter, Generation Y is well suited to questioning the answers and adapting to change, and these traits should be embraced in the fire service. If the generations arriving at work today do not have the courage to speak up when they see something going wrong—a tenet of good crew resource management—then errors made on scene will continue to haunt the profession. The question training officers must ask themselves is how can they empower Generation Y employees, who are a group of independent-minded inquisitive future leaders, to find the courage to speak up as needed during fire training scenarios? The ability to empower this group will only occur if the culture of the fire service does not suppress their willingness and ability to speak up. In order for cultural change to occur, Generation Y's willingness to share their thoughts and opinions must be supported in real time by their trainers and supervisors. This behooves trainers and company officers to use an inclusive style of leadership.

Training officers should provide their younger fire fighters with the tools to make those challenging statements, especially when their knowledge will prevent a tragic occurrence. They must also teach these recruits how to tactfully communicate their ideas. If the method of communicating is not effective or respectful, then they will not be taken seriously or listened to. As a result, the goal is not met.

Tying it all Together

The fire service needs to embrace the diversity of ages and generations. Fire service is, after all, a team effort every time the big red truck rolls out the door. The expectations of the everyday American are not any lower, and for more informed citizens, the expectations of service are higher than ever. The differences are what tools the workforce is using to achieve those goals and in the leadership styles that will be successful in the future.

The leadership styles utilized today are those of inclusiveness, open and frequent communication, and the embracing and understanding of generational cultures. These approaches are the key to leadership accomplishing organizational goals. The old-school theory styles may work in the short term, but the new Millennials, as well as other generations, do not view a purely autocratic style as the correct method to use in managing people when they are not involved in an emergency situation. Training for new fire officers must include learning how to use a variety of situational leadership styles, once they have diagnosed the developmental levels of their crew. As demonstrated in the study conducted by Hersey and Blanchard in Northouse's, *Leadership: Theory and Practice* and in the courses at the National Fire Academy, it is now more important than ever for company officers to be flexible in their management styles in order to get the most from their multigenerational employees.

A training officer who is aware of the learning styles of his or her students will have a distinct advantage, in that they can target the learning experience to best reach the variety of learners. To assess learning styles, training officers can use an online learning style inventory questionnaire. Once the surveys are completed and the results are compiled, you will find it much easier to develop training that factors in the learning styles of your students. As we discussed earlier, there is no quicker way to alienate the new Millennial fire fighters than to subject them to a day or two of "death by PowerPoint." If the material being taught is strictly informational and available online, a good strategy would be to assign this reading to students to complete at their own pace and on their own time. Online materials can be followed up with hands-on skills training conducted in groups with the instructor coaching members on how to master the skills. It is this blended format—a combination of online and live training—that will reach all generations and move the profession forward. When information and skills can be presented in multiple formats, the skills and talents of training officers can be focused on the repetition needed by the student to be a master of those skills.

Within a few years, most of the Traditionalist or Veteran generation will have retired and a large portion of the Boomer generation will be facing retirement as well. Currently, Generation X promotions are underway, with the future looking bright for many of these members who are positioning themselves to become chief officers. This generation is left with the responsibility of embracing new training techniques, not just for the Millennials, but also for whatever the next generation will bring when they enter the workforce.

Howard Cross, a National Fire Academy adjunct instructor, encourages fire service leaders who are dealing with generational issues to ask questions that evaluate how well they embrace the changing dynamics of their departments. Some of the questions he suggests to ask include: *Does the organization tolerate some failures and risk taking? Does the organization capitalize on the strengths of their diverse department?* The differences in a team are what make it a more strongly-functioning unit.

Leaders today need to make an effort to get to know the team and utilize their strengths. A leader may understand generational differences, but unless the workforce is educated on what makes these different generations tick and relate, it will be missing a key piece. Generational differences can be used to enhance the ability to deliver service, and although it is important to learn the characteristics of each generation and use this knowledge to train and lead departments, caution is urged to not overgeneralize employees based on what studies say about their particular generation. Not everyone fits neatly into a category, especially when many people have characteristics from two or more generations.

Case Study

You have a two new training officers, 27 and 28 years old, who both hold the rank of Lieutenant. Although they are relatively young, they have approximately 8 years of fire service experience and therefore are not rookies. Given their ages, these two training officers are from Generation Y. However, before you assume that these training officers have all the traits of a Millennial, consider some of other factors. Both of their parents were in the fire service, as long-time volunteers and career, and are very active. The officers' group characteristics (Gen Y) are tempered not only by their upbringing in the values of the older Boomer generation group, but also in their immersion in the fire service as a culture at an early age. This upbringing has given them a distinct edge in understanding multiple age groups. They have had the opportunity to witness multiple fire service generations and relate well to those in the younger group. The training officers share their experience with the students in their recruit programs. Their insight into how each generation thinks and works helps students understand how they can respect other generations and fit into the fire service department.

One challenge for the training officers in this next decade will be to teach the older group to be tolerant of the differences in attitude and work styles demonstrated by the younger group and to respect their aptitude for technology and their ability to work as a team. It is important to remember that we cannot train someone in either technique or knowledge just one time and then expect them to perform well under pressure. Skills and knowledge have to be ingrained through repetition and practice to the point where they become routine and automatic. The question is: What is the best way to accomplish this with a different generation in the class? Talented and creative training officers will make the difference.

Key Points

- Today we have the greatest generational diversity in the workforce than we have had at any other time in history.
- We should invest the time and effort to teach students, especially those from Generation Y, the principles and practicals. We should do this in such a way that it builds their confidence.
- Generation Y wants to make a correct decision. They seem to favor getting as much information as possible and obtaining a general consensus from their peers before making any decisions.
- Generation X members rated having job satisfaction as being more important than getting promoted. They tend to focus on life outside the organization. They are not willing to be turned into the workaholics their parents were.
- Training officers will have to learn how to train individuals who may be from four different generations on skills. To prepare for this new type of coaching, training officers will need to explore the characteristics of each generation to understand how best to motivate and train each one and ensure that they work together. This knowledge will affect the development of training content and how instruction is delivered.
- A key responsibility of any officer is to teach skills and pass on his or her knowledge about the fire service to lower-ranked, newer fire fighters, as well as further his or her own skills and knowledge by learning the job above.
- In order for cultural change to occur, a student's willingness to speak up must be supported in real time by the trainers and supervisors.
- The leaders of today need to make an effort to get to know the team and utilize their strengths.

References

Alsop, R. *The Trophy Kids Grow Up: How the Millennial Generation is Shaking Up the Workplace.* San Francisco, CA: Jossey-Bass; 2008.

Arsenault, P.M. Validating generational differences: A legitimate diversity and leadership issue. *Leadership & Organization Development Journal* 2004:*25*(2): 124–141.

Alyn, K. Developing future leaders in the fire service. *Firehouse Magazine* 2010:*35*(8):82–83.

Behrstock-Sherratt, E. and Coggshall, J.G. Realizing the promise of Generation Y. *Educational Leadership* 2010:*67*(8):28–34.

Brokaw, T. *The Greatest Generation.* New York, NY: Random House; 1998.

Butler, P.I. Succession planning in Homeland Security: How can we ensure the effective transfer of knowledge to a new generation of employees. *Compilation of Thesis Abstracts March 2010.* Monterey, CA: Naval Post Graduate School; 2010. https://www.hsdl.org/?view=docs/theses/10Mar_Butler.pdf. Accessed September 7, 2014.

Brunacini, A.V. *Essentials of Fire Department Customer Service.* Stillwater, OK: International Fire Service Training Association; 1996.

Cline, K. Develop new officers for organizational success. *Fire Engineering* 2010:*163*(7):107–110.

Covey, S.R. *The Seven Habits of Highly Effective People: Restoring the Character Ethic.* New York, NY: Simon & Schuster; 1989.

Deloitte. *Generation Y: Powerhouse of the Global Economy.* New York, NY: Deloitte Development LLC., 2009. http://www.deloitte.com/assets/Dcom-UnitedStates/Local%20Assets/Documents/us_consulting_hc_GenerationY_Snapshot_041509.pdf. Accessed September 7, 2014.

Dries, N., R. Pepermans, and E. De Kerpel. Exploring four generations' beliefs about career: Is "satisfied" the new "successful"? *Journal of Managerial Psychology* 2008:*23*(8):907–928.

Feyerherm, A. and Y.H. Vick. Generation X women in high technology: Overcoming gender and generational challenges to succeed in the corporate environment. *Career Development International* 2005:*10*(3):216–227.

Gasaway, R.B. Trained to fail. *Situational Awareness Matters.* April 22, 2012. http://www.samatters.com/2012/04/22/. Accessed September 7, 2014.

Glass, A. Understanding generational differences for competitive success. *Industrial and Commercial Training* 2007:*39*(2):98–103.

Hart, J. Understanding today's learner. *Learning Solutions Magazine* September 22, 2008: 1–11. http://www.learningsolutionsmag.com/articles/80/understanding-todays-learner. Accessed September 7, 2014.

Larton, D. Dispatch: The Next generation. *9-1-1- Magazine: Managing Emergency Communications* 2005:*18*(8):22–24.

Ludwig, G. Generation Y: Why you can't connect. *Journal of Emergency Medical Services* 2009:*34*(11):30. http://www.jems.com/article/administration-and-leadership/generation-y-why-you-cant-conn. Accessed September 7, 2014.

Northouse, P.G. *Leadership: Theory and Practice, Fourth Edition.* Thousand Oaks, CA: SAGE Publications, Inc.; 2007.

Parow, J. A Case for not putting your difference aside. *Fire Chief* 2011:*55*(3).

Prensky, M. Digital Natives, Digital Immigrants, Part I. *On the Horizon* 2001:*9*(5):1–6. http://www.marcprensky.com/writing/Prensky%20-%20Digital%20Natives,%20Digital%20Immigrants%20-%20Part1.pdf. Accessed September 7, 2014.

Taylor, M. Teaching Generation NeXt: A pedagogy for today's learners. *A Collection of Papers on Self-study and Institutional Improvement, 26th Edition.* Chicago, IL: The Higher Learning Commission; 2010.

Tippett, J. *Leading the Millennials.* A Presentation to the Georgia Fire Chief's Spring Conference, Savannah GA, April 17, 2011.

Theil, A. Are we doing it all wrong? *Fire Chief Magazine* 2010:*54*(9):16.

Yu, H.-C. and Miller, P. Leadership style: The X Generation and Baby Boomers compared in different cultural contexts. *Leadership & Organization Development Journal* 2005:*26*(1/2):35–50.

Zemke, R., C. Raines, and B. Filipczak. *Generations At Work: Managing the Clash of Veterans, Boomers, Xers and Nexters in your Workplace.* Boulder CO: AMACOM; 1999.

CHAPTER

13 Special Topics in Training

Scott Maddison, Brian P. Kazmierzak, and Brian Ward

Training in Combination Fire Departments

One of the most important jobs in any fire department is that of the training officer. The training officer is usually the bridge between the current direction of the fire department and the line fire fighters. The training officer's attitude, skill, and dedication usually set the tone for training in the department. While it may not be the case in every department, generally, the quality of the training officer is what can make or break a fire department, especially in the areas of fire-ground operations and fire fighter safety.

This is especially true in a combination fire department, where the training officer is the common link between the career members of the department and the part-time/volunteer members. Due to the intermixing of career and volunteer members, the training officers are usually the one constant, as they must pass the message of fire fighter safety via training to the entire department. The United States military has made this concept work for years using career and reserve military personnel, requiring both to meet the same training standards. So, can the same be done in the fire service? In this section, we will look at the challenges of being a training officer in a combination fire department and offer some suggestions for developing a training program for this department type.

Training Standards and Expectations for Career Versus Volunteer Members

A common complaint and struggle in a combination fire department is the lack of consistency in training between the career and volunteer members of the department. However, most training officers will find that no matter how hard they try, they cannot have the same exact training standards and expectations for career and volunteer members. The reason for this is that career fire fighters, on average, work 56 hours per week, train about 1–2 hours each shift, and typically run more calls than the volunteer members. Due to the added exposure to firefighting tasks performed on the job, the skill level of career members will increase at a greater rate. There is just no way to overcome this. On average, a career fire fighter will achieve approximately 24 hours of fire training per month, while a volunteer achieves only about 9 hours. One way to understand these figures is that the Insurance Services Office (ISO) only credits one-third of an hour to a volunteer for every hour of firefighting training they complete, whereas a career fire fighter is credited the full hour. It should be noted, however, that in some departments the volunteers achieve 20 or more hours of training per month, exceeding those achieved by career members, because of the hard work and efforts of some very dedicated members and training officers.

Ultimately, when determining training standards for career members versus volunteer members, the training officer must find a happy medium and allow for the maximum use of each group's skill level on the fireground.

The best way to achieve balance when establishing training standards is most likely through performance-based training or job performance requirements. Training officers have a limited amount of time to train and observe all members, but this is especially true with volunteers. When training officers train with volunteers, which is typically in the ballpark of 1–4 times per month, they must make that 2–4 hour block of training as productive as possible. This means training officers must have volunteers hitting hydrants, dragging hose, squirting water, searching structures, and, in general, acquiring as many interior and exterior firefighting skills as they can. In doing so, training officers hope to ensure as high a skill level on the fireground as possible.

Training Volunteer Fire Fighters

You may be wondering that if all training time is spent on honing skills, how do training officers address other training requirements, such as those mandated from Occupational Safety and Health Administration (OSHA), local entities, and so forth? For the volunteer member, it is best that these courses are completed at his or her leisure through the use of Internet-based courses or other types of self-study. There is no good reason to spend valuable and limited hands-on training (HOT) time on these subjects on an annual basis, especially when this material can be easily accessed outside of a classroom environment and requirements can still be met. In order to do this, however, you must have a system in place that documents the completion and knowledge retention of each of these courses.

Another concern you may have is how can training officers provide specialty training, such as driver training, officer training, Emergency Medical Services (EMS) training, and weapons of mass destruction (WMD) training, to volunteer members when their availability is so limited? The answer is actually quite simple. If these fire fighters are not drivers or preparing to be drivers, for example, do not spend time on this training. Instead, establish a mentoring program. Put new drivers with seasoned operators who can mentor them during hose and stream exercises, or let them drive to the drill site. Furthermore, the driver/operator certification course is one that all members should have completed prior to driving/pumping, in addition to the Emergency Vehicle Operators Course (EVOC). To teach strategic and tactical decision-making skills, you should place prospective officers in command roles at drills. Prospective officers should take some initiative to get the proper certifications prior to the promotional or election process. When training prospective officers in the combination fire department, it is important to consider the fact that promoted career officers and elected volunteer officers often do not mix well, unless the requirements for each position are the same. In other words, an officer should not fill a command position unless they are qualified. Make sure the requirements are the same so there is never a question on a scene about who is qualified to hold that command position.

With respect to WMD training, while it is a necessity in today's fire service, do not spend all your training time preparing for this event if your fire fighters do not have the basic structural firefighting skills mastered. As Billy Goldfeder once said, "We are not killing fire fighters with nuclear or chemical weapons; however, we are still continuing to kill fire fighters due to a lack of knowledge of building construction, fire behavior, and a lack of basic fireground skills."

Holding All Members to the Highest Standard

What all fire fighters need to remember is that firefighting is a profession, regardless of whether you are a career or volunteer fire fighter. We must all hold ourselves to the highest standard and remember that people's lives are at stake, including those of our brothers and sisters in the fire service. It easy to forget this key point, especially when firefighting requirements and standards vary so greatly from state to state (province to province). As it was pointed out many years ago by Fire Chief Alan Brunacini, although a barber in most states is required to have 1500 hours of training to cut hair, some states require no training for fire fighters, while other states require 240 hours of training for recruits and mandatory requirements for officers. It does not make sense that we would not allow anyone without a license and 1500 hours of training to cut hair, but we would allow a poorly trained fire fighter to enter a burning building or an untrained incident commander to run a fire scene. As leaders in the fire service, it is our duty to change this and hold all fire fighters to the same standards we expect from other professions. It is the only way we can expect to decrease the number of fire fighter line-of-duty deaths (LODDs).

Training in Small Rural Fire Departments

It is a well-known fact that being a training officer is one of the hardest and most challenging positions in the fire service. It does not matter whether you are training members from a career or volunteer department, the challenges are there, even more so with small, rural fire departments. Some of the challenges you will face with training members in rural departments are the same as those encountered in career and large urban fire departments. However, there are also those challenges that are truly unique to the rural department training experience. We hope that these tips will help address some of these issues.

■ The Rural Department Fire Fighters

The most important factor with training in small, rural departments is training decisions! In career and larger departments, the majority of the fire fighters are young individuals with a love for firefighting and the fire service career. It is a challenge in and of itself to provide training fast enough to them. The small rural department, however, is often made up of individuals ranging in ages from 18 to 80 years, and they typically are the types of fire fighters who just want to help out their neighbors and communities. Depending on the demographics of the community, these people are mostly farmers, ranchers, and other hard-working individuals whose focus is on making a living and taking care of their families through their full-time jobs. For training officers, taking the time to get to know and understand the personal lives of the fire fighters they train can greatly aid in planning training. Keep in mind, though, that they deserve the right to have the best training possible, just as career fire fighters do.

■ Developing a Training Schedule for Rural Departments

When developing the training schedule, the first thing you need to do is meet with all the members and get a majority consensus on training days and how long each training session should be. Including members' input in the establishment of a schedule will increase attendance, and it lets them know that you are dedicated to providing them with training to keep them safe and ensure their ability to perform the tasks needed on every call. Once the schedule is established, though, stick to it! You will find a major complaint of rural department members is that training is frequently canceled or rescheduled for trivial reasons. If a major holiday or emergency conflicts with the training schedule, do everything possible to try and reschedule at a date and time that works for everyone. Furthermore, canceling training should only be considered as a last resort. Another consideration that needs to be made with scheduling is the alignment of appropriate topics for certain times of the year. It makes no sense to teach winter operations in June or heat-related injuries in January. In addition, if your fire fighters make their living as farmers or ranchers, it does not make sense to have a training session that requires their attendance during peak busy times, such as harvest, planting, and calving seasons. Last, the experience and certification level of the fire fighters should dictate the types of training you conduct. If everyone has their Fire Fighter I certification, then it would make sense to provide training for Fire Fighter II certification for your department. Alternately, another type of training you could consider providing is apparatus driver/operator training, as your members would have already met the Fire Fighter I certification requirement. Local and department requirements should also be factored into your training schedule.

■ Maintaining Training Records

At the very least, you should have a training record for each individual on the departmental roster and a copy of the certifications they have received on file. This makes it easier for you to confirm that a fire fighter has met certain prerequisites before enrolling him or her in outside training. It will also help when your records need to be reviewed by outside agencies, such as the ISO. If these records are kept up to date, they can be very useful in deciding what areas of training you need to work on. Another great training record tool is to post lists of the major certifications required by your state/province and your department for each department member on a large bulletin board or white board. At a simple glance, this board can clue you in on the training needed by your departmental members. At the end of the year, your training records will also help you determine how successful your program was and help you evaluate the needs for next year.

Assessing Training Needs for the Rural Department

When assessing training needs for your department, consider your geographical location and the demographics of your territory. It makes no sense to spend a lot of time training on high-rise fires when the tallest structure in your area may be a two-story farmhouse. Tour your department's response area to determine the types of structures and occupancies you will need to service and base the majority of training on the types of calls your department responds to. Keep in mind that you may still have to train members on topics that do not always seem particularly relevant to your department due to state/provincial certification requirements. If this is the case, train members on these areas to ensure they understand them, but do not allow it to make up the bulk of your training schedule. Also, do not discount training opportunities that may arise because the chances of a particular incident occurring are very low. Any situation has the potential to involve the teaching of skills and strategies that could be transferred to other incidents. For example, when touring your area, you may see a new gas station has been constructed. While it is true that you may never respond to the gas station, it would be a good idea to be prepared in the event an incident did occur there, and conducting a training involving a fire at this gas station would enable your department to explore several firefighting topics, such as building construction, fire-ground tactics, hoseline deployment, and ventilation. Although the focus of the training would be on how to respond to a fire at this location, the tactics and strategies discussed could be applied to other situations as well.

Training Topics and Delivery Methods for Small Rural Departments

A huge tip in training is to never miss a training opportunity. In other words, if a member comes to you with a question, address it and use it as an opportunity for a 15-minute training session. You may be out on the floor and a member may come up to you with a question about a piece of equipment on an apparatus. Do not tell this person that we will cover that in next month's training. Instead, grab the piece of equipment and give a short demonstration right then and there. If the member asked the question, his or her brain is most receptive at that moment in time, not a month later. To keep members interested, another option for training may be to allow them to request training topics. Simply ask your class what topics they would like to train on, or take a training day, walk out on the apparatus floor, and ask them what equipment they would like to practice with. Generally, the members will pick a topic that needs to be covered anyway, and this allows them to have a say in the training being provided, increasing the chances for a buy-in.

In contrast to training officers who instruct larger career departments with an ample amount of time and resources to schedule training during shifts, the rural department training officer must balance the right ratio of classroom and practicals to optimize the students' time. In addition, training must not be achieved at the expense of your department. You will need to ensure that you are not taking so much of the members' time that you are depleting the resources needed for incident responses.

When faced with a limited training schedule, you may find that HOT will be the most effective type of training, as this will ensure the greatest chance of knowledge and skill retention. Remember that HOT increases muscle memory, so that the tasks and decision making the fire fighters need to perform will become second nature to them. However, this does not mean that you do not impart key facts and information as you conduct HOT. It is important to mix in key informational points while conducting the HOT. In addition to being able to perform a drill automatically, fire fighters need to know why they are using a specific method, as well as the types of situations that may require them to adapt that method. In order for this training to flow smoothly and offer the greatest amount of benefit, a well thought-out lesson plan should be constructed. This lesson plan can be borrowed from the traditional classroom setting; however, it needs to be altered for the specific topic that will be covered and for the allotted time. As to the use of HOT versus the use of traditional classroom sessions in rural fire departments, most members will likely tell you they learn best when they are actually performing the task. This can be an excellent way to use the "See it, Do it, and Teach it" training method. It will not only build the members' confidence in themselves, but it will also promote team work. It is up to us, as training officers, to give fire fighters the tools they need to succeed.

Using Outside Resources

When faced with a limited amount of time for training, it can be almost impossible to provide all the training the department needs or desires. It is no shame to have some of the training accomplished outside of your department. In today's world, the amount of resources available is almost unlimited. The training officer's role has almost changed from being the main instructor of a department's training to being the guide who helps fire fighters find the training they need. Do yourself a favor and investigate and explore these resources. To begin with, the National Fire Academy provides on-campus, off-campus, and online courses. Your local state/provincial fire marshal's office may also provide some training. The state/provincial fire academy can also provide you with numerous resources. There may be local and regional weekend fire schools you could send your fire fighters to. For example, a group of fire fighters from the Atlanta area host the Metro Atlanta Firefighter's Conference (MAFFC) every May. Here, all classroom sessions are free and the practicals are only $25 each. The MAFFC has numerous sponsors, and the Georgia instructors agree to teach sessions for free, which is the mark of true brotherhood. Seminars and conferences are great options for weekend training and for departments that have limited budgets. Another option is to explore partnerships with local colleges, as they may offer fire science classes that your members are interested in. Additionally, a neighboring department may have training and may be more than willing to let your people attend. There are several trade publications that have training articles and drills in them. Simply placing the publication on the training table and saying, "Let's discuss how this type of incident can happen here," is a great way to start a discussion, and it encourages critical thinking. A lot of these same publications have websites that offer online training and drills that can be performed during a training session, or reviewed by fire fighters according on their own schedule.

In the past, fire officers were encouraged to have a good reading library of firefighting books. Today, you should encourage even the rookie fire fighter to start a personal library. The recent advancements in technology with Kindle and Nook eReaders have made it easier and more convenient than ever to access these resources. Many fire-related titles can be downloaded onto these readers. For individuals who have to spend a lot of time in their vehicles for work, another option that you could consider is assigning and recommending audio books for them to listen to while driving. Lastly, do not forget that you would also benefit from using outside resources to develop your training and lesson plans. There are many different resources you can use that have published ready-made lesson plans and PowerPoint presentations. The International Society for Fire Service Instructors (ISFSI) has an excellent networking program for this. In addition, there are numerous, reliable training classes and materials available on the Internet that anyone can download for free. However, as is the case when using any resources downloaded from Internet sites, always reference the material before presenting. Remember, when planning lessons, you do not have to reinvent the wheel, just keep it moving around.

■ The Rural Department's Training Budget

In the past, most of our budgets were focused on classroom materials and training aids. Today, the rural fire department may find need the majority of their training budget going toward travel expenses, registration fees, and tuition. However, members of rural fire service departments may have additional skills, such as carpentry, welding, and metal fabrication, that can be utilized when developing training materials. Most members are willing to help the department build props and acquire materials if they are asked. This has helped many small career and volunteer departments gain valuable props without having to deplete their budget. Although there seem to be more and more grants available for training props and classroom materials, smaller rural departments tend to be overlooked when they apply for these grants. One excellent way to overcome this problem is to submit a joint grant proposal with a neighboring department to increase the chances of receiving the grant. You may want to extend this effort even further by forming some type of training association that could not only pursue these types of grants, but also share training resources among each other. In the state of Georgia, several areas have banded together and created regional training officers' and chiefs' associations for the purpose of sharing information and resources. These training associations allow instructors to meet, share ideas, brainstorm, and provide classes to neighboring departments. In a time of decreasing budgets, we have to think outside of the box and consider ways we can optimize our resources. Your department may have a great fire attack program, while another has a great ventilation program. Both departments can benefit from each other's expertise if they combined their resources and skills into one regional training department.

Rewarding and Recognizing Rural Department Fire Fighters

Rewarding and recognizing the rural department's fire fighters for their hard work and dedication is very important. You must motivate your fire fighters! Always remember your fire fighters have put forth a lot of time and effort to train, and they deserve to be recognized for it. When fire fighters complete a certification program, do not just hand the certificate to them, present it to them at a department meeting. Post it on the departmental bulletin board or publish it in a local newspaper or in your department newsletter, if your department has one. If you have an awards banquet or ceremony, hand them a certificate of achievement for the amount of training they have accomplished in the past year (yet another reason for maintaining good training records). Another good way to reward fire fighters is to feed them. Provide a meal or snack at each training session. Or, as one small department does, host a dinner every fifth Monday before training that includes immediate family to say thanks for their service and to cultivate the feeling of family among members of your department. An expression that comes to mind is, "If you feed them, they will come." It will bring a smile to your face when you find your department members asking not what this month's training is, but rather what they will be eating at training. Regardless of what you do, do something to show that you appreciate them and you care. Giving thanks and recognizing the efforts of your small but valuable department will motivate them.

Re-Evaluating the Rural Department's Training Program

Evaluation should be an ongoing process. Invite your fire fighters evaluate how your department is doing. This assessment should also include the evaluation of department members' progress, as this will also give you some insight as to how well your training division is performing. Evaluate every aspect of your program. If something is not working, you must be willing to adapt and change it. If something is working, look for ways to improve it and make it even better. If need be, compare your program to other fire departments. Do not be afraid to ask other departments what does and does not work for them.

Although fire departments are not required to conduct evaluation on an ongoing basis, doing so has helped many small rural fire departments. You may choose to use some or all of the suggestions we have made, but always remember the purpose of your training program is to help your fire fighters perform efficiently and safely, and ensure they go home to their families after every call.

Training in the Industrial and Private Sector Emergency Services

As fire and rescue is still fire and rescue regardless of the employer, there are numerous similarities in emergency services between the public and private sector. However, differences between the two do exist. Most of these differences revolve around the standards and regulations that govern the two entities. When emergency services from the public and private sector have to work together, one challenge that may arise is that they tend to use different terminology. In this section, we will discuss the industrial section requirements as well as the importance of public and private sector emergency services working together.

Even if you are not involved in private sector/industrial emergency services, it would be a good idea to familiarize yourself with this branch of the fire service industry, as it may help you eliminate any confusion that arises when responding to incidents inside an industrial facility as part of a mutual aid response. In addition, the public sector should be aware of the Lock Out, Tag Out (LOTO) system, which is not taught or practiced in most public emergency services, but is used in the private sector. LOTO is generally thought of in the electrical manner; however, LOTO should be applied any time there is an energy source regardless of its nature, whether it is mechanical, hydraulic, electrical, pneumatic, chemical, thermal, or some other type of energy. The private sector can assist with neutralizing these sources and confirm that the LOTO is complete prior to a fire attack or entry into a confined space. At the very least, they should be able to provide information on how their LOTO system is applied.

As an additional note for the public sector, even if your state does not have an OSHA state plan, the same regulations could be held against the public sector during a lawsuit, similar to an NFPA standard. In addition, most of these standards are not difficult to comply with and they provide added protection to all levels of emergency services responders, regardless of whether they are public or private. States with Approved Occupational Safety and Health Plans include (asterisk [*] indicates public employee plans only):

- Alaska
- Arizona
- California
- Connecticut*
- Hawaii
- Illinois*
- Indiana
- Iowa
- Kentucky
- Maryland
- Michigan
- Minnesota
- Nevada
- New Jersey*
- New Mexico
- New York*
- North Carolina
- Oregon
- Puerto Rico
- South Carolina
- Tennessee
- Utah
- Vermont
- Virgin Islands*
- Virginia
- Washington
- Wyoming

Terminology in the Industrial or Private Sector Emergency Services

Typically, an organization that provides emergency services in the industrial or private sector either has an Emergency Response Team (ERT), Fire Brigade, Incipient Fire Attack Team, or an Emergency Operations section. These terms can have several meanings and be interpreted in different ways. It should be noted, however, that the difference between the Fire Brigade and the Incipient Fire Attack Team is the use of a 1½-inch (38-mm) hose line versus a 1¾-inch (44-mm) hose line or larger. Each organization should develop standard operating procedures (SOPs) or standard operating guidelines (SOGs) for their organization. The SOPs/SOGs should outline the parameters of what is expected of the organizational members. Below we will list some of the operations/expectations that may fall under the different emergency services categories.

- **Emergency Response Team (ERT).** Examples of services this team may provide include first aid, confined space rescue, rope rescue, or incipient fire attack.
- **Incipient Fire Attack Team.** These teams are trained and authorized to fight incipient fires. They generally are allowed to use up to a 1½-inch (38-mm) hoseline or a fire extinguisher.
- **Fire Brigade or Industrial Fire Department.** These types of departments or organizations may be strictly limited to providing fire protection only or they may provide a variety of services. They generally use 1¾-inch (44-mm) hoselines and larger, and they provide interior structural firefighting.
- **Emergency Operations.** Examples of services this team may provide include, but are not limited to, first aid, confined space rescue, rope rescue, hazardous materials, fire attack, and incident command.

Fire Brigade Versus Incipient Stage Fire

As a training officer, it is very important to understand the difference in levels of service provided, especially if you are not accustomed to the private sector. If you review OSHA's definition, it will help you understand and define the level of service that your department will be providing. However, there is a large amount of information left up to interpretation, and as a training officer, you will help define exactly what is expected of an employee who is knowledgeable, trained, and skilled. Another option is to only use 1½-inch (38-mm) hose lines and fire extinguishers; however, employees still need to have a certain level of knowledge and skills in order to safely perform these methods as well.

Requirements

The same general requirements exist whether you are a public or private sector organization. The largest difference comes from understanding OSHA state plans versus federal plans. Also, there are numerous references in the 1910 OSHA regulations that cite other standards or best practices from organizations such as American National Standards Institute (ANSI), National Fire Protection Association (NFPA), International Society of Fire Service Instructors (ISFSI), and National Fire Academy (NFA). Before implementing any training regimen, it is best to review all of the referenced materials in order to produce the most accurate and up-to-date product. The diagram listed in Appendix D, "Fire Department Required Training Self-Assessment Guide," and the chapter, *Guiding Requirements*, can be used to assist organizations with fulfilling their training requirements and satisfying all applicable standards. It is also beneficial to check your organization's corporate standards and regulations, as they can be more stringent than the minimum OSHA standards.

Establishing Mutual Aid Training

As in all sectors, public and private, adequate staffing and equipment are a constant struggle. In the private sector, emergency response agencies generally rely on mutual aid to supplement their emergency services. In addition, most public sector emergency services have no idea of the hazards, escape routes, layout, water supply, or any other necessary pieces of information regarding the property normally protected by private sector services. It is a good idea to have public and private emergency service agencies and organizations train together to plan for potential emergencies and to minimize the loss of life and property. To coordinate cooperative training between private and public agencies, the training officer can develop

Creating a Lateral Fire Academy (Fast Track Program)

In his article for *Fire Engineering*, "Hiring Experienced Firefighters During Tough Times,"[1] Brian Ward explains that one of the most difficult tasks is to attract the right personnel to your department. We all wish to have the top recruits and fire fighters knock on our door instead of us actively searching for them. One of the best programs that can be implemented that will allow this to happen is the creation of a Lateral Academy, or Fast Track Program, for experienced fire fighters. If the program is structured appropriately and held to a high standard, it will provide numerous benefits to your department, including:

- Attracting experienced fire fighters who are looking to relocate, but are not interested in repeating recruit school
- Immediate monetary savings. The more individuals you move through the fast track program, the less expensive it becomes and the more you would save over sending them through a traditional fire/EMT recruit school.
- Expediting turnaround time after the departure of a fire fighter. As is the case in most departments, the chief is not allowed to hire a replacement until the fire fighter retires or leaves the department. In some cases, it may take up to 6 months to train and certify a fire fighter off the street in fire and EMS practices. The fast track process can expedite the turnaround time with training a replacement after the departure of a fire fighter.
- As noted in the article, it is important to not trade quality for quantity. Regardless of how badly we may need trained fire fighters or how much money can be saved through the use of a particular training program, if one fire fighter gets hurt, then such a program is not worth it. As the training officer, you must validate your program based on NFPA standards and document everything involving your program. The standards and documentation will allow you to justify your program and help to keep it objective-based. During the development of this process, it is important to consider the schedule, content, qualifications of applicants, and transparency with new hires.

[1]Data from: Ward, B. Hiring experienced firefighters during tough times. *Fire Engineering* 2010:163(8):97. http://www.fireengineering.com/articles/print/volume-163/issue-8/Features/hiring-experienced-firefighters-during-tough-times.html. Accessed September 17, 2014.

a simulated scenario and invite the local responders to participate in the drills. These drills may consist of any number of emergency response-type incidents, including plant-wide evacuations that include a fire department response, confined space rescues, rope rescues, medical emergency responses, hazardous materials release responses, and fire attack drills. The one constant of these drills is that they would require the experience, expertise, and resources of both entities.

The location of these drills may be at the fire department's training center or simulated at a local structure, such as a local business, school, office space, or commercial building. In some cases, private sector organizations may have already established an agreement between the public and private sectors to share training centers, burn buildings, props, and classroom space. This tends to be the most cost-effective route for both organizations.

An example of a simple joint drill is a plant evacuation due to a fire located on the B side of a commercial structure in the railcar loading area. This drill may include the response of fire and medical emergency organizations. Before conducting the drill, an action plan should be developed that outlines when, where, how, and who will be involved. A map or liaison should be provided to each fire apparatus that provides directions around and through the facility. Locations of hydrants and standpipes should also be identified by the liaison or map. This drill is timed throughout, and the training officer should complete a tactical worksheet and note the benchmarks achieved during the drill. As plant employees evacuate, emergency response personnel respond to the simulated fire and perform LOTO for involved equipment. Meanwhile, the training officer should have the local fire department and EMS respond to the front gate for staging assignments. A unified command system should be established and assignments given based on the information provided by the initial emergency responders. After the drill is complete, the training officer should conduct a critique of the entire drill that includes local responders and plant personnel in the discussion.

Summary

The industrial sector may seem difficult to understand if you operate in a non-OSHA plan state. However, whether you are using NFPA, OSHA, or ANSI, the bottom line is to protect our personnel through proper and effective training. Researching and learning the different standards will broaden your overall knowledge as a training officer, safety manager, or chief (according to the title used in your organization) and allow you to develop relationships with other responding services. This will also affect the resources that may be available to your organization during an emergency incident.

Key Points

- The training officer's attitude, skill, and dedication usually set the tone toward training in the department.
- The best way to achieve a happy medium with establishing training standards is most likely through performance-based training or job performance requirements. Training officers have a limited amount of time to train and observe all members, but this is especially true with volunteers. While the number of training hours may vary, certification levels should not.
- The small, rural department generally consists of individuals ranging in ages from 18 to 80. They typically are the types of fire fighters who just want to help out their neighbors and communities. Depending on the demographics of the community, these people are mostly farmers, ranchers, and other hard-working individuals whose focus is on making a living and taking care of their families through their full-time jobs.
- It is important to remember that the members of the rural fire department have put forth a lot of time and effort to train, and they deserve to be recognized for it.
- When emergency services from the public and private sector have to work together, one challenge that may arise is that they tend to use different terminology.

References

Goldfeder, B. personal communication, August 2012.

Occupational Safety and Health Administration (OSHA). 29 CFR 1910 OSHA General Industry Regulations. Davenport, IA: Mancomm Mangan Communications Inc.; September 2011.

Ward, B. *Hiring Experienced Firefighters During Tough Times*, Fire Engineering, Glencoe, IL: Penwell, August 2010. http://www.fireengineering.com/articles/print/volume-163/issue-8/Features/hiring-experienced-firefighters-during-tough-times.html. Accessed September 17, 2014.

CHAPTER

14 Investigations

Brian Ward

Introduction

Depending on the structure of your department, the training officer may be required to assist or lead investigations concerning accidents that occur during training, emergency responses, and in other instances. In some departments, the training officer also fills the role of the department safety officer. Regardless of whether you are leading or assisting in an investigation, there are several items to consider in order to properly analyze the accident and prevent future occurrences. The number one priority is to prevent the loss of life and limb, followed by the need to conserve money and equipment.

According to the National Safety Council, in 2012, the average comprehensive costs, listed by injury severity, were as follows:

- Death: $4,538,000
- Incapacitating injury: $230,000
- Non-incapacitating evident injury: $58,700
- Possible injury: $28,000
- No injury: $2500

Accidents

An accident is the final event in a series of unplanned events that results in harm, such as injuries or property damage. All accidents are preventable to some degree, and most accidents occur due to human error. People can prevent accidents caused by poor judgment and human error by using situational awareness and anticipating them. If we train on and make an effort to understand the incidents with which we are faced, we can deepen our situational awareness and prevent 99 percent of all accidents from occurring. In the fire service, the top three killers of fire fighters are cardiac arrest, fire-ground operations, and operating apparatus. If we take care of ourselves, eat right, and continue to train, the incidence of injuries and line-of-duty deaths (LODDs) will decrease.

Situational awareness consists of three distinct points of reference and is a continuous process that should never cease. The first component of situation awareness is simply identifying that a problem or a hazard exists. For example, if you were practicing situational awareness at an incident, you would note hazards such as power lines on the ground, or you may note that the structural fire is producing products of combustion and smoke. The second component to developing situational awareness is to make sure that you understand exactly what the action or item is telling you. If you are one of the first response units on scene, it is very likely that the lines on the ground are live wires; upon arrival at a fire involving a large commercial structure, you may notice that the smoke is turbulent and has a tremendous amount of velocity. The final part to developing situational awareness is to make an attempt to foresee or predict what could occur as a result of a certain action. In doing so, you may consider changing the current course of action or a decision previously made. For example, you can easily predict that if you step on the downed power lines, it is likely that you will be electrocuted, which will create a secondary incident at the structure fire. With regard to the high-velocity, turbulent smoke rising from the commercial building, you may conclude that, based on the smoke's characteristics and the size of the structure, flashover could be very near, and entering the structure would endanger the fire fighters' lives. The ability to make sound and accurate decisions is based on your situational awareness capability level. As your capability improves, the time that it takes to make these sound judgment decisions should improve as well.

A key phrase that should be used within any changing culture is the term *accidently successful*. This is when we intentionally or unintentionally take a shortcut or miss a clue, yet the outcome remains positive. Because the outcome was positive, we decide that this course of action or decision is a viable option at other similar incidents, and we fail to recognize the flaw in our decision-making process. As time goes on, we continue to use the faulty approach at similar incidents, and we even teach other fire fighters, either directly or through their observations, to use it. It is only a matter of time before we will learn the error of our ways. It takes just one bad incident to realize we were only successful by accident, and the misguided approach results in an injury or an LODD statistic. This outcome is unacceptable.

Accident Investigation

The Occupational Safety and Health Administration (OSHA) defines the term *accident investigation* as a process in which investigators and supervisors determine the how and why of an accident occurring and who is liable for it, whereas *accident analysis* is a process in which the accident is analyzed to determine system weaknesses. While determining liability in the court system is extremely important, it is equally, if not more, important to understand our system's weaknesses. As you begin your investigation, be careful to not place blame on anyone or anything; the objective of the investigation is to determine what caused the accident and how we can correct the issue.

Process

After an accident has occurred, there are several steps that should be taken to start the accident investigation process. If your department has a policy on investigations, refer to it. Inside of the policy, you should find the following guidelines for conducting an accident investigation:

- The parties that should be notified in the event of an accident
- The individual(s) who are authorized to speak on the department's behalf in response to questions from the victim's family and the media
- The individual who will lead the accident investigation
- Chain-of-custody for the reports and any evidence
- General timetables for completing the report

Once these specifics are confirmed, the actual collection of data can begin. A three-step process can be used to conduct the investigation in a systematic and methodical manner after the emergency has been handled:

Step #1: Preserve and document the accident scene.
Step #2: Establish a sequence of events and determine a cause.
Step #3: Recommend improvements and finalize reports.

The 5 Why Concept

- Good to use when it involves human interaction.
- Questions each response with a simple "why" in order to dig deeper than just the first level response.
- Assists in determining the root cause, not just the symptom.
- 5 "Whys" is a rule of thumb—some situations may require fewer or additional questions.
- Simplest form of investigation and easy to learn.

Step 1: Preserve and Document the Accident Scene

Step one is the stage in which the investigator is responsible for collecting and gathering information. The investigator should begin to analyze the accident and try to determine why the situation occurred. One concept that could be used is the "5 Why" concept, which is a tool used for determining the root cause of failures. Once you determine why the accident occurred, ask yourself what caused the next component or failure to occur. However basic, the information you receive from applying this concept is directly related to the effort and level of detail you put into it. As you finish the initial documentation and collect whatever physical evidence there may be, be sure to take photos and film any relevant footage. Remember to also record interview statements. It may be helpful to sketch a drawing of the accident scene, noting dimensions and distances.

The interview process will allow you to obtain fresh facts from all involved individuals. By getting a detailed account from all involved parties, a clearer picture of the overall scene should begin to develop. Again, be sure to not accuse anyone during this process, and be sure to explain the investigation process and your role in it. Though it will take some practice to develop interviewing skills, be specific with your interview questions and try to use open-ended questions; this will encourage the interviewee to explain what happened instead of responding with a "yes" or "no" answer. If possible, it is best to conduct interviews privately, away from other individuals who may be involved, but still at a location near the accident. Generally, private interviews will help put both parties at ease. This process will also allow you to compare accounts of the different individuals without the facts being skewed from how one individual perceived the incident. Additionally, interviews should be held as soon after the incident as possible, but at a time that is still appropriate for the situation.

Tips for Interviews

- Set an appropriate time and place.
- Choose a quiet environment without distractions.
- Use a casual environment without an interrogation approach.
- Ask questions that require a response and allow the individual to explain/discuss.
- Listen.
- Prepare questions ahead of time.
- Always be respectful and professional.

Step 2: Establish a Sequence of Events and Determine a Cause

During this step, you begin to analyze the collected facts, interviews, and evidence. Based on this information, you should be able to establish a timeline of when and how the events occurred. Be careful not to jump to conclusions or assume that one particular action or event is the underlying cause. The initial cause may be an issue; however, it may only be a symptom and not the main concern. It is also important to determine if the accident was a cause of several failures occurring simultaneously.

As the sequence of events is arranged and a timeline that aligns with the information and evidence gathered is established, you will need to determine the primary cause and any secondary causes associated with the accident. The primary and secondary causes of the accident may be due to a number of things, including errors made by personnel and/or equipment failures. One commonly used method for determining the causes of accidents is the Root Cause Analysis (RCA) process through Cause Mapping. The Cause Map helps illustrate the relationship between the cause and effect of each event, and it provides a visual representation of the accident. The Cause Map also compels you to drill down into the root cause of a problem.

As you start to develop the Cause Map, remember that every cause must have evidence to support it and any factors of the incident that are missing evidence could potentially indicate a cause. Be specific in your responses and ask "Why did this happen?" after each cause until you reach the root cause. The root cause(s) is the point where there is no further evidence to support additional investigation. All Cause Maps start with an impacted goal. For example, if the goal was to ensure that the training was conducted without injury, a way in which that goal might be adversely affected would be if one of the members broke his or her index finger during the training. From this incident, you would work backward to determine what caused the injury to occur. If it appears that there were multiple causes, each cause should be listed in a vertical line. For each effect, there should be a cause determined with an evidence block supporting that correlation (**FIGURE 14-1**).

Step 3: Recommend Improvements and Finalize the Report

Once the primary and secondary causes have been determined, recommended improvements should be established for future reference. These improvements should be related to National Fire Protection and Association (NFPA) standards, Occupational Safety and Health Administration (OSHA) regulations, National Institute for Occupational Safety and Health (NIOSH) research, best practices, departmental standard operating procedures (SOPs), and other guidelines from similar organizations. As always, be sure to be very clear and concise when stating the necessary improvements and do not place blame. You are nearing the end of the investigation process, and the words you write and the comments you make will become public record and can be used against you or your employer, even if they are not true. You should be truthful, upfront, and respectful to all parties. In addition, there are three types of controls to consider when developing recommendations:

1. Engineering
2. Administrative
3. Personal protective equipment (PPE)

Engineering controls are implemented to mitigate hazards, and this is accomplished by either engineering a correction to the faulty equipment to eliminate a hazard or by adding a safety feature as a standard component to the equipment. Engineering controls are the most effective way of mitigating hazards. When re-engineering, if a component is not practical or possible, administrative controls may then be implemented. An administrative control is the development of SOPs or standard operating guidelines (SOGs) to dictate how the fire department will handle a particular situation or hazard in the future. The last defense is the application of PPE controls. The PPE control involves dictating what protective equipment the responder will need to wear when responding to a specific hazard or incident. Although this control does not prevent a hazard from occurring, it will ensure that the individual performs the work with an added layer of protection.

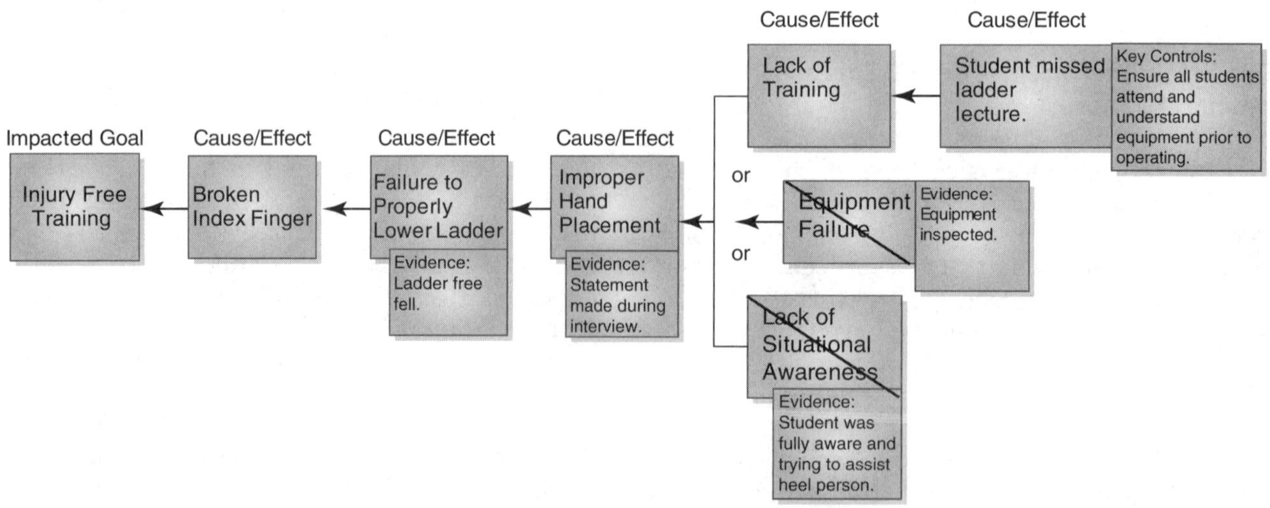

FIGURE 14-1 A Cause Map showing the primary and secondary causes of a member sustaining a broken index finger during training.
Courtesy of Brian Ward

The last part of an investigation is to write the final report. The final report is where the investigator puts all of the information regarding the accident down in a logical manner. Generally, there are five to seven topics that need to be addressed. If you become unsure or are unfamiliar with how a final report should look, review a few of the NIOSH reports that are available online. NIOSH has conducted several investigations, and their reports of these investigations serve as excellent examples because the NIOSH provides documentation from an unbiased position, without any blame or naming, and the reports are presented in an acceptable format. The final report should include the following:

- **Background.** Details leading up to the incident. Why were personnel where they were or performing a certain task?
- **Description of the accident.** Details explaining how the accident occurred. What was the final outcome of the accident?
- **Findings.** Details explaining why the accident occurred. Discuss the root cause(s) and the sequence of events that led up to the accident.
- **Recommendations.** Provide your recommendations and cite your reference materials.
- **Summary.** Summarize the investigation.
- **Review and follow-up actions.** Explain any timetables for reviews of recommendations and for the owner's corrective actions.
- **Attachments.** Attach any documents, such as pictures, statements, and/or drawings, that contain pertinent information to your report.

Once you have completed your final draft, have a colleague proofread it to ensure accurate information and data. This will also ensure professionalism of the document.

Summary

Accident investigations can be extremely tough, especially if you attempting to conduct one inside of your own department. It may be necessary to solicit outside help, or at a minimum, you may need to form an internal committee to assist. Regardless of which method you choose, ensure that the investigation identifies the root, or underlying, cause of the accident, and results in the development of guidelines or measures that will prevent additional injuries and equipment failures. Always remember, as you are conducting the investigation, that everything you say and do as part of an investigation can be used in a court of law.

Key Points

- Human error is the leading cause of accidents.
- All accidents are preventable at some level.
- Situational awareness and foreseeability are our two most valuable assets when preventing accidents.
- An accident is the final event in a series of unplanned events that results in harm, such as an injury or property damage.
- Be careful to not place blame on anyone or anything; the objective of the investigation is determining what caused the accident and how to correct the issue.
- There is a three-step process that can be used to conduct the investigation in a systematic and methodical manner after the emergency has been handled.
  - Preserve and document the accident scene
  - Establish a sequence of events and determine a cause
  - Recommend improvements and finalize reports
- It will require some practice, but be specific in your interview questions and try to use open-ended questions.
- The initial cause may be an issue; however, it is a symptom and not the underlying problem.
- Recommended improvements should be related to NFPA standards, OSHA regulations, NIOSH research, best practices, department SOPs, and other guidelines from similar organizations.

References

National Safety Council. *Estimating the Cost of Unintentional Injuries*. 2011. http://www.nsc.org/news_resources/injury_and_death_statistics/Pages/EstimatingtheCostsofUnintentionalInjuries.aspx. Accessed April 27, 2012.

Occupational Safety and Health Administration (OSHA), Accident Investigation. 2012. http://www.osha.gov/SLTC/accidentinvestigation/index.html. Accessed April 27, 2012.

CHAPTER

15 Physical Fitness Training for Recruit Academies

Ian Cassidy

Overview

Many newsworthy stories have developed over the past several years concerning recruits and instructors being injured or suffering from heat-related illnesses. Some of these were due to a lack of adequate training on the instructor's part concerning workloads, heat stress, and pertinent medical information from those participating. There are times when the occurrence of incidents during training is unavoidable. One example is the 28-year-old fire fighter candidate from Clayton County, Georgia, who collapsed after scoring a 100 on a hiring physical agility test. Although the candidate was cleared to participate in the drill, it turned out that he had an unknown condition. Such incidents are rare, however. A vast majority of injuries or illnesses that occur during or after training could have been prevented if the instructor possessed the proper knowledge and had been adequately prepared.

Unfortunately, training injuries are not uncommon and districts across the country have been considering or making strides to overhaul their physical training programs. One district that has made an effort to make their physical training program safer and more effective is the Northwest Fire District from Tucson, Arizona. This district, too, had several fire fighters who experienced injuries and illnesses during the recruit training program—although not to the magnitude described in the scenario above—and decided that they needed to make some changes. They decided to create a physical training program based on Orange County Fire Authority (CA) standards. To design the fitness program, they enlisted the help of paramedic Brian Castillo, a Crossfit Level 1 instructor and Aerobics and Fitness Association of America (AFAA)-certified personal trainer, and Captain Ian Cassidy, the academy coordinator. Upon the program's completion, they identified an across-the-board increase in the strength, endurance, and stamina of their recruits. Furthermore, zero injuries or heat-related incidents were recorded during its implementation.

The success of the physical training program created by Northwest Fire District prompts us to reflect on how we prepare fire fighters for the duties in the field. As we learn more about functional movement exercises and the importance of proper lifting techniques—all of which help reduce job-related injuries—we realize that the way we have been doing business does not prepare recruit fire fighters for the long career ahead of them.

Since certain conditions, such as geographic location, climate, and available resources, vary from one department to another, a one-size-fits-all program does not exist. That being said, the program created by the Northwest Fire District was ultimately very successful, and so, for the purposes of this text, we are using it as a model physical fitness training program. The idea is that you, as the training officer, can use this program as a foundation for your department's physical fitness training program, adapting it to fit your organization's needs.

Recruit training is the point where newly hired fire fighters are molded into the kinds of fire fighters that the fire department wishes to have: a department of long-living and healthy fire fighters. With 50 percent of fire fighter fatalities occurring from cardiac-related issues, finding a program that fire fighters will continue to use after recruit school is more valuable than the physical training conducted in the recruit school itself.

Introduction to Academy 2010-01

Academy 2010-01 was the first academy that the Northwest Fire District ran that successfully tracked data regarding recruit physical fitness in an attempt to develop a standard for training future recruits. The overall goal of the fitness program in Academy 2010-01 was to improve the overall fitness of the recruit fire fighters and prepare them for on-the-job experiences, while decreasing the average number of injuries typically seen at previous Northwest Fire District academies. At the beginning of the program, the academy had 22 recruits, 21 male and 1 female, performing at various levels of physical fitness (**TABLE 15-1**). By the program's completion, 20 of the 22 recruits (19 males and 1 female) successfully met all the requirements and graduated.

As the physical training program was being created, its developers knew that they needed some guidance, particularly since this was the first time they would track data on the performance, injuries, and illnesses of their recruits. Rather than starting completely from scratch, they decided to create a training program that was similar to that of the Orange County Fire Authority (OCFA). Centered on high-intensity functional movements performed at various intervals, the OCFA's program is intended to prepare the participant for a variety of tasks and provide the skills necessary to be able to perform each task exceptionally. In this case, high-intensity workouts are defined as those that are above 85 percent of the maximum heart rate of the individual. This program is classified as a General Physical Preparedness (GPP) program.

In order to assess the recruits' performance, the recruits had to track data concerning their fitness level, such as body weight, blood pressure, and heart rate. From day one, recruits were provided with a blank fitness log and an Ekho Fit-18 Heart Rate Monitor. Recruits were tasked with taking an accurate pulse rate upon waking the next morning, before getting out of bed. Each recruit was then given time in class to work out their heart rate percentages (i.e., 90, 80, 55). The recruits would update this data on a weekly basis, each Thursday. Additionally, recruits were monitored during fitness activity to make sure that they were operating in a safe, yet effective, manner. Within 15 minutes of every workout, recruits recorded data on heart rate, body weight, blood pressure, and pulse. Although not a specific data aggregate, recruit blood pressures dropped significantly during the course of the academy, specifically the diastolic pressures. This fact, along with the decrease in resting heart rates and oxygen consumption (VO_2), helped to validate incorporating high-intensity interval training into a physical training program. In addition, post-workout pulse rates gave the instructors an idea of whether or not recovery times were improving by participating in the shorter, yet higher intensity, workouts.

In terms of the types of workouts performed, there were several key differences between previous academies and this one. One key difference was the way in which running exercises were performed. In past academies, it was customary to have recruits run long distances on a regular basis. In this academy, the recruit class only took part in three 3.1-mile (5-kilometer) runs at the beginning, middle, and end of the academy, as a means of assessing their cardiovascular endurance (which, incidentally, increased significantly). Instead of running long distances on a regular basis, this academy had the recruits run in short bursts up to 800 meters at any given time to more closely simulate fireground activities.

For the Academy 2010-01 training program, the movements recruits performed as part of their exercises were all-functional in nature. In other words, they were all movements that are regularly performed on the fireground. In addition to these activities, recruits were

Table 15-1 Academy 2010-01 Recruits

	Male	Female	Total
Number of Recruits who entered Academy 10-01	21	1	22
Number of Recruits who completed Academy 10-01	19	1	20
Average Age of Recruits			29.5

Source: Courtesy of Ian Cassidy.

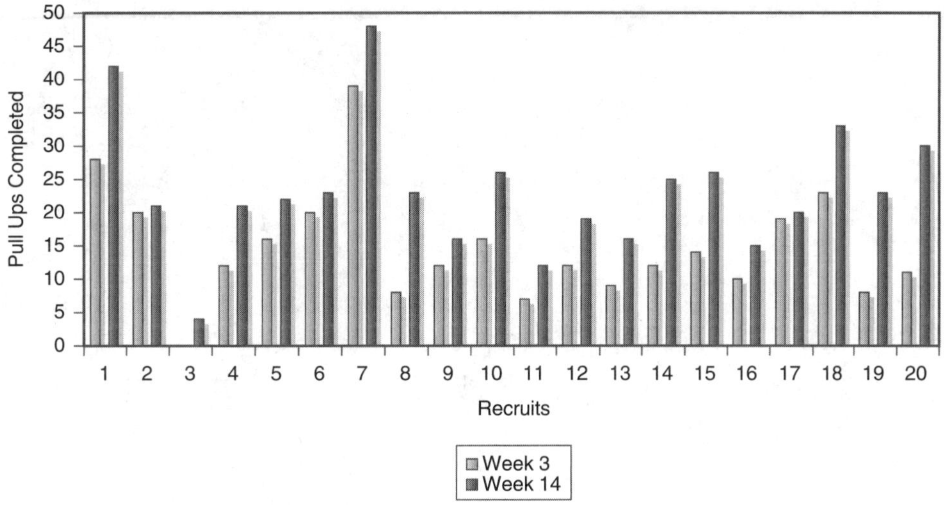

FIGURE 15-1 Academy 2010-01: Number of pull ups each recruit completed for weeks 3 and 14.
Courtesy of Ian Cassidy

given time to practice "new skill work" after each physical training session. During these training sessions, each recruit would use a 5-foot-long piece of ¾-inch PVC pipe as a barbell. The instructors would teach the recruits proper lifting techniques in the form of squats, deadlifts, shoulder presses, cleans, lunges, and overhead squats. By repeating these sessions on a regular basis, the recruits developed muscle memory on how to perform these movements. Then the instructors would teach them to transfer these skills over to the manipulative drills (i.e., ladder carries, loading hose, etc.). Through the use of this instruction, the program developers noted good lifting techniques throughout the academy and no back injuries were recorded.

Academy 2010-01: Physical Fitness Results

■ Pull Ups

An assessment commonly used to measure upper-body strength and endurance, the results of the pull-up exercises showed that the Academy 2010-01 recruits had marked improvement across the board (**FIGURES 15-1** and **15-2**). The smallest increase noted during the

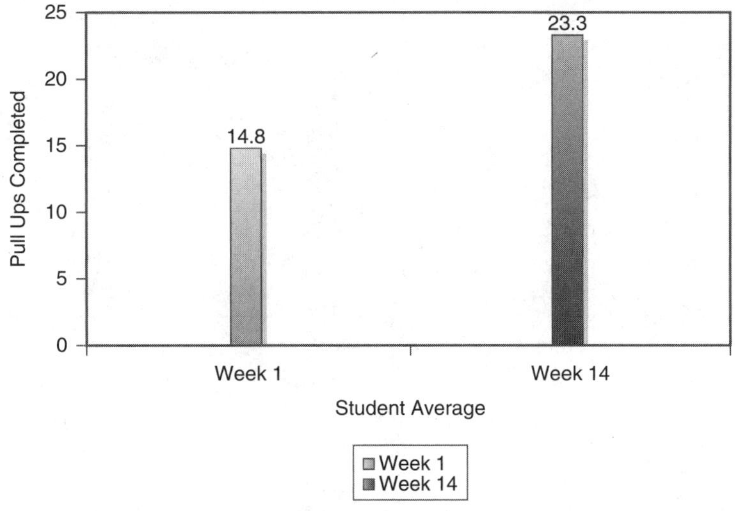

FIGURE 15-2 Academy 2010-01: Average number of pull ups completed by the academy for weeks 1 and 14.
Courtesy of Ian Cassidy

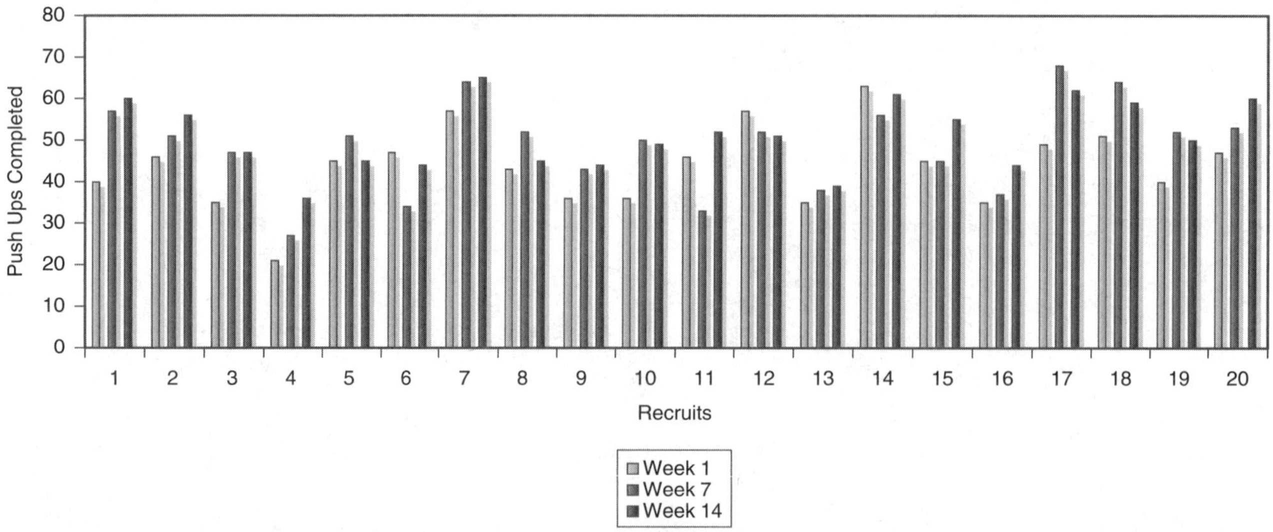

FIGURE 15-3 Academy 2010-01: Number of push ups completed by each recruit for weeks 1, 7, and 14.
Courtesy of Ian Cassidy

academy's term was 5 percent, while the largest was 400 percent.

Push Ups

Academy 2010-01 recruits also demonstrated consistent improvement in the push-up category. The push-up scores provide a picture of the recruits' upper-body strength. Collectively, the recruits averaged 51 push ups, a score that was well above the district standard for full credit on the FIT test. Individually, scores ranged from 21 to 68 push ups (**FIGURE 15-3**). The recruits made improvements ranging from 4.7 percent to 71.4 percent (**FIGURE 15-4**).

Fire Fighter VO_2 Max Assessment: The 3.1-Mile (5-Kilometer) Run

The 3.1-mile (5-kilometer) run was a core component of the academy's fitness assessment, and it is a valid test of estimated maximal oxygen uptake (VO_2 max). Typically, the 1.5-mile run is used and recommended

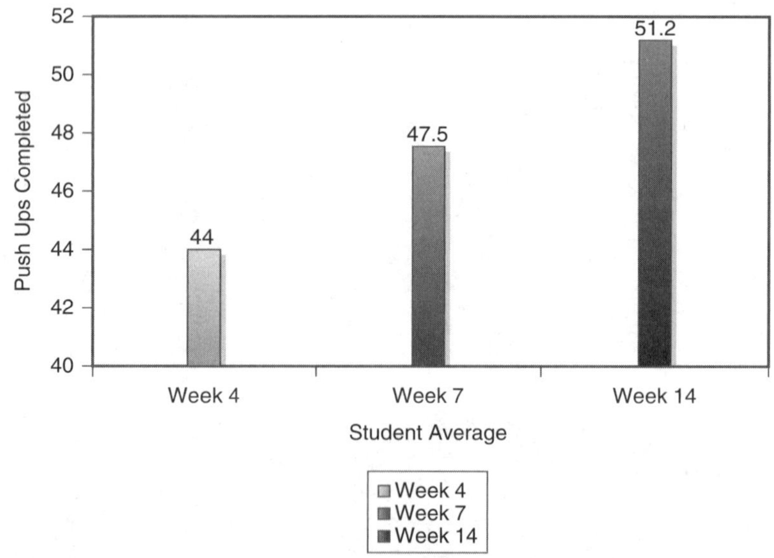

FIGURE 15-4 Academy 2010-01: Average number of push ups completed by the academy for weeks 4, 7, and 14.
Courtesy of Ian Cassidy

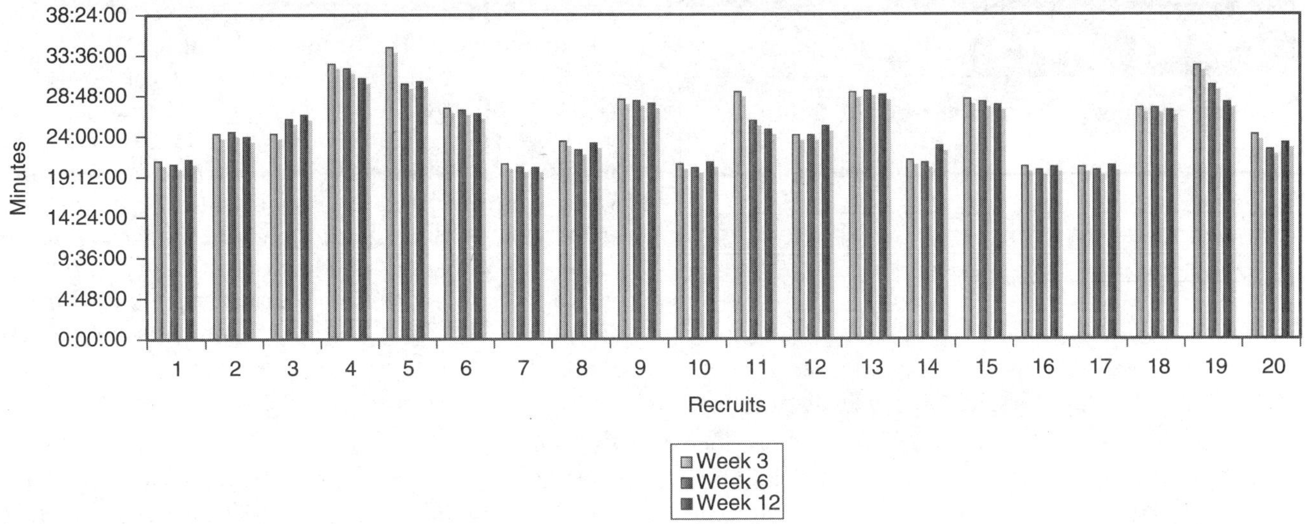

FIGURE 15-5 Academy 2010-01: Length of time it took each recruit to complete the 3.1-mile (5-kilometer) run for weeks 3, 6, and 12.

Courtesy of Ian Cassidy

by the Academy of Sports Medicine and National Fire Protection Association (NFPA) 1500, *Standard on Fire Department Occupational Safety and Health Program.*

Predictably, the results from Academy 2010-01's run show that 75 percent of the recruits successfully reduced their respective pace times (**FIGURE 15-5**) and collectively decreased the class average by 1 minute and 15 seconds (**FIGURE 15-6**). Half of the class could complete the run in less than 25 minutes. The fastest time was 19 minutes, 56 seconds, while the slowest time came in at 30 minutes, 50 seconds. Overall, the class demonstrated a 5.1 percent pace-time improvement without running long distances on a regular basis.

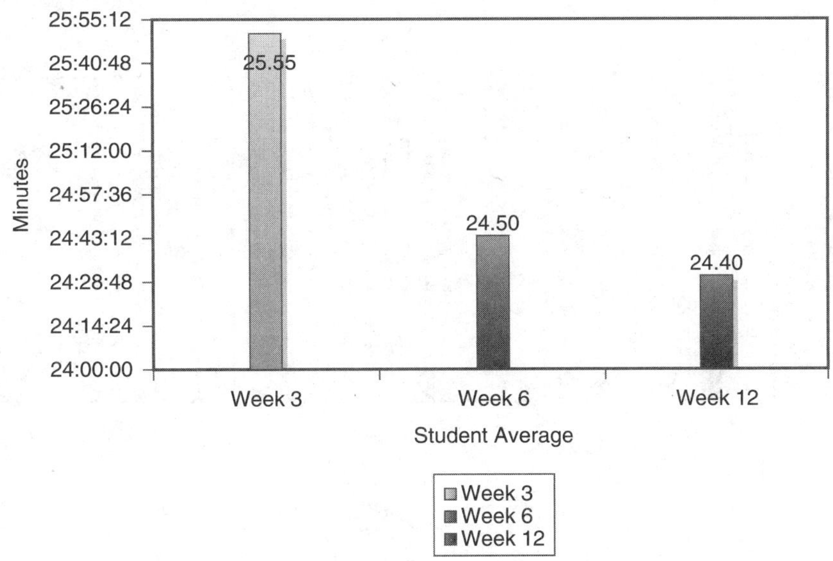

FIGURE 15-6 Academy 2010-01: Average length of time for the academy to complete each 3.1-mile (5-kilometer) run for weeks 3, 6, and 12.

Courtesy of Ian Cassidy

Table 15-2 FIT Score	
Week Number	**Average FIT Score**
Week 1	52.3
Week 7	55.7
Week 13	63.3

Source: Courtesy of Ian Cassidy.

FIT Test: Standard District Fitness Evaluation

To ensure that the academy program would produce results that were consistent with local district standards and practices, the recruits were subject to typical department fitness tests (FIT tests) in weeks 1, 7, and 13 (**TABLE 15-2**). A Golder Ranch Fire District certified physical fitness trainer (PFT) administered the tests and tallied the results. Overall, each recruit demonstrated a dramatic increase in his or her FIT scores (**FIGURE 15-7**), and the entire academy demonstrated marked improvement by week 13 (**FIGURE 15-8**).

Work Capacity Test

The work capacity test was designed to measure the amount of work each recruit could complete while breathing air from the district-issued self-contained breathing apparatus (SCBA). The test is very basic in nature; it does not measure workloads, which change on the fireground, but nonetheless, it provides a starting and ending point for how each recruit improved his or her ability to manage air while working (**FIGURES 15-9** and **15-10**). The test involved an 800-meter circular track that recruits walked around with purpose while breathing air from their SCBA. The number of laps was counted as work completed, and the total time each recruit was able to manage his or her air was also recorded (**FIGURE 15-11**). Time stopped when the recruits ran completely out of air.

Body Weight

Recruit body weights were taken each week on Thursday. It should be noted that while not all recruits lost weight during the academy, their FIT test scores, as well as the pull-up and push-up aggregates, reflected an increase in muscle mass (**FIGURE 15-12**). Some recruits began with a low percentage of body fat and gained muscle mass during physical training, which added additional weight. Furthermore, since muscle mass is heavier than fat, recruits who may have lost fat but gained muscle may also show an increase in weight.

Heart Rate Data

As explained previously, during Academy 2010-01, recruits were issued an Ekho Fit-18 Heart Rate Monitor.

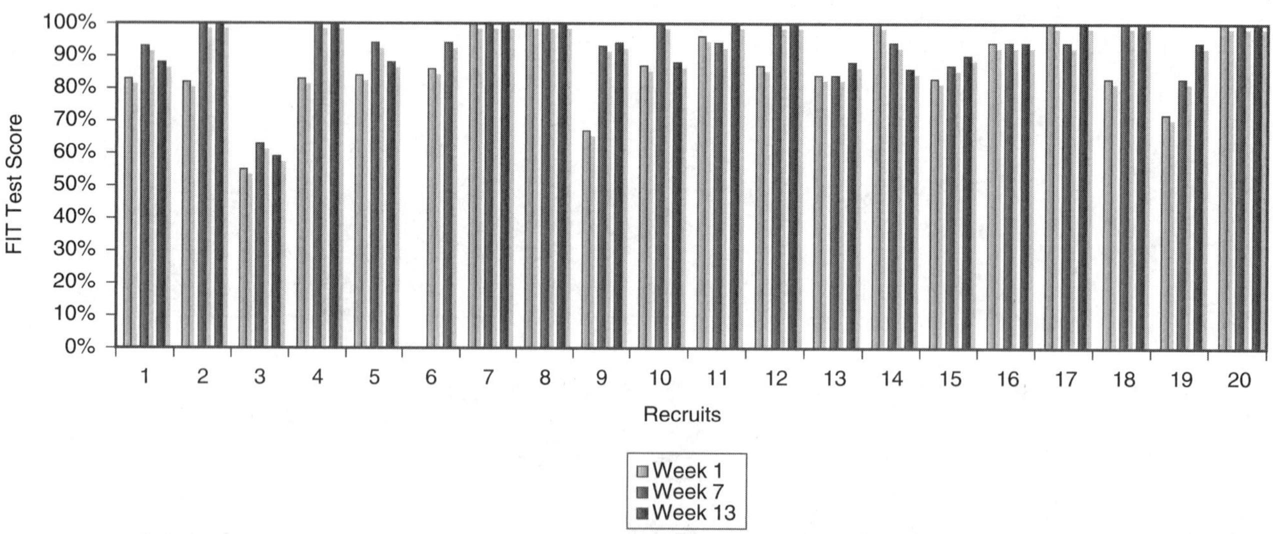

FIGURE 15-7 Academy 2010-01: A comparison of the FIT test results for each recruit for weeks 1, 7, and 13.
Courtesy of Ian Cassidy

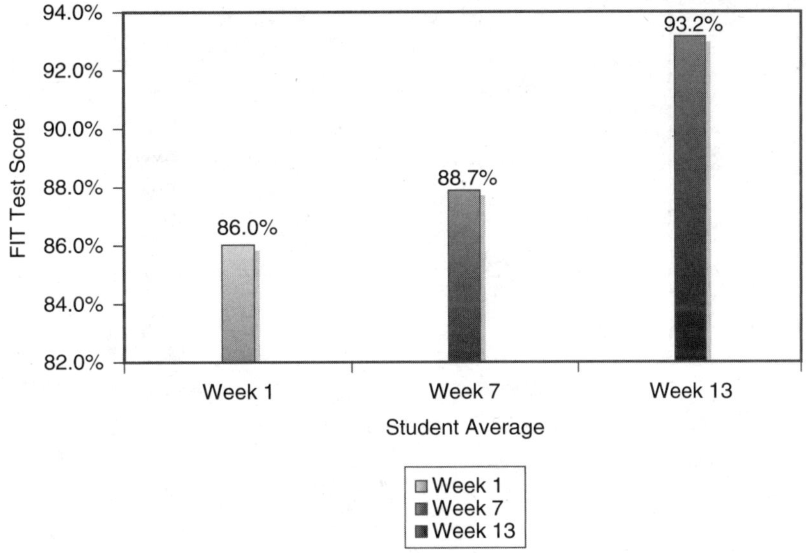

FIGURE 15-8 Academy 2010-01: The academy's average FIT test score for weeks 1, 7, and 13.
Courtesy of Ian Cassidy

The recruits were tasked with determining their maximum heart rate, and the 90 percent, 80 percent, and 55 percent of maximum rates, with PFT supervision. The recruits performed physical fitness workouts designed to mimic fireground activities. The data were then checked during all physical fitness workouts to ensure that each recruit was achieving the maximum performance of a given workout.

It was found by the developers that during both the live manipulative skills training (e.g., live fire simulations) and physical training sessions, the recruits were reaching the desired heart rates. As a result of the

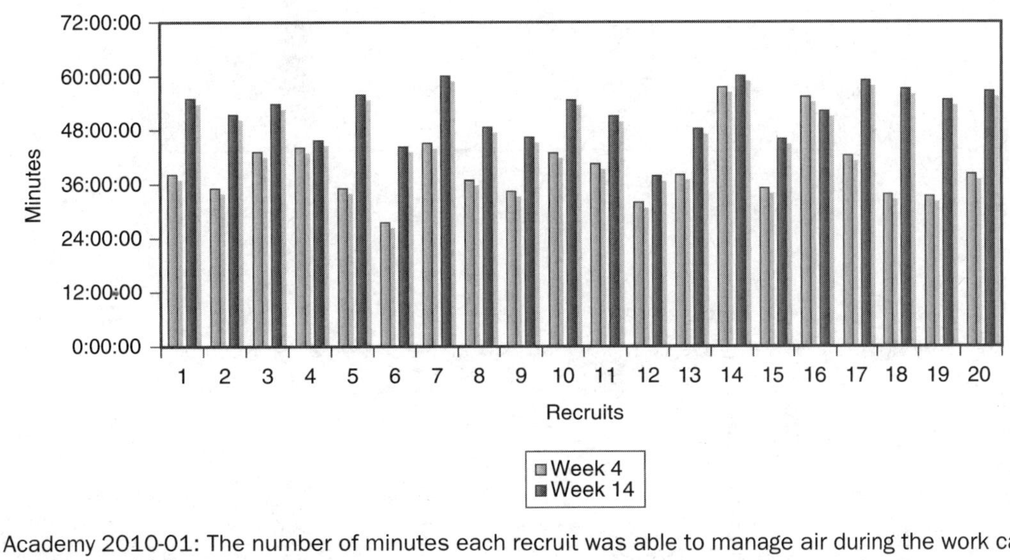

FIGURE 15-9 Academy 2010-01: The number of minutes each recruit was able to manage air during the work capacity tests performed in weeks 4 and 14.
Courtesy of Ian Cassidy

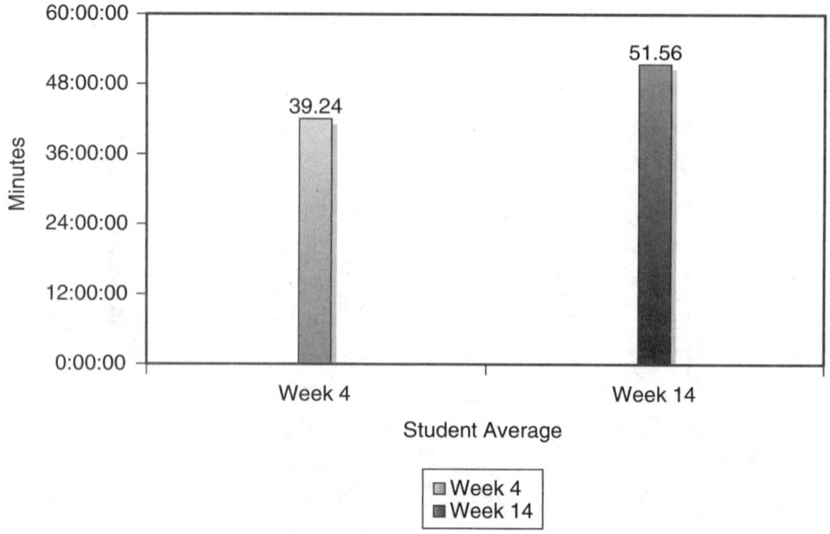

FIGURE 15-10 Academy 2010-01: The average number of minutes the academy was able to manage air during the work capacity tests performed in weeks 4 and 14.
Courtesy of Ian Cassidy

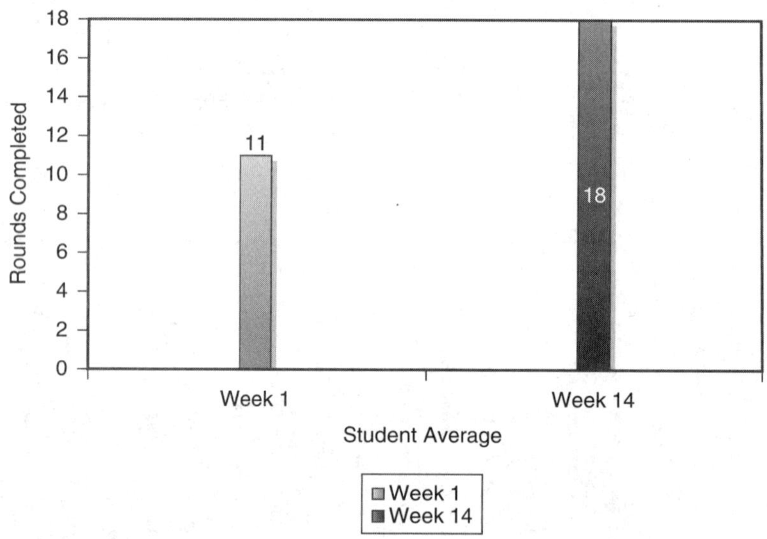

FIGURE 15-11 Academy 2010-01: The average number of laps the academy was able to complete for the work capacity tests performed in weeks 1 and 14.
Courtesy of Ian Cassidy

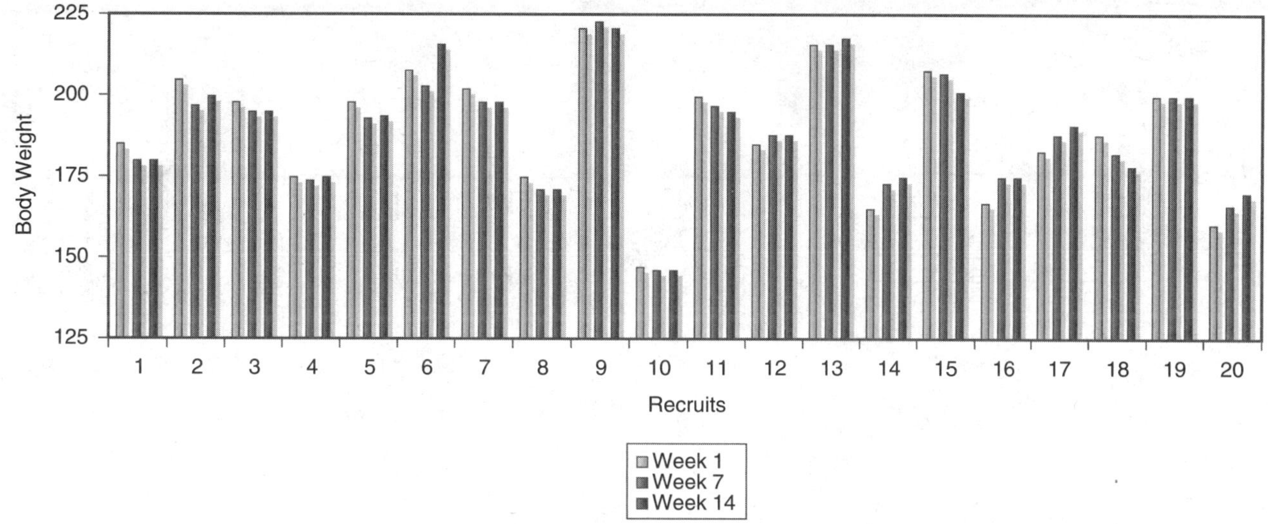

FIGURE 15-12 Academy 2010-01: Weight measurement in pounds per recruit for weeks 1, 7, and 14.
Courtesy of Ian Cassidy

physical training program, the academy's resting heart rate average decreased by *three* beats per minute over 14 weeks (**TABLE 15-3**).

Benchmark Workouts

Each of the four benchmark workouts was meant to test the recruits in areas that would most closely mimic fireground activities. When they enter the field as a probationary fire fighter, recruits are required to have a high level of upper and lower body strength as well as a stable and strong core. A notable increase in overall fitness was achieved through these workouts and others like them. When looking at these results, you should be aware of the fact that recruits may have begun the program using resistance bands in week 1 and finished using their full bodyweight in week 13. Although these details were recorded on their fitness logs, they are not included in this report. Therefore, there are recruits who did not appear to improve during the course of the training, when, in fact, they increased the amount of resistance over time, thus showing improvement in overall strength.

Benchmark Workout #1

For the first benchmark workout, recruits were evaluated based on how many rounds they could complete in 10 minutes when performing the following exercises:

- 5 pull ups
- 10 push ups
- 15 squats (using one's own body weight for resistance)

The recruits were permitted to use resistance bands for the pull-up exercises, but only once they had

Table 15-3 Academy 2010-01: Average Resting and Target Heart Rates

Week Number	Average Resting Heart Rate	Average Target Heart Rate (80%)
Week 1	57.6	
Week 14	54.8	162

Source: Courtesy of Ian Cassidy.

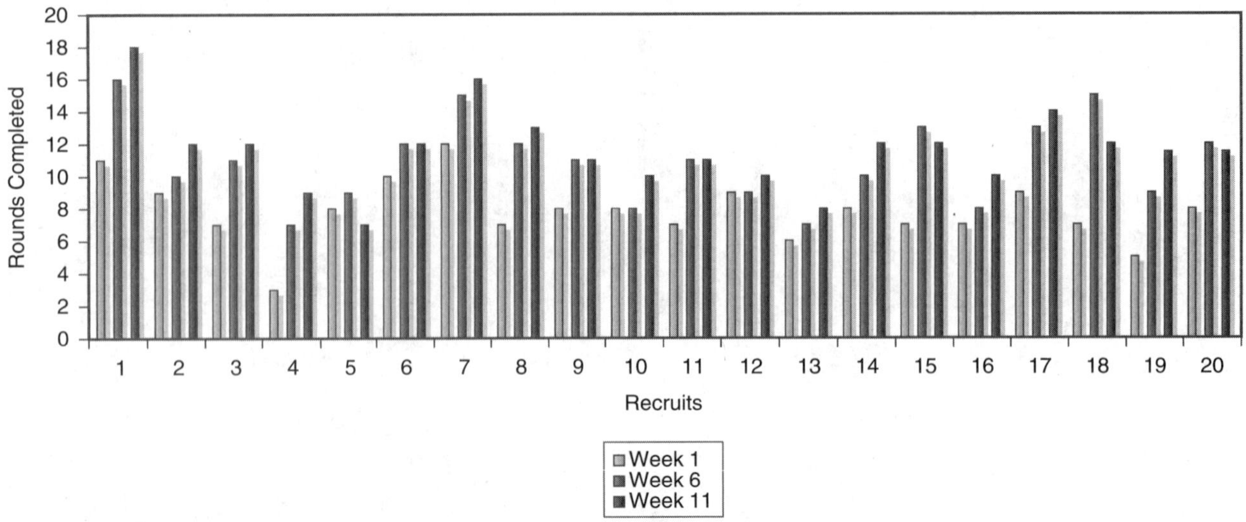

FIGURE 15-13 Academy 2010-01: Number of rounds each recruit was able to complete for the Benchmark Workout #1 performed in weeks 1, 6, and 11.
Courtesy of Ian Cassidy

demonstrated that they could not pull up using their own body weight. The program developers looked for an *increase* in total rounds completed. **FIGURES 15-13** and **15-14** compare the number of rounds completed for the Benchmark Workout #1 performed in weeks 1, 6, and 11.

Benchmark Workout #2

For the second benchmark workout, the recruit was evaluated based on how much time it took for him or her to complete the workout described below (**FIGURES 15-15** and **15-16**). Short rest periods were included in this workout to simulate the rehab/recovery period typically used

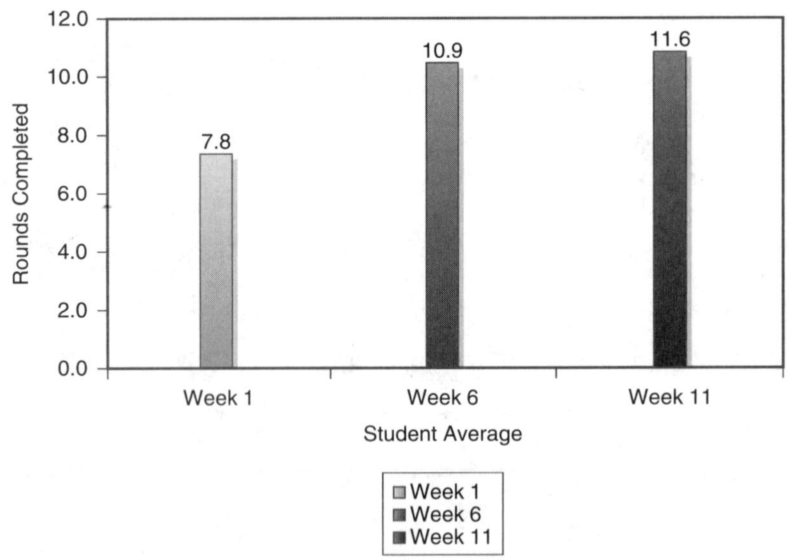

FIGURE 15-14 Academy 2010-01: Average number of rounds the academy completed for the Benchmark Workout #1 performed in weeks 1, 6, and 11.
Courtesy of Ian Cassidy

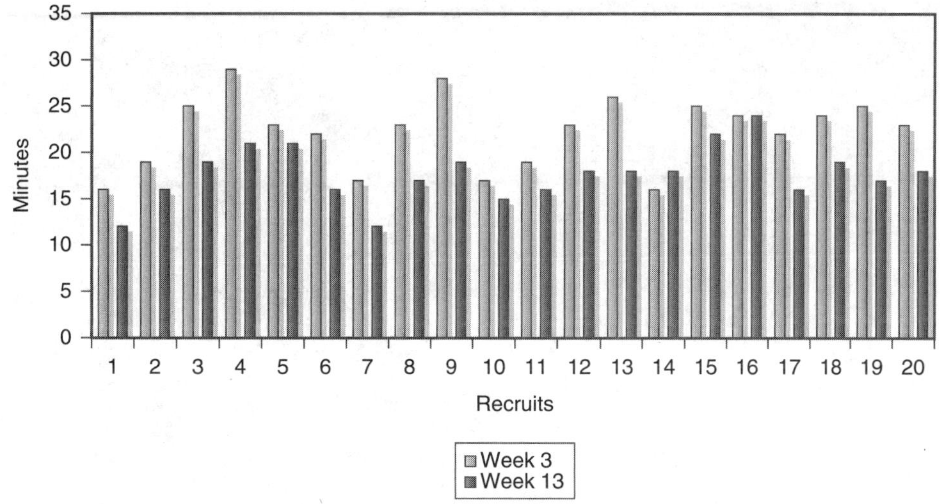

FIGURE 15-15 Academy 2010-01: Average length of time it took each recruit to complete the Benchmark Workout #2 performed in weeks 3 and 13.
Courtesy of Ian Cassidy

on the fireground. The recruits were permitted to use resistance bands for the pull-up exercises. For this benchmark workout, the developers looked for a *decrease* in time to complete the work. This workout included one set of the following exercises:

- 20 pull ups
- 30 push ups
- 40 sit ups
- 50 squats

Benchmark Workout #3

In the third benchmark workout, recruits were required to complete three sets of the exercises listed below in the shortest amount of time possible (**FIGURES 15-17** and **15-18**). For the first set, recruits performed the highest number of repetitions, 21, and this number was reduced with each subsequent set, with 15 repetitions for set 2 and 9 repetitions for set 3. The cadre used this workout to evaluate the recruit's ability to work for a relatively short period of time, but at a maximum

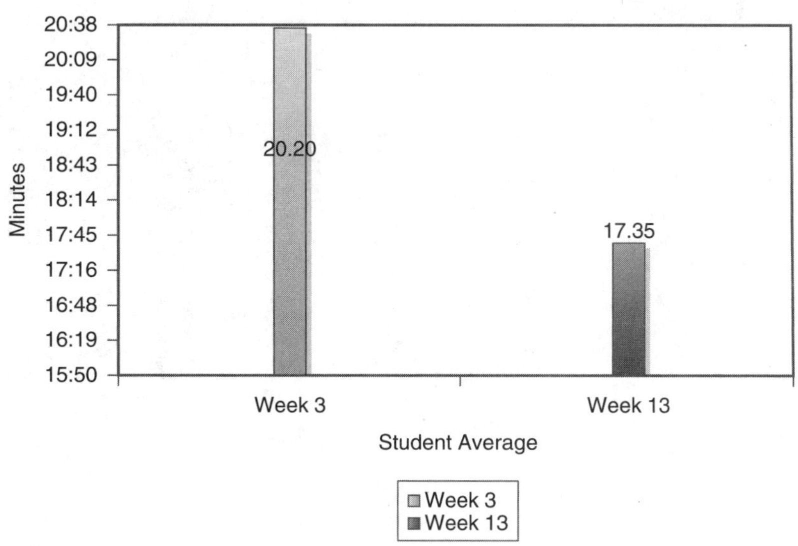

FIGURE 15-16 Academy 2010-01: The academy's average time to complete the Benchmark Workout #2 performed in weeks 3 and 13.
Courtesy of Ian Cassidy

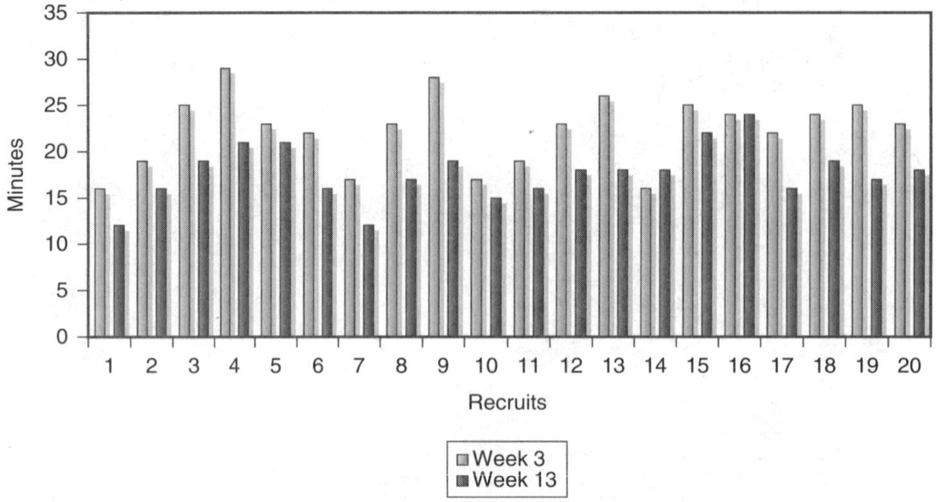

FIGURE 15-17 Academy 2010-01: The length of time it took each recruit to complete the Benchmark Workout #3 performed in weeks 3 and 13.
Courtesy of Ian Cassidy

output, measuring equally upper and lower body strength. The exercises for this benchmark included:

- Thrusters (up to 75)
- Pull ups

Recruits were able to scale the weight down to promote safe lifting, but a minimum required weight of 40 pounds overhead was required, as this measurement is consistent with the weight of district hose packs and other tools commonly used on the fireground that fire fighters are regularly required to lift over their heads. For an indication of success, the developers looked for a *decrease* in the amount of time it took to complete this benchmark workout.

Benchmark Workout #4

In the final benchmark workout, the recruits were required to complete one round of the exercises

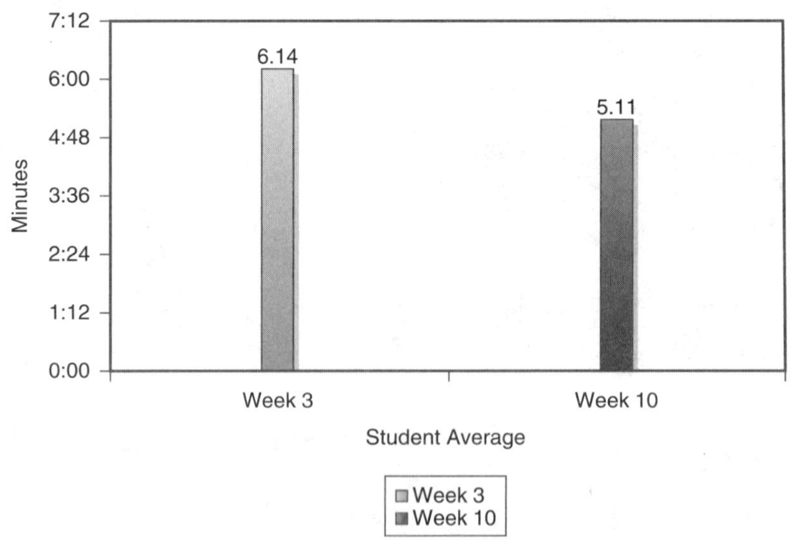

FIGURE 15-18 Academy 2010-01: The average length of time it took the academy to complete the Benchmark Workout #3 performed in weeks 3 and 10.
Courtesy of Ian Cassidy

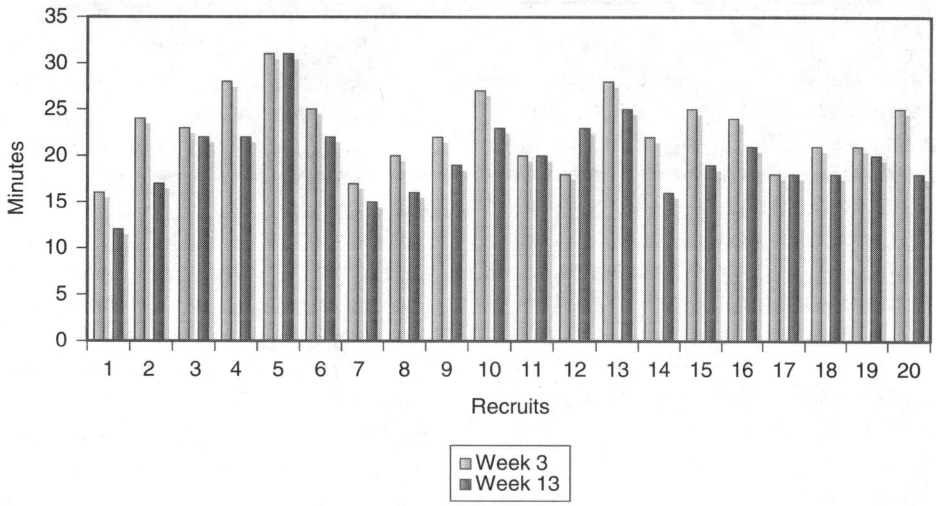

FIGURE 15-19 Academy 2010-01: The length of time it took each recruit to complete the Benchmark Workout #4 performed in weeks 3 and 13.

Courtesy of Ian Cassidy

listed below as quickly as possible (**FIGURES 15-19** and **15-20**). The recruits were not allowed to move onto the next exercise until the one previous had been completed. The recruits were permitted to use resistance bands for their pull ups. As a measurement of success, the developers looked for a *decrease* in time to complete the workout. This benchmark workout included one set each of the following exercises:

- 100 pull ups
- 100 push ups
- 100 sit ups
- 100 squats

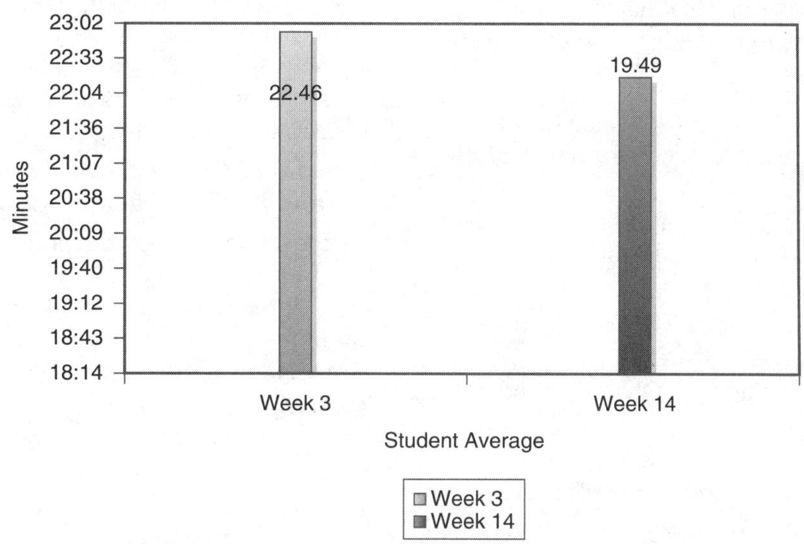

FIGURE 15-20 Academy 2010-01: The average length of time it took the academy to complete the Benchmark Workout #4 workout performed in weeks 3 and 14.

Courtesy of Ian Cassidy

Post-Academy Questionnaires

Upon completion of the program, each recruit was given a 10-question post-academy questionnaire to evaluate the effectiveness of the physical fitness training provided. The questions also prompted recruits to provide feedback on ways in which they could improve the program. In general, the recruits rated the training favorably and reported that they felt prepared and physically fit for the firefighting activities that they would be performing. Many reported that they would continue using the program, or some version of it, after the academy. Below we present a discussion of the recruits' responses to each question.

Question 1: Do you feel the workout log provided you with some "ownership" over your academy physical training? Was the log worth the time you put into it? Why?

Each of the 20 recruits reported that the logging of physical training each day gave them a sense of ownership in the physical fitness training and helped them track their progress from start to finish. Furthermore, they stated that the log provided them with an exercise plan that they could continue to use after the academy to continue with their fitness goals.

Question 2: What could we add or subtract from the physical training program to make it more beneficial to you? Is there anything that you would absolutely change for the next round of incoming recruits?

The majority of recruits reported that they would like to see the following changes made to the physical fitness program:

- The provision of a detailed schedule of the next day's workout to help them prepare for the training
- The daily addition of a warm-up that includes 3 sets or rounds of 5 pull ups, 10 push ups, and 15 bodyweight squats
- The addition of a short 1600-meter run to the warm-up
- More strength training (i.e., Olympic weight lifting, dead lifts, etc.)
- The addition of mandatory weekend workouts adhering to the "3 on, 1 off" Crossfit regime
- Introducing "as prescribed" workouts earlier than week 7
- Including more flexibility training after the prescribed workout
- Adding more time for running and swimming
- Reducing the number of squats performed
- Adding more team-building activities that include problem solving
- Including more "new skill" work into follow-up workouts
- Adding more YMCA visits for physical training

Question 3: Did the movement preparation (a.k.a. dynamic stretching) before each workout prepare you for the PT/drill ground? How so? If not, what would you add or take away from this exercise?

Overall, the recruits thought that the dynamic stretching prepared them for the workout and for any manipulative drills they completed during the academy. As noted on the FIT test scores, the majority of recruits reported that, from weeks 1–14, they noticed a significant increase in their overall flexibility. Many recruits reported that they planned to include dynamic stretching in their personal routines after the academy.

Question 4: Do you strongly agree/agree/disagree/strongly disagree that the intense workouts helped condition you for firefighting activities? Did the program prepare you for your career in the fire service? If not, what would help prepare you better?

Of the 20 recruits, 85 percent reported that they "strongly agreed," and the remaining 15 percent reported that they "agreed," with the statement that the physical training program prepared them for firefighting activities in the field. All of the recruits agreed that the stretching exercises helped them reduce the risk of incurring potential injuries during the academy.

Question 5: Are you now more or less physically fit than you were before starting this academy?

Of the 20 recruits, 100 percent reported that they were more physically fit after completing the academy. A few recruits noted a change in their physical fitness from high levels of strength only to much more stamina and cardiovascular endurance.

Question 6: What did you like most about the physical training program? What did you like least?

The aspects of the fitness program that the recruits liked the most include:

- The as many rounds as possible (AMRAP) workouts, because it was easy to track progress

- Quick, intense, and effective workouts (recruits responded that they felt more fatigued and sore after completing these short workouts than they would if they had worked out at the gym)
- An obvious improvement in muscular endurance and multisystem capacities
- That the workouts pushed recruits to their maximum potential
- The Crossfit-style workouts
- The team effort/concept
- High-intensity workouts
- That exercises varied on a day-to-day basis
- The engine company relays

The parts of the physical fitness program that recruits disliked:

- Workouts were too short
- Burpees
- The weights were too light and would have preferred Olympic-style lifts
- The 3.1-mile (5-kilometer) runs
- Slightly more variation
- That the schedule did not indicate what the next day's workout would be (recruits reported that they wanted to know the workout exercises planned for the next day so they could perform supplemental PT without impacting the next day's activities)

Question 7: What portion of the PT program best prepared you for the job of a fire fighter?

Recruits reported that the following items prepared them best for the job of a fire fighter:

- High-intensity workouts with short rest breaks
- Short, intense workouts with quick recovery periods that pushed them to their maximum potential
- Tire flip relays
- The cardio portion of the training
- That the training revealed both strengths and weaknesses
- That some of the drills prepared them for working with explosives—the same type of work found on the fireground
- The program in its entirety
- The AMRAP workouts
- The strength and endurance training
- The mental aspect of the workouts
- The dynamic stretching
- The core body work
- Pull ups, push ups, and squats (recruits reported that they felt they felt their upper and lower body strength was much stronger due to performing push ups, pull ups, and squats)

Question 8: Did you suffer any injuries while performing any of the training program's workouts that you did not report? If so, which workout(s)?

Although no injuries were reported to the training officers during the academy, the academy knew that injuries sometimes go unreported, so they asked this question to give each recruit the opportunity to report anonymously whether they experienced an unreported injury. One hundred percent of the recruits denied experiencing any unreported injury during the academy.

Question 9: Do you plan to continue a similar type of fitness program after the academy? Why or why not?

One of the program's major objectives was to create a physical fitness program that recruits would continue to use after the academy was over. To evaluate whether the physical training was something that they would continue as a whole or in part, each recruit was asked if they would continue with a similar-style workout plan. One hundred percent of the 20 recruits stated that they would continue a Crossfit-style training program either in whole or as part of a weightlifting regime after the academy was completed.

Question 10: Has your energy level/recovery level increased over the course of this program?

Of the 20 recruits, 95 percent stated that their energy levels and, more importantly, recovery levels have increased significantly. Only 5 percent of the recruits stated that their energy levels had not increased significantly.

Academy 2010-01: Conclusions and Recommendations

As seen throughout the report, the physical results of Academy 2010-01 show a significant improvement in all areas assessed. The 10 evaluations conclude:

- FIT test scores with a 6.85 percent improvement on average
- VO_2 increase of 11 points on average over 14 weeks

- A 57.4 percent improvement in the number of maximum pull ups completed
- A 16.4 percent improvement in the number of maximum push ups completed
- Work capacity test with a 33 percent improvement in pace time and 63.6 percent improvement in work completed
- The resting pulse rate decreased 3 beats on average
- An increase of 48.7 percent increase in rounds completed during Benchmark Workout #1
- A 15 percent decrease in the time required to complete Benchmark Workout #2
- A 16.6 percent decrease in the time required to complete Benchmark Workout #3
- A 13 percent decrease in the time required to complete Benchmark Workout #4

In conclusion, the Northwest Fire District-Fire Academy Physical Fitness Program either met or exceeded all of the objectives it set out to achieve. Not only did the recruits feel that they were prepared for the rigors of day-to-day firefighting tasks, they unanimously stated that they would pursue career-long fitness goals by the example set in the academy. They maintained that the ownership that they felt over the fitness program in the recruit training also motivated them to continue the program.

As previously noted, no injuries occurred during the physical training—a major goal of the program—and only one contusion occurred during a manipulative drill. Not only is the lack of injuries incurred during the course of the program a testament to the program's fitness program, it is also proof that the program was a success in its day-to-day operations. The developers were able to produce 20 healthy, fit-for-duty fire fighters upon completion of the academy, thus protecting the taxpayer's investment and allowing the district to more accurately plan for future hiring.

Recommendations for Continuous Improvement

The Northwest Fire District has committed itself to providing its members with a safe and healthy environment. With the emphasis on health and wellness over the past 5 years, it is long overdue that we begin to effect change at the beginning stages of a fire fighter's career path. Instilling a strong sense of duty in recruit fire fighters to remain physically fit and healthy throughout their careers should be one of the primary focuses of a district.

Based on the findings of this academy and the recruit questionnaires, it was recommend that the Northwest Fire District continue with this fitness program "as is" with some modifications made based on the results from the questionnaires and lessons learned. The Northwest Fire District should also continue to track predetermined benchmarks from academy to academy to acquire measurable results and evaluate the readiness of its recruit fire fighters. Furthermore, it is believed that the health and safety division should include that the district PFTs take a larger role in upcoming academies. The work required in advance to make a fitness program succeed the way that this one has requires individuals to be proactive at least 3 months prior to the academy's start. The district PFTs should work hand-in-hand with the training developers to accomplish its goals. It is vitally important that the developers and PFTs participate in physical fitness each day with the recruit fire fighters to build credibility for the program's success.

Based on Orange County Fire Authority's (OCFA) information, the district may also want to explore the possibility of building a Fitness Fair or Pre-Academy Fitness Workshop into the selection process. This would give each selected recruit the opportunity to prepare for the specific physical fitness activities that are conducted during the fire academy and give the health and safety division an opportunity to flag those members who may be at risk of injury prior to their entering the academy. Identifying these recruits pre-academy and giving additional support in their weakest areas would further protect the investment made in training them.

Finally, it is believed that the district should continue purchasing heart rate monitors for each recruit who enters the academy to promote safety and to help recruits evaluate the effectiveness of each workout and manipulative drill. The Northwest Fire District should also consider investing in software that will automatically track heart rate data from the personal monitors to a laptop computer. This will provide real-time data during both fitness and manipulative drills, which would allow trainers to determine the exertion levels that the fire fighters are operating at during exercises such as live fire drills or hose pulling operations.

Key Points

Although there were many lessons learned throughout the implementation of Academy 2010-01 and areas that would benefit from improvement, the Northwest Fire

District agreed that several key changes needed to be made. These include:

- The academy must have a level 1 Crossfit instructor, as these instructors deliver better instruction on new skill movement and form. At the very minimum, the instructor should at least be highly familiar with Crossfit training, as it was proven that use of Crossfit methodology was a success.
- The academy needs to purchase 4–5 rowers to increase the number of cardio workouts that recruits can perform and continue to keep injuries at a minimum.
- The academy must set aside 3 hours on the first day for FIT testing, or they must coordinate with the district physician to perform FIT tests in his or her office prior to academy start. On average, FIT test days take 3–4 hours to complete for a class of 22 recruits.
- The academy should make sure benchmark workouts do not follow an extremely hard PT day, as this was shown to have negative effects to tracking. For accurate tracking, it is best to schedule all benchmarks to occur on Mondays.
- When working through new skill work, the academy should break up teams into smaller groups for more one-on-one time. To achieve more personal attention, at least four training members with knowledge of the movements would be required to attend.
- New skill work should be put into practice during the days following a PT session instead of the following week or two. By practicing new skill work immediately after the day of instruction, the movements will still be fresh in the recruits' minds. They also tend to develop better muscle memory over the long term.
- During FIT testing, recruits should be encouraged to continue push ups and sit ups until they can no longer perform them, not until they reach the 100 percent mark on the exam. For this reason, none of the recruits were told what constituted passing or what the highest score achievable was. Maximum push ups and sit ups should be recorded this way (metronome).
- Maximum pull ups can also be completed post-FIT test. No resistance bands should be used during the evaluation. Repetitions only count when the chin breaks the plane of the bar and the body is lowered until the arms are completely straight.
- Thursday's were Engine Company Relays (ECR-team-based PT session). This could be expanded to include some problem solving or true relay-type drills. This was key in fostering the team environment.
- Because ECRs are typically shorter workouts, time would be better spent on teaching recruits rehabilitation or injury prevention techniques instead of spending time on new skill work. These workouts may include additional stretching, instruction on self-massage techniques, use of rollers, etc. The academy should coordinate with the district physician's office for the arrangement of a certified trainer to attend ECR days.
- Add more Tabata, or 4-minute workouts, into the program.
- Add in "Griff," or another workout of the day, to the named benchmark workouts for additional lower body strength evaluation.
- The program should remove the 1-ton overhead ECR workout, because it was too quick and ended up being too easy for the academy, forcing the addition of PT on this day.
- All benchmark workouts should be completed toward the beginning, middle, and end of the academy. Although the midterm evaluation could be completely removed, it does provide an indication of how effective the program has been up to that point. It can also be used to identify potential areas of weakness before the academy ends so adjustments can be made.
- The academy should add additional gymnastics work (i.e., rings) for core strengthening. It was also suggested by the district physician's office that specific core workouts be added, possibly on the ECR days.
- Although the work capacity test was simple and effective, it was recommended that the developers use the District's consumption course in the future, as it seems to be more in line with what the district is doing and it provides more specific fireground application to the test. This evaluation needs to be completed during the first week or two of the academy and then again at the end.
- The academy's fitness schedule should *not* be given to each recruit despite the comments received on the evaluations. The reason a fitness schedule was not supplied was because the academy was trying to provide an environment that would closely mimic the fire service lifestyle.

A fire fighter should be able to perform a task well at any given moment, no matter what it is. Since fire fighters never know what the next call may be, it does not make sense to provide a schedule on the tasks that they will perform beyond the day's training. Even though recruits were unable to prepare mentally for the next day's workout or were concerned that performing supplemental workouts would fatigue them, the physical training program was still highly successful. There is no reason why it should not be continued as-is.

- With the addition of district PFTs to the program, it was recommended that the academy invest in 4 or 5 barbells, as well as some bumper (rubber) plates, for Olympic-style weightlifting. This workout should be strictly supervised by PFTs. The addition of weights will significantly increase strength and prepare recruits for fireground activities.

References

11 Alive News. *Fire Recruit Dies Hours after Fitness Test*. Atlanta, GA: 11 Alive News: September 22, 2008. http://www.11alive.com/news/local/story.aspx?storyid=121444&catid=3. Accessed October 8, 2014.

National Fire Protection Association (NFPA) 1500, *Standard on Fire Department Occupational Safety and Health Program*. 2013. http://www.nfpa.org/codes-and-standards/document-information-pages?mode=code&code=1500. Accessed October 8, 2014.

CHAPTER

16 Promotional Assessments

Steve Pegram and Brian Ward

Introduction

In various departments throughout the country, training officers are charged with the development and implementation of their department's promotional assessments. It can be quite a challenge preparing the right materials for the right assessment and ensuring that they are valid. In this context, the term *valid* refers to the assessments that only contain questions that are pertinent to the skill or knowledge being assessed for and that the answers do not vary among people who have similar fire service knowledge and experience. Ensuring the assessments are fair and valid is absolutely essential, as there has been legal case after legal case where employees have submitted grievances or filed lawsuits against departments for administering assessments they deemed were discriminatory against particular individuals.

One example of a lawsuit is the 2009 case of *Ricci vs. DeStefano*, in which a group of white fire fighters sued the City of New Haven on the basis of race discrimination. The fire fighters maintained that their civil rights were violated when the city decided to dismiss the results of the promotion tests and not promote the fire fighters who had passed because not enough minority applicants had performed well on the test. After 7 years of fighting this case, the fire fighters won the lawsuit and the city had to pay approximately $2 million in damages.

As you can see from this example, the consequences of administering a poorly conceived promotional assessment can be quite severe, with the costs of a lawsuit far outweighing the cost of hiring a third-party to properly develop these assessments. When deciding whether to develop and administer promotional assessments internally or hire a third party to do so, departments must weigh the risks and benefits as well as the resources and funds available. Departments that choose to develop and administer their own assessments must do so with painstaking care and attention.

Use of Internal Versus External Contractors for Assessment Development and Administration

Some departments have chosen to completely remove themselves from administering the promotional process and hire a third-party contractor to develop the assessment and compile the data. Most contractors have employees who hold degrees in assessment development and are certified to provide physiological examinations. However, contracting such individuals to develop and administer assessments can be extremely costly. Furthermore, contractors are unable to operate with complete autonomy, as they require some input from the department. These subject matter experts (SMEs) will need a list outlining the specific topics required for your department, but the contractor will establish the guidelines and administer the assessment.

Several assessment companies are available to provide whatever services you may require. However, for those of you whose departments choose to develop and administer their own assessments, we have provided a few items you should consider before carrying through with your promotional assessment.

Developing Promotional Assessments

■ The Preparation Stage

If your training department develops promotional assessments for your organization, there are several items you will need to consider before you can begin to develop materials. First, you will need to determine the specifics of the position that you are trying to fill. The education, certification, responsibilities, and experience the position requires will dictate the terms and content of the promotional assessment. For example, if you are looking to fill a fire fighter position, what level fire fighter are you looking to hire? Will the position require additional certification(s)? For example, would the fire fighter also need to be a certified emergency medical technician (EMT), or will the fire fighter need to be trained in hazardous materials release? If such certifications are required, your promotional assessment will need to be designed with them in mind. You will also need to consider any prerequisites or certifications that may be mandated by local or state/provincial ordinances, as these will also need to be factored into the development of these assessments.

You must also consider whether you are filling a vacancy for a position that already exists or a newly created position, and if the hire will be an internal or external process. If this is an external hire, you may choose to include more questions about the candidate's personal history and experience in your assessment. If you are hiring for a preexisting position, you may already have a detailed job description on hand that you can update as needed. You will also need to consider the reference sources for the assessment materials. Your department may have standard operating procedures (SOPs) that dictate the testing or promotional process, so those should be referenced before and during the development of the assessment to ensure you are in compliance with them.

You will also need to consider how much time you have to develop the promotional assessment—when the position needs to be filled often dictates how much time you have to create it. Regardless of how much time you have, however, you should never rush through the development of a promotional assessment. As the case of *Ricci vs. DeStefano* demonstrated, the stakes of using a rushed, ineffective assessment are too high.

■ SMEs and Assessment Development

Once the training officer begins to answer the questions described in the section above, he or she can start to develop a plan surrounding how, who, and when you want to conduct the assessment. These questions are vital to the entire process, and each step must be documented along the way. To aid in this effort, the training officer should select a group of SMEs who are certified to either the level required for the position or beyond to help develop the assessment plan. They can guide the department in the best possible direction by ensuring that the department is creating a test that is reliable and fair.

One task SMEs are charged with is the discussion of the process. They will also perform a job analysis to determine exactly what the position entails. For example, one question they will seek to answer is, "What are the daily duties for this position?" By offering their expertise on the position, they can help the training officer ensure that a job description and assessment that cover all of the necessary requirements have been developed. The SMEs should also assess the written test, to ensure it is valid and reliable, by taking the test via a random sampling of the assessment items. For the assessment item to be considered valid, it should roughly receive a 70 percent passing rate among the SMEs who answered it. Once the assessment is determined to be valid and reliable, it will also be checked for spelling and grammar errors and cross-referenced. Finally, the assessment items should be ranked for the level of difficulty. Using this information, the training officer determines the adequate number of questions, the difficulty level of each question, and the passing score.

■ Ensuring Integrity in the Assessment Process

The integrity of a promotional assessment is paramount to selecting the right individual(s) for the position and the department and for preventing lawsuits. Each individual SME and all candidates should sign a statement declaring that they will not show, discuss, or provide specific examples of any assessment-related items with anyone else. Furthermore, all SMEs should be removed from training or the perception of training any individual who will be taking the assessment. All individuals should be completely clear that if they have any concerns about an instructor or department having a conflict of interest, they should discuss this up front with the training officer to avoid potential problems at a later time. With their signature and an understanding of the rules, any individual caught in violation could face disciplinary action, up to termination, for cheating. All assessors should follow these same procedures and sign the same statements.

Creating the Job Description

If you are filling a vacancy of an already existing position, many questions as to what the job will entail and what certifications it will require should already be answered. The training officer should quickly review the job description, consider the necessary qualifications, and review the SOPs used in the past. For many departments, a job opening can present the opportunity to review and rewrite the SOPs for promotional assessments. When performing these revisions, you should consider the changes that occurred in your organization since the position was last filled. Certifications, prerequisites, and department SOPs may need to be updated as a result of changes in the community's expectations, changes in technology, and differing levels of service. The direction of the department may have also changed, precipitating a change in the department's responsibilities. You may now provide levels of service you did not before (i.e., EMS and specialty teams). The job description must be current, thorough, and exact to ensure that the candidates you select for this vital position have the right knowledge, skills, and abilities to lead your organization forward.

Internal Versus External Candidates

The next consideration is to determine if you currently possess the individual(s) to fill this role. Do you have the right candidate in your organization or is it time to look outside and see what a regional job search produces? The decision to hire from within or go outside will have a lot to do with the size of your organization, the availability of qualified candidates, union contracts (if they exist), and your current policy on promotions or hiring processes. There are many advantages and disadvantages to hiring internal candidates versus an external candidate. With an internal candidate, they are already familiar with the organization, how you operate, various personalities, the department's policy and procedures, and daily operations. This could also be a disadvantage, as the person may have become indoctrinated to a certain ideal or way of thinking and training, and a leadership position can benefit from a fresh face and someone able to introduce new ideas to the organization.

If your department has a culture or a system you want to change, then looking outside the organization may be the best option. By opening up your process to outside candidates, you will have the opportunity to assess your internal candidates who apply, but also put them up against other candidates from around the region and elsewhere who are looking for a new challenge. With an internal candidate, you know the employee and they know you—which can be both an advantage and a disadvantage—and with an external candidate, you are relying more on your testing/hiring process, references, and candidate interviews to choose the best candidate for the job.

If your assessment is validated and a high level of integrity is built around the assessment, this will create the best chance for selecting the right individual. To aid in this process, most departments use outside assessors who are completely removed from your department and are not from nearby adjoining communities. The assessors will come in for 1 to 2 days to receive training on several items including SOPs, department organization, key points to identify, resources used, and typical strategy and tactics scenarios. This allows the assessors to review the content and ask questions concerning the department and grading scale. This system generally works best when testing internal candidates, removing any potential for subjectivity or favoritism.

Promotional Assessment Study Guide

Prior to any candidate being given the assessment, it is only fair that each candidate understand the rules and reference sources for the assessment. A promotional assessment study guide should be developed that outlines several items for the candidates. This not only provides them with direction, but it also protects the department. It is also a good idea to set up a question and answer session in the event there are any unclear or ambiguous statements. The areas covered in the study guide should typically include the following items, such as has been determined by Industrial/Organizational (I/O) Solutions:

1. Process components and development
2. Information on the format and structure of the exam
3. Advice on the importance of reading specified study materials
4. Advice on how to study for the written examination
5. The development of a study plan
6. General preparation guidelines
7. Rules for the test
8. Test-taking principles
9. Use of the answer sheet
10. Posttest review and appeal process

Sample Interview Question and Answer

Interviewer: Tell me a time when you demonstrated initiative. What was the task and what was the outcome?

Interviewee: I noticed that we had completed all of our mandatory, minimum-standard training for the year; however, I felt that there was more that we could do to better prepare ourselves for the challenges that we might face. In between calls and other station duties, I developed a simulation program, which allows each fire fighter to practice situational awareness, decision-making opportunities, and sharpen skills around best practices for the given situations. After about a month working on the project, it was completed and implemented throughout the department as a training initiative.

Assessments and Interviews

As an organization prepares to hire any position, it should consider using multiple assessment avenues. These typically include a written portion, a scenario "Hot Seat," a personnel role play, and an interview. A well-rounded combination of assessments will provide the best candidate. Review the chapter, *The Evaluation Step,* for additional information on types of tests and testing procedures.

Another method to consider during the interview is to use the "portfolio" approach to the process. A portfolio process consists of several documents and other items the applicant must prepare before any test, interview, or assessment can be conducted. The contents of the portfolio are then typically submitted with the application and resume and will show competency in several required job tasks.

Examples of items to include in a portfolio could be reports, articles, presentations, executive summaries, dissertations, or any other work already produced and/or performed by the candidate that is relevant to the position he or she is applying for. Having the portfolio prior to the interview will allow the interviewer to prepare questions and clarify any information. This is also a good time to have the interviewee provide subjective based outcomes, or SBOs. The interviewer will provide open-ended questions that allow the applicant to use past experiences to answer them. It has been said by some that the best opportunity to assess if someone will succeed is to review how they have performed in the past. With this format, the interviewee is able to apply their knowledge and experiences during the interview process.

Some departments have begun to videotape all scenarios, so that if there are any discrepancies between the assessors and the candidate, the videotape can be reviewed for accuracy. This video becomes part of the assessment process and should be stored for record-keeping purposes.

Appeals Process

After the assessment process has been concluded, an appeals process should be provided in case there are any discrepancies. This should be developed prior to the assessment, and the rules for the appeals process should be provided in the promotional assessment study guide. All rules should be covered adequately and upheld. Typical items covered in the study guide under appeals include:

- Timeframe
- Appeals process
- Documenting the appeal
- Predetermined timeframe for reviewing the written assessment with reference materials after initial grading. Generally, it is a 60-minute window, and you are not allowed to leave the assessment room.
- Candidate format for justification: specific to question, state/province concern, and provide proof in referenced materials.

■ Selection Process

At this point, generally the process is complete for the training officer. All results and data are compiled and delivered to the command staff for selection. As with any course or program, constant evaluation is necessary to develop the best program or assessment. Also, even after the assessment is complete, be sure to not release any results or discuss the assessments with anyone. Last, ensure that privacy acts and laws are upheld throughout the process for all candidates.

Key Points

- It can be quite a challenge to prepare the right materials for the right assessment and ensuring that they are validated.
- Most contractors have employees who are degreed in assessment development and providing physiological examinations; however, they can be very expensive.
- The SMEs' responsibilities include discussing the process and performing a job analysis to determine exactly what this position entails.
- The integrity of a promotional assessment is paramount to selecting the right individual(s) and avoiding lawsuits.
- Who you select for this vital position must have the right knowledge, skills, and abilities to lead your organization forward, and the job description must reflect these needs.
- In order to help accomplish validity and reliability (in a nonbiased environment), most departments use outside assessors who are completely removed from the department.
- As with any course or program, constant evaluation is necessary to develop the best program or assessment.

References

The Associated Press, New Haven Firefighters Settle Claims of Racial Bias. *The New York Times*, July 28, 2011. http://www.nytimes.com/2011/07/29/nyregion/new-haven-firefighters-settle-race-discrimination-claims.html. Accessed October 8, 2014. (Note: A version of this article appeared in print on July 29, 2011, on page A22 of the New York edition with the headline: "Firefighters In New Haven Settle Claims Of Race Bias.")

I/O Solutions, Florida DOT Sergeant Promotional Examination. http://www.iosolutions.org/uploadedFiles/IOS/IO_Solutions/Research_and_Resources/Resources_for_Promo_Processes/Florida_DOT/Sergeant%20Promotional%20Study%20Guide.pdf. Accessed October 8, 2014.

CHAPTER

17 Sample Training Drills

Brian Ward and Bryan Altman

Introduction

This chapter is dedicated to all of the training officers who are constantly looking for new skills or a different way to teach drills. One of the greatest difficulties as a training officer is finding the perfect drill that challenges the student, is meaningful, and provides the student with something to walk away with. For a volunteer department training officer, this may be the most important chapter; even with the time constraints placed on volunteers, they still need to be as knowledgeable and capable as any other fire fighter.

Here we have supplied you with numerous drills of various backgrounds, and as a training officer, you should be challenged to create your own as well. Each department and each territory provides for its own set of obstacles and challenges, and these situations should be trained on in order to reduce the amount of surprise once the incident occurs. Developing great drills requires a certain amount of time, diligence, and creativity; however, the better the drill, the more the student will retain. When developing your own drills, review different situations and ask the following questions:

- When was the last time these situations were faced?
- What could the severity be if a similar incident occurred?
- Do we know what to do if this situation occurred?
- We may know what to do, but do we understand why we perform certain operations? If our first option does not work, do we know how to adapt to the current situation?

Thermal Imaging Camera Familiarization

The first step is to show the student proper operation of the thermal imaging camera (TIC), including on/off, image interpretation, white-out, scan time, surface temperatures, temperature modes, and equipment specific technology (**FIGURE 17-1**). Once the students have been shown how the TIC operates, set up three rooms, each as a different station so students can familiarize themselves with the operations and image interpretation. This drill can be done in two stages, the first with cold smoke and second under live fire drills (National Fire Protection Association [NFPA] 1403, *Standard on Live Fire Training Evolutions*, should be followed where applicable for live fire drills). Students should be divided into groups of 3 to 4 to allow for maximum hands-on time.

■ Room One Set Up

The first station ("You tell me") is designed to allow the students to take turns with the TIC and show how to properly scan a room from high-level, mid-level, and low-level (**FIGURE 17-2**). As they scan the room, they should be telling the instructor where they are scanning and what they see at each level. The low-to-high scan is taught by first directing the camera in an area that is most likely to reflect the outside ambient temperature. The fire fighter will stop the camera at two points (left and right) on each level as the camera is raised. This is called the six-point scan. This drill will help the student gauge the speed with which they should move the camera, along with image interpretation. Items to acknowledge during the scan include:

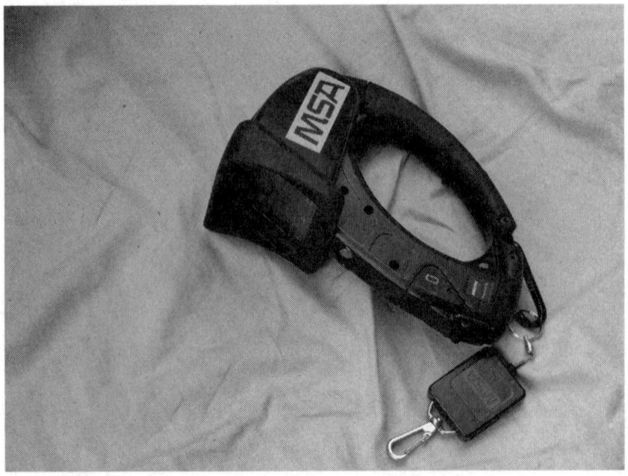

FIGURE 17-1 A thermal imaging camera (TIC).
© Joe Nedder/Jones & Bartlett Learning

- **High-level scan:** Fire, superheated gas, and collapse potential.
- **Mid-level scan:** Egress routes and other openings leading to different areas, and also any victims not on the floor. Structural integrity issues from racked windows and doors should also be noted.
- **Low-level scan:** Holes, obstacles, and victims.

■ Room Two Set Up

The second station ("What is it?") is designed to have the group take turns identifying different objects in a room, anything from a person to pieces of furniture.

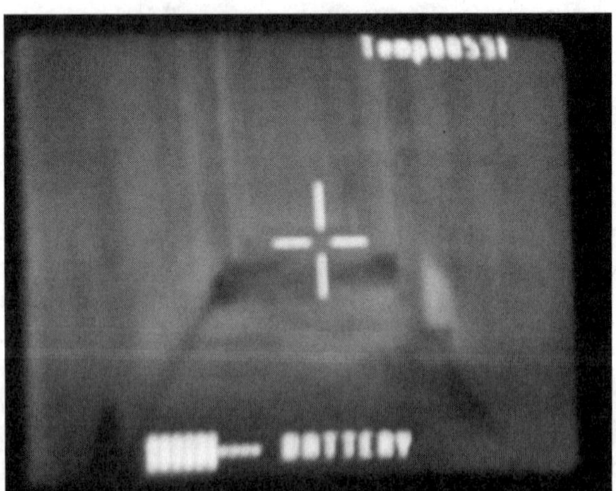

FIGURE 17-2 TIC scan of the hallway in a non-immediately dangerous to life or health (non-IDLH) environment.
Courtesy of Brian Ward

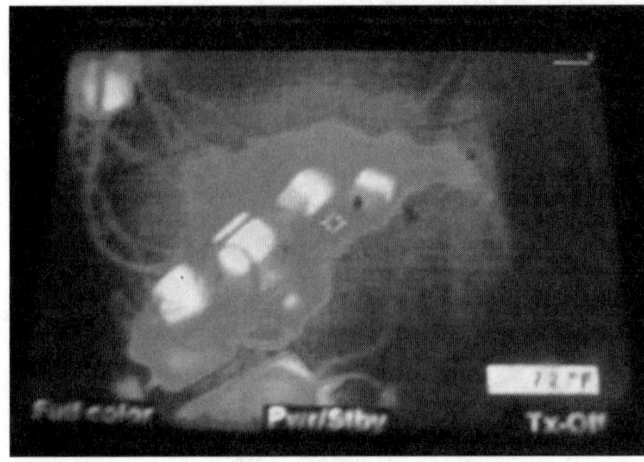

FIGURE 17-3 TIC scan of a fire fighter in an entanglement drill in a non-IDLH environment.
Courtesy of Brian Ward

Using adult and pediatric manikins, place a blanket over the manikins and guide the student in how to interpret the image on a couch, bed, or in a corner. It is important to show how images may be different from the perception that may be in our heads (**FIGURE 17-3**).

■ Room Three Set Up

The third station ("Follow the leader") is designed to allow the students to take turns at the doorway with the TIC and to lead the team through the room to different objects, around objects, and to locate victims. This drill allows the student to develop the skills from stations one and two and build upon them by leading a team of fire fighters through a structure. Although fire and heat are great training tools, this drill can be conducted in a completely dark room if resources are limited and live fire is not an option (**FIGURE 17-4**). Stations one, two, and three are progressive stations, such that if the students are not proficient with a station, they should not move to the next one.

Air Consumption Drill

There are several air consumption drills available in the fire service, and we will discuss a few variations here. The level of prior training will dictate the complexity of this drill, and this can also be combined with gear acclimation drills.

Students are put in groups of three to four, and the group starts by quickly dressing together. Once they are all finished and breathing air, they then jog to a cone placed 30 to 40 yards away. Once the team is at the cone, they then drop to their hands and knees and crawl to another cone 60 yards away. When the team is at the second cone, they should all stand and walk a

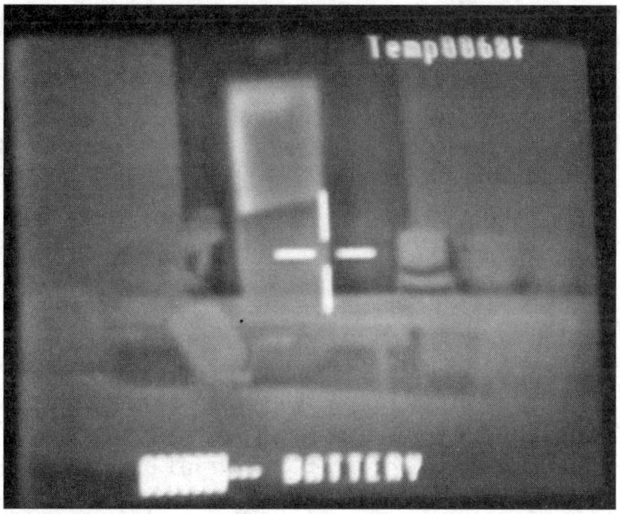

FIGURE 17-4 Image of the TIC scan used to navigate the crew through a blacked out room.
Courtesy of Brian Ward

route back to the starting point and then repeat. The team should stay together the entire time and also have an instructor with them at all times. The instructor needs to remind the students about conserving their air. One method to teach is the Riley Breathing Technique (RBT), which consists of inhaling normally, but humming while exhaling. As documentation is always important, the student's number, quick dress time, starting air, finishing air, and total time need to be recorded. This drill can also be performed with the rapid intervention crew (RIC) Pittsburgh drill obstacle course.

A second version of this drill is to have the students quickly dress and simply walk in gear while breathing air for 15 minutes, and then take a 5-minute break (**FIGURE 17-5**). After the break, the students return for a 10-minute walking drill with a 10-minute break.

FIGURE 17-5 Students walking in full gear with air on.
Courtesy of Brian Ward

The third rotation is 10-minute walk with a 10-minute break. This drill is geared more towards the first part of recruit school, where students are learning personal protective equipment (PPE)/turnout gear for the first time. The students should be reminded to pay attention to how the self-contained breathing apparatus (SCBA) reacts once the low air alarm activates, as well as how they react to the low air alarm and once the alarm discontinues. Students should be instructed to stop walking, get low, and finish breathing air from the SCBA. If the bypass valve releases air, the student did not complete the drill. The goal here is to become familiar with SCBA operations and to completely breathe down an air cylinder to become familiar with emergency operations.

The third version is to have the students reduce air from within their SCBA to 1000 pounds. All students should be packed out and sitting down. The key is to show the students how long 1000 pounds of air can last a fire fighter if needed. The drill involves using various breathing techniques and learning to minimize the air coming out of the SCBA. The student should practice cracking the valve slightly (but enough to take a breath) so that the Heads-Up Display (HUD) or personal alert safety system (PASS) device does not activate. The key here is to show the students how long an air cylinder can last if a fire fighter becomes trapped, but remains calm during that time.

SCBA Familiarization Drills

Once a fire fighter becomes comfortable with gear acclimation and the use of their SCBA, the complexity of the drills should increase. The next drills are simple in concept, but allow the fire fighter to gain confidence in their skills and equipment. One of the greatest advantages that we as fire fighters have is the ability to train beyond the minimum standards.

■ Hot Cylinder Swap

The next step is to perform a hot cylinder swap. This means starting with all PPE on, including face piece and breathing air. The fire fighter should time it correctly to turn off the valve, disconnect the cylinder-holding strap, take a breath, and quickly disconnect the air cylinder from the pack. Replace this cylinder with a fresh cylinder, while holding your breath and slowly exhaling. After the cylinder is reconnected, turn on the valve and take a breath. Reconnect the cylinder-holding strap to complete the drill (**FIGURE 17-6**). As the fire fighter becomes more comfortable with this drill, it should be performed with gloves on and then blindfolded. Once fire fighters have

FIGURE 17-6 A drill to practice swapping out the cylinders.
Courtesy of Brian Ward

accomplished these drills, they should feel completely comfortable with their skills and equipment. Fire fighters should also feel comfortable with their body's reaction to limited air and the stress that is created. Other areas to consider include what actions should be taken if something happens to the air supply and it cannot be reconnected. With several SCBAs, a fire fighter can place the airline into their coat and filter breath; it is difficult, but possible. Also, a discussion should be had regarding carrying an extra hood or a crown royal bag, which tends to fit well into certain face pieces.

Maze and Confidence Builders

Mazes come in many different forms and, along with entanglement drills, are typically considered confidence builders and should be practiced in non-immediately dangerous to life or health (non-IDLH) environments. Depending upon the maze design, these drills may be set up for an individual or a team (**FIGURE 17-7**).

FIGURE 17-7 A maze drill.
Courtesy of Brian Ward

As with many drills, this can be set up as a progressive level of drills: first, the fire fighter simply wears all of their PPE and SCBA; second, the fire fighter's mask is blacked out or simulated smoke is added; and third, heat would be added. This drill can also be performed in the station without taking equipment out of service or requiring tremendous resources.

Students are brought into the maze and placed on either a right- or left-hand search. The students are then instructed that this is an IDLH/hostile environment, and they need to evacuate the structure. They should practice all of the knowledge and skill sets that they have learned up to this point. The maze drill is designed to teach them how to negotiate through different areas that they may encounter in a structure, and they should be feeling for egress points, holes, and victims. If they fall, become entrapped, or if their low air alarm activates, they should go through their Mayday procedures. Students should remember to practice the use of LUNAR (Location, Unit, Name, Air, Resources).

For simple disorientation drills inside the station, place a student packed out with wax paper inside of their face piece. Their task is to find the where the PASS alarm activation is located and bring it to safety at a point of egress. This drill works well for team drills—the key point being to not become disoriented in a structure, even while attempting a rescue. The student should develop a plan to stay in contact with a landmark or the wall while attempting this rescue (**FIGURE 17-8**). Simple enough, but it could prove difficult for even the experienced fire fighter. Simply blacking out a fire fighter's mask and adding a rescue manikin can be quite stressful if it is not practiced.

FIGURE 17-8 Fire fighters performing a right-hand search following the wall and identifying the window as an egress point.
Courtesy of Brian Ward

Disentanglement/Memphis Drill

Drill Requirements and Set Up

The disentanglement or Memphis drill requires an open floor area where entanglement/collapse props can be constructed. The materials needed include rope and accessory cord in several sizes, survey cord or kite string, and salvage cover or shipping pallets for collapse simulation. To set up for this drill, tie up rope, cord, and string in such a way that fire fighters, using a standard search technique, will become entangled in the loops that are hanging in their paths. Salvage cover or shipping pallets should be arranged to simulate a collapse situation.

This drill requires two instructors, who will ensure that the disentanglement/collapse scenario is properly constructed and that thorough instruction and demonstration are provided to students. Both instructors will serve as safety officers during the drill and will assist students in case of panic or other problems. The instructors will also ensure that all drill objectives are met with safety and efficiency in mind.

Drill Objective

Review the basic information surrounding the 1994 incident at 750 Adams Avenue in Memphis, Tennessee(see **Disentanglement/Memphis Drill Motivation** box), emphasizing the following key points necessary to successfully complete this skill:

- Keep in constant communication with your partner.
- Limit anxiety.
- Make contact with the entanglement.
- Identify the entanglement hazard by feeling resistance early.
- Immediately stop forward motion when resistance is perceived.
- Initiate swim technique until contact is made with entanglement.
- Back up slightly until tension against entanglement is relieved.
- Lift entanglement with hand and proceed underneath.
- Communicate with your partner and pass entanglement to his/her hand.
- If unsuccessful, have partner assist in disentanglement.
- Remove the entanglement.

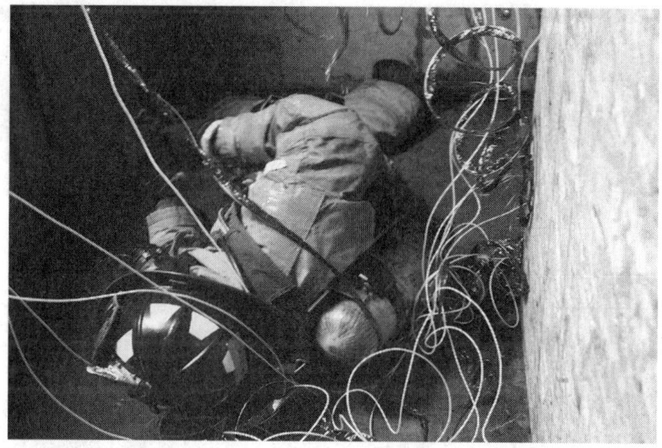

FIGURE 17-9 Fire fighter moving through a disentanglement drill.

Courtesy of Brian Ward

To successfully complete this skill, each student, while remaining calm and maintaining proper communication with other team members, will use the swim technique and direct contact to disentangle themselves from entanglement/collapse props (**FIGURE 17-9**). Students should navigate this drill in two-person teams, and, when finished, they should observe other students attempting the drill.

Disentanglement/Memphis Drill Motivation

On April 11, 1994, a fire broke out at the Regis Towers Apartments at 750 Adams Avenue in Memphis, Tennessee. William E. Bridges, a private assigned to Engine 7, was searching the 9th floor hallway for a fellow fire fighter when he became entangled in fallen wiring. Bridges became trapped by the wiring and was later found unconscious and in respiratory arrest. William Bridges's life was claimed in an open hallway, 10 feet from the stairwell door, by what was later determined to be television cable.

Tool Lab

This drill is designed to show students the pros and cons of different hand lines and nozzles, standpipe operations, and hose extension/burst hose replacement. Students are divided into three groups and rotate through three stations.

- **Station 1: Hoseline and nozzle familiarization.** The hoseline and nozzle station familiarizes

FIGURE 17-10 Students using different size hoselines and nozzles.
Courtesy of Brian Ward

students with the use of different size hoselines (1¾-inch [44-mm], and 2½-inch [65-mm]) and nozzles/streams (fog, straight stream, smooth bore). This is also a teaching station; students should examine the reach; distance, force, nozzle pressure, and handling techniques (**FIGURE 17-10**).

- **Station 2: Loading standpipe packs.** In this station, students learn to load standpipe packs (either standard flat shoulder load or the Cleveland load). Next, they should learn where and how to connect to standpipe systems, at grade or multistory (**FIGURE 17-11**). Empty parking lot decks are a great way to utilize these drills if a training tower is not available.
- **Station 3: Replacing and extending hoselines.** In this station, students learn how to replace burst sections of lines and extend hoselines (**FIGURE 17-12**).

FIGURE 17-11 Standpipe system.
Courtesy of Brian Ward

FIGURE 17-12 Extending hoselines.
Courtesy of Brian Ward

All Inclusive (Non-Fire) Drills

Using an all-inclusive scenario allows the students to practice numerous techniques and teamwork. To start the drill, the students should be broken into engine, truck, and squad companies, depending on the number of students. Simply pulling the apparatus out of the bay and using the fire station for this scenario are the quickest methods for this drill. However, several other options can be used to increase the realism of the drill.

As the students graduate from the basic levels, a live fire scenario can be used inside the burn building. A scenario should be dispatched to the incident commander, who is allowed to run the incident. According to the scenario, ensure all components of the command objectives Rescue, Exposure, Confinement, Extinguish, Overhaul, Ventilation, and Salvage (RECEO-VS) are covered. Based on the new National Institute of Standards and Technology (NIST) and Underwriters Laboratories (UL) research, the components Size-up, Locate the fire, Identify and control the flow path, Cool the fire from a safe distance, Extinguish, Rescue, and Salvage (SLICE-RS) have been developed as tactical objectives. Additional information on SLICE-RS and its application can be found on the International Society of Fire Service Instructors (ISFSI) website.

As the facilitator, the training officer should announce when a simulated apparatus has arrived and is available for assignment by the incident commander. Typically, the first apparatus is assigned as either fire attack or search. Manikins should be placed inside the structure for rescue, and hoselines should be charged for the full effect. Ladders are raised for secondary means of egress and/or ventilation, RICs and secondary search teams are established, and back-up hoselines are placed in service.

As the drill nears the end, the incident commander transfers command to the training officer. This confirms the acknowledgment of accountability and the tracking of crews based on availability. After the scenario is completed, all equipment is broken down and returned to service on the apparatus. The crews come together and discuss what the scenario was, what the assignments were, and how they thought the drill went. In addition to the hands-on training, the discussion around learning from our experiences (right or wrong) is the most important aspect of the training.

After the discussion, all crews should rotate and be assigned to different apparatus, along with a different scenario and incident commander. These drills generally take 30 minutes from dispatch to discussion, and allow everyone to be involved at the same time. If you are a volunteer training officer, these types of drills may deliver the biggest bang for the buck concerning the involvement of personnel and different tools.

Standpipe Drill

This drill is based inside of an empty six-story parking deck. We start with a scenario that simulates a fire inside of a low-rise commercial structure. As the facilitator, you can decide if occupants are still inside or if all are evacuated based on the objectives of the scenario. The students should be divided into groups based on your typical assignments for this type of structure and assign them as appropriate. Other items to consider are simulated damage to the Fire Department Connection (FDC) or the location in which the pumping apparatus hooks to a building's standpipe and or sprinkler system on the outside of the structure. This requires the fire fighter to think about what to do next, and to possibly connect to the first floor standpipe versus the exterior.

During this time, one apparatus is assigned fire attack and ascends the stairwell with a high-rise pack. The third team member in this group should be the "gate man," carrying the high-rise connection bag. This bag contains all the necessary tools, including a 2½-inch (65-mm) wye, elbow, and an inline pressure gauge (**FIGURE 17-13**). The apparatus assigned to water supply should obtain positive water. Communication among all involved is a vital point to emphasize during this drill—especially among the fire attack group, the pump operator, and command. After the first line is deployed, send a second line (preferably a master

FIGURE 17-13 High-rise bag contents include various adapters, spanner wrenches, a mallet, a pipe wrench, a 2½-inch (65-mm) elbow, webbing, and inline pressure gauge.

Courtesy of Brian Ward

stream, not a typical line to pull inside a structure; however, this depends on your scenario). A second high-rise connection bag with an additional 2½-inch (65-mm) hose high-rise pack should be brought to the sixth floor. The portable master stream is assembled and deployed. Considering the volume of water needed to supply the master stream and the 2½-inch (65-mm) hoseline smoothbore, a second 3-inch line is connected to the Siamese on the first floor to help boost volume and reduce friction loss at the standpipe. The Engine 2 driver calculates and adjusts the Pump Discharge Pressure (PDP) and the attack crews extinguish the fire.

Engine 2 Driver's Calculations

80 psi (552 kPA)—Smoothbore Master Stream: 13/8 inch (35 mm)
75 psi (517 kPa)—150ft(46 m) of 2½-inch (65-mm) supply line to the Portable Master Stream
10 psi (69 kPa)—Appliance
25 psi (172 kPa)—Standpipe
25 psi (172 kPa)—Elevation
215 psi (1482 kPa)—Total psi at standpipe

12 psi (83 kPa)—Two 3-inch(76-mm) lines from Engine 2 to Standpipe
227 psi (1561 kPa)—Pump Discharge Pressure (PDP)

There may be several obstacles faced during this scenario. The purpose of this drill is to learn to navigate the obstacles and train for adaptability. After the drill is complete, topics to be discussed by all involved should include what they saw in the drill, what tools would make their job easier, and what the difficulties were. Important points to consider would include engine calculations, communications, and preplans for the various structures inside your territory.

Tactical Decision Games

Tactical Decision Games are timed assessments in which a student reads a scenario and then must answer questions based on his or her evaluation of the situation. Normally a student would be given 2 to 8 minutes to read each of the scenarios below and answer the corresponding questions.

■ Tactical Decision Game #1: RIC Assignment

You are the Lieutenant assigned to E10. E10 is dispatched to a reported structure fire in CO. E14 is first-in district. The fire is located in a restaurant. The fire building consists of a large one-story, Type II, commercial building (**FIGURE 17-14**). The fire is located on side D of the building. E14, E26, and E21 are conducting fire attack operations inside the building. Battalion 3 is the incident commander and assigns E10 RIC responsibilities. Your crew consists of a 4-year fire fighter, a 5-year relief driver, and yourself. Delegate responsibilities as you see fit.

- What equipment will you need?
- What verbal communications will you give to your crew?
- What radio channels will your crew monitor?
- What are your responsibilities as a RIC Group Leader?

FIGURE 17-14 Large one-story building with active fire.
Courtesy of Brian Ward

■ Tactical Decision Game #2: Improvised Explosive Device (IED)

You are a Lieutenant assigned to E7. This area is a heavily congested area with both vehicular and foot traffic. It is 1230 hours when your Engine is dispatched to a rubbish fire. The location of the rubbish fire is on a narrow one-way street. This street is a lunchtime favorite for the many civilian workers in the area. The sidewalks are very wide, with many places to eat and shop. As you respond down this street you observe the rubbish fire in a wire mesh trash can. It is a small fire and your crew goes to work on it. A civilian approaches you and states that the fire is the result of a small explosion in the trash can. Your nozzle fire fighter gets your attention and points out a cell phone strapped to the bottom of the wire mesh trash can. While you are inspecting the cell phone from a distance, you hear two separate minor explosions down the street. Two more trash cans have now burst into flames. There are no injuries to nearby civilians. A few seconds later, a large explosion rips apart what appears to be a parked car. This explosion takes place approximately 200 yards away from you and your crew. You estimate that there are 50 people injured to some degree (**FIGURE 7-15**). There is no additional fire problem.

- List priorities and note concerns for responders.
- Describe civilian safety procedures.
- List additional resources requested.
- Indicate placement of additional resources.

■ Tactical Decision Game #3: Mass Casualty With Triage

Part I

You are a Lieutenant assigned to Engine 14. It is approximately 1200 hours on a Saturday. Engine 14

FIGURE 17-15 Result of an IED in a busy shopping area.
© David Silverman/EdStock/Getty Images/iStockphoto

FIGURE 17-16 A vehicle crashed into a building.

and Medic 14 are dispatched to an "accident with injuries" on Mall of Georgia (MOG) Blvd. MOG is crowded with both vehicular and foot traffic. Temperatures are hovering around 100 degrees. Your fire fighter is a first-house probationary fire fighter and your Engineer is a 10-year veteran. Both units arrive on-scene at the same time. You observe that one car is involved in the accident. The vehicle crashed into the building, and there are many civilian injuries (**FIGURE 17-16**). A citizen reports that there were approximately 25 individuals in the area of the building hit by the ear.

- Discuss what individual assignments you will give to your members.
- What are your expectations of Medic 14?
- List additional resources requested.
- Indicate placement of additional resources.

Part II

Twenty-two victim injuries have been identified through triage and are listed below. These were the only victims who could be extricated during the first operational period due to the instability of the structure. Technical Search and Rescue are on-scene clearing debris and searching for additional victims.

33-year-old pregnant female with tibia/fibula fracture
10-year-old male with small laceration to forehead
17-year-old female complaining of headache
19-year-old male with crushed pelvis
78-year-old female (A×3, blood coming from nose and ear)
52-year-old female (A×0, not breathing, no pulse)
55-year-old male in cardiac arrest
2-year-old female with difficulty breathing (10 breaths per minute)
34-year-old male complaining of chest pain (pain = 8)
9-month-old male infant (possible loss of consciousness, 20 breaths per minute)
13-year-old male with profuse bleeding from left arm.
Five victims have walked up to you with minor complaints wishing to be transported.
Six victims have been declared dead by an on-scene doctor.

- How will you assign the resources requested in Part I with the additional patient information?
- List any additional resources requested.
- Describe additional considerations while operating at this scene, including National Incident Management System (NIMS) requirements for extending into a second operational period (24-hour operational periods).

Multiple Training Scenarios

The next several scenarios were conducted at the same time during the Gwinnett County (Georgia) Leadership and Safety Conference. In order to pull these off at the same time, a Type 3 Incident Management System was built. Due to the high hazard of some of these drills, the following positions were filled: Incident Commander, Safety Officer, Assistant Safety Officer at each site, Lead Instructor, and Assistant Instructors. A 5-to-1 student-to-instructor ratio was implemented as well. All standard PPE was required, along with additional PPE for some drills, and safety systems were built for the safety and survival techniques, such as a belay for the fire fighter bail out. During this conference, there were four drills being conducted simultaneously. These drills were the heavy lifting and extrication, vehicle extrication, safety and survival techniques, and hoseline operations.

The planning process for all of these drills began approximately 8 months prior to the conference. This timeline is very important, as you will see in the first drill, where a school bus was acquired. This required contact with the local school board and waiting for a school bus to be taken out of service. Thankfully, the local school board came through about 1 month before the conference, with the only stipulation of returning it so that they could use the scrap metal. In addition, several vendors that had the tools needed for these drills were contacted and the vendors customized a class, not to showcase their tools, but to deliver a training objective. As the training officer, bringing in outside ideas and techniques allows fire fighters to walk away with more tools in the toolbox. However, setting the tone early is imperative so that the training remains just that; most vendors understand that it is a promotion of their tools and will donate the time and tools to help teach the class.

All props were delivered 2 weeks prior to the drill date, and all instructors arrived 1 day early with equipment to reposition props in the proper orientation for the scenarios. For this, a local heavy rescue tow truck company supplied a heavy-duty rescue tow truck and skid steer. The drill sites were designed as a several-evolution rotation, but the class was purposefully kept to a limit of 10 to 15 students, so that all received ample hands-on training time. As well, most of these scenarios were designed as all-hands evolutions.

Vehicle Versus Bus Extrication

The first scenario consisted of a vehicle under a bus (**FIGURE 17-17**). This is a relatively simple drill, if you have the appropriate tools to lift and crib the bus. Once this is complete, extrication of the patient or removal of the vehicle to allow extrication to be completed should be accomplished. One key is to stabilize the bus and vehicle and not allow fire fighters under the bus to extricate the vehicle first. This drill was designed to have the fire fighters focus on the various tasks and to learn the tools that they were using for specific task. As the day progressed, scenarios became more complex.

Vehicle Versus Tanker Extrication

The second scenario was a vehicle that had flipped over and landed under an 18-wheeler tanker (**FIGURE 17-18**). The first objective is for the students to decide what hazard the tanker contains by locating the placards or any other means of identification. Because the objective of this training was not based on hazardous materials but lifting techniques, we decided that it was an empty milk truck; however, we wanted the students to go through the mental process of determining any potential hazardous situations. The keys to success in this scenario are teamwork, communication, correct lifting techniques, adequate cribbing, and the safety of crew members. As the fire fighters started extrication on the vehicle, stabilizing the tanker was extremely important. One of the concerns with this process was that as we compromised part of the vehicle by removing doors and other structural components, the weight of the tanker could possibly collapse the vehicle further. In order to stop this process, the tanker was stabilized by chocking the wheels and ensuring that the brakes were applied, after which the suspension system in the tanker was cribbed (**FIGURE 17-19**). As this was completed, extrication began on the vehicle by one crew and lifting of the tanker was initiated by a second crew.

FIGURE 17-18 A flipped vehicle under an 18-wheel tanker scenario.

Courtesy of Brian Ward

FIGURE 17-17 A vehicle under a bus scenario.

Courtesy of Brian Ward

FIGURE 17-19 Cribbing and air bags placed under the tanker.

Courtesy of Brian Ward

FIGURE 17-20 Vehicles in a clamshell collision.
Courtesy of Brian Ward

FIGURE 17-22 Rescue struts are used as a secondary safety to support the weight of the rear of the vehicle.
Courtesy of Brian Ward

■ Vehicle Head-On Collision (Clam Shell) Extrication

The third scenario involves two cars colliding head-on, with one vehicle landing on top of the other (**FIGURE 17-20**). This is a very hazardous drill for students if they do not pay attention to details. The objective is to safely secure and stabilize the vehicles in a clamshell collision while extricating the victim. The first part of this scenario is to locate and assess the patients, and the second part is to stabilize these vehicles and develop a plan for extrication. While one crew is placing cribbing and rescue jacks in place, the second crew should be developing an action plan. After the rescue jacks are set, the vehicles are married together to further reduce slipping in either direction (**FIGURE 17-21**). For this, heavy-duty (10,000 lbs [4536 kg]) ratchet straps were used and placed in the locked position.

Relief points were then cut into each side of the floorboard in order for the "clam shell" operation to operate properly—similar to rolling a dash. Due to the damage in rear of the vehicle and limited access through the rear, it was decided to raise the rear of the vehicle. First, the catalytic converter had to be removed and folding points were cut into the floorboard. Second, the C-post was cut and secured with a Come-A-Long to the opposite side of the vehicle. Rescue struts should be used as a secondary safety to support the weight of the rear of the vehicle (**FIGURE 17-22**). A Come-A-Long was secured to the axle, and cribbing placed under the cable in order to keep if off of the vehicle. After this, the C-post was cut on each side; the Come-A-Long stabilizes the vehicle in preparation for the C-post cuts. As the Come-A-Long is ratcheted, the rear of the vehicle closes down onto itself (**FIGURE 17-23**). During this process, all nonessential

FIGURE 17-21 Vehicles are married together to further reduce slipping in either direction.
Courtesy of Brian Ward

FIGURE 17-23 As the Come-A-Long is ratcheted, the rear of the vehicle closes down onto itself.
Courtesy of Brian Ward

personnel should stand clear and for those remaining, all PPE should be double-checked. The remaining metal floorboard still attached causes this to be a slow process, as it does not simply break over onto itself. This process provides two advantages: (1) there is full exposure to the car on top, and (2) it reduces the hazard by lowering the center of gravity.

Bus Rollover With Heavy Tow Wrecker

The fourth and final drill is an all-hands scenario with the assistance of a heavy tow wrecker to perform extrication on a standard school bus that has rolled over (**FIGURE 17-24**). The students' main objective is to safely extricate victims involved in a school bus using all possible resources. The first priority is to ensure that the bus is stabilized. Due to the orientation of the bus, this was already taken care of. Next, the students should identify any hazards, such as ignition potential and fuel spills. After the hazards are identified, victims should be assessed as best as possible, and any victims who can be moved or extricated with minimum effort should be removed. Any other victims should be covered or protected from falling debris.

Making entry through the back door could be a viable option; however, in this scenario, the rear door is not an option due to a second vehicle colliding with the back door. We wanted the students to think outside the box and use the resources available. The second option for entry was to roll the roof. Using the heavy tow cables as a stabilizer, the students cut all connection points along the rear and roof with standard cutters. If this process is used, two sets of cutters will expedite this process. After the roof is rolled, full access to the entire interior of the bus is gained (**FIGURE 17-25**).

FIGURE 17-24 School bus rollover accident scenario.
Courtesy of Brian Ward

FIGURE 17-25 Once the roof of the school bus is rolled, full access to the entire interior is gained.
Courtesy of Brian Ward

Important points to emphasize include communication, foot and body placement, and overall safety.

Bailout Drills

Several bailout drills can also be conducted during training. The drills discussed here are the ladder bailout, the hoseline bailout, and the rope bailout. The common thread among all of these drills is the use of a belay safety system (**FIGURE 17-26**). The belay is a vital point to ensure when performing these drills. Fire fighters should be taught to stay low and remain calm. The objective for all these drills is to bailout of a building, simulating a last chance effort of survival using experienced personnel and applicable safety systems.

Ladder Bailout

In the ladder bailout drill (**FIGURE 17-27**), the fire fighter needs to remain low, use one hand to guide himself

FIGURE 17-26 A belay safety system for the fire fighter bailout drill.
Courtesy of Brian Ward

out the window head first, and stabilize himself at the second or third rung. Using the opposite arm (with that hand in a fist), hook your arm at the elbow under the second or third rung—do not grab the rung. Once your elbow is hooked, rotate your body in the opposite direction, and slide down the ladder with your feet on the outside of the rungs.

Hoseline Bailout

In the hoseline bailout drill, the fire fighter needs to stay low and remain calm. Slowly ease yourself over the window, wrapping one leg in front and the other leg around the back of the hose (**FIGURE 17-28**). Once your feet are out, squeeze the hoseline to your chest. In order to slow down more, squeeze the hoseline harder to create additional friction.

Rope Bailout

In the rope bailout drill, the fire fighter needs to stay low and remain calm. Slowly ease yourself out over the window, and as you are doing this, a tool or some object should be in your inside hand. Your life safety rope should be secured to the tool and the tool placed in the bottom corner of a window. The rope should follow the corner straight over the ledge. The fire fighter should have the rope wrapped around their backside with one end of the rope in each hand (**FIGURE 17-29**). As the fire fighter eases out, bringing the ropes tight and close to the chest will slow down the descent, whereas opening up the arms will increase the speed of the descent.

FIGURE 17-27 Setup for the ladder bailout drill with an anchor below and the belay system above.
Courtesy of Brian Ward

FIGURE 17-28 The hoseline bailout drill with belay system anchored above.
Courtesy of Brian Ward

FIGURE 17-29 The rope bailout drill with belay system.
Courtesy of Brian Ward

Various Fire Fighter Packaging and Retrieval Methods

Fire fighter packaging and retrieval methods should also be practiced during training. Drills for packaging and retrieving a downed fire fighter discussed here include the 2:1 mechanical advantage drag, and two different methods for moving a downed fire fighter up or down a flight of stairs.

■ Creating a 2:1 Mechanical Advantage

For this drill, students will need an adequate length of rope with two carabiners. One carabiner is attached to yourself (you become the anchor), and the second carabiner is attached to the fire fighter (typically the air pack) (**FIGURE 17-30**). The student should use friction or a doorway, if possible, to brace himself for the pull (**FIGURE 17-31**). This method is used when extra strength is needed to remove a downed fire fighter.

FIGURE 17-30 The 2:1 mechanical advantage drag utilizing rope and two carabiners.
Courtesy of Brian Ward

■ Moving a Downed Fire Fighter Up Stairs

The drill to move a fire fighter up stairs requires two students and at least one piece of webbing looped together, typically with a water knot. The first part of this lift is to package the downed fire fighter by placing the waist strap between their legs. This will keep the air pack from sliding up and displacing the face piece. The student at the head of the downed fire fighter loops the webbing over his or her back and then places the opposite loops under the downed fire fighter's air pack (**FIGURE 17-32**). This allows the student at the downed fire fighter's head to use their legs instead of back or

FIGURE 17-31 The student bracing himself to pull the downed fire fighter.
Courtesy of Brian Ward

FIGURE 17-32 Looping the webbing under the downed fire fighter's air pack.
Courtesy of Brian Ward

arms to lift under the downed fire fighter. If webbing is unavailable, the student at the downed fire fighter's head can loop his or her waist belt through the downed fire fighter's shoulder straps in order to assist in the lift. The second student then gets as low as possible to the stairs and places the legs of the downed fire fighter over his or her shoulders and provides an up and forward movement. With limited help, students should use the handrails, if available, to assist during the lift (**FIGURE 17-33**).

FIGURE 17-33 Using handrails to assist during the stair lift.
Courtesy of Brian Ward

■ Second Downed Fire Fighter Carrying Method

The second fire fighter carrying method described here can be used to carry a downed fire fighter up or down stairs or over debris spread across the floor. This requires that the downed fire fighter is wearing his SCBA and that the RIC is using two pike poles. One pike pole is placed through the downed fire fighter's shoulder strap and waist strap, and then the downed fire fighter's leg is placed over the pike pole. This process is repeated with the second pike pole on the opposite side. Both students should coordinate the lift together, lifting with their legs and not their back (**FIGURE 17-34**).

FIGURE 17-34 Carrying a downed fire fighter up stairs using pike poles.
Courtesy of Brian Ward

Team Drills—Shift Challenge Day

Various team drills can be used in a friendly competition. In this example, the shifts are selected via draft at the end of the course by the Engine Officers: A, B, and C.

Round 1: Academic Quiz

- Ten random questions
- One shift member
- Classroom

1st: 7 points 2nd: 5 points 3rd: 3 points

Round 2: Hose-Roll Race

- Straight roll/donut roll
- Two shift members
- Station bay

1st: 5 points 2nd: 4 points 3rd: 3 points

Two hoses will be laid out from one end of the bay to the other. The first fire fighter will begin at one end of the bay and will roll the hose in a straight roll. Then, the fire fighter will carry the hose to the opposite end of the bay, at the beginning end of the other hose, where fire fighter two is waiting. Fire fighter one tags in fire fighter two, who then does a donut roll. Both then advance to the end where fire fighter one begins and the first team done wins.

Round 3: Rapid Dress

- First finisher wins
- One shift member
- Station bay

1st: 5 points 2nd: 3 points 3rd: 2 points

Round 4: Knots Shuttle Relay

- Five different knots
- Five shift members
- Station bay

1st: 7 points 2nd: 5 points 3rd: 3 points

Stage 1: Clove hitch on an object horizontal
Stage 2: Bowline around an object
Stage 3: Figure 8 on a bight
Stage 4: Connect two ropes together
Stage 5: Clove hitch around a vertical object

Round 5: Salvage Cover Race

- First finisher wins
- Two shift members
- Station bay

1st: 5 points 2nd: 3 points 3rd: 2 points

Both fire fighters must demonstrate folding a salvage cover for a one fire fighter roll as done per skill sheet.

Round 6: Rapid Dress Relay Race

- First team to finish together wins
- All shift members
- Outside the bay area with marked (coned) areas, or cordoned off enough room for all students.

1st: 10 points 2nd: 7 points 3rd: 4 points

Continued

Round 7: Kieser Sled Race

- Timed event
- Two shift members
- Fireground simulation or on the drill ground

 1st: 7 points 2nd: 5 points 3rd: 3 point

One fire fighter drives the sled all the way to the end of a marked area, such as 10 feet, and then fire fighter two drives it back. Best team time wins.

Round 8: Hoist Equipment Drill

- Timed event
- Three shift members
- Fireground simulation or on the drill ground

 1st: 5 points 2nd: 3 points 3rd: 2 point

Fire fighters start on the ground. One advances up to the third floor with the rope and lowers it to the other two fire fighters on the ground. There, the chain saw is attached and tag line put in place. The fire fighter then raises the saw, unties it, and carries it to the third floor door, where the judge calls time. Best time wins.

Round 9: High Rise Pack Carry

- Timed event
- One shift member
- Fireground simulation or on the drill ground with any set of stairs that equals six floors.

 1st: 5 points 2nd: 3 points 3rd: 2 point

Fire fighter must ascend all six floors and then return. Fire fighter must hit every step on the way down.

Round 10: Dummy Drag Race

- Timed event
- One shift member
- Mark off a predetermined distance, such as 50 feet (15 m), with start and stop lines.

 1st: 4 points 2nd: 3 points 3rd: 2 point

Fire fighter must demonstrate large dummy drag on the "track" for two laps.

Round 11: Hose Pull Race

- Timed event
- Three shift members
- Mark off a predetermined distance, such as 100 feet (30 m), with start and stop lines.

 1st: 10 points 2nd: 7 points 3rd: 5 points

Fire fighters must all start at the determined starting point. All must move hose to the designated area without running. The lead and backup fire fighters must be in proper place before the flow of water begins. The team must knock down three cones. Time stops when the last cone is knocked down.

Round 12: Roof Ladder Deployment

- Timed event
- 4 shift members
- Ventilation simulation on roof of the fire station.

 1st: 7 points 2nd: 5 points 3rd: 3 points

Fire fighters must demonstrate this as it was shown in class. The fourth fire fighter will climb the ladder and place the roof ladder. Time will stop when the judge determines the ladder is properly in place.

Continued

Round 13: Burst Hose Drill

- Timed event
- All shift members
- On the drill ground

1st: 10 points 2nd: 7 points 3rd: 5 points

All fire fighters start from a designated starting position. Fire fighters must demonstrate the replacement of a burst hoseline properly as demonstrated in class. Time ends when the hoseline is flowed with a proper backup and a cone is knocked down.

Round 14: Hose Deployment Drill

- Timed event
- Four shift members
- Mark off a predetermined distance, such as 150 feet (46 m), with start and stop lines.

1st: 7 points 2nd: 5 points 3rd: 4 points

All fire fighters start from a designated starting position. Fire fighters must unroll, place appliances, and completely extend 150 feet (46 m) of 1¾-inch (44-mm) hoseline to knock down three cones at different distances. Time ends when the hoseline is flowed with a proper backup and all cones are knocked down.

Round 15: Hose Load Drill

- Timed event
- Three shift members
- Engine or a simulated hose bed.

1st: 5 points 2nd: 4 points 3rd: 3 point

Fire fighters will demonstrate an accordion hose load (200 feet [61 m] of 3-inch [76 mm] line). Time stops when judge determines the hose load is complete.

Round 16: ***Bonus Event*Race**

- One shift member
- Drill ground

1st: 10 points 2nd: 7 points 3rd: 3 points

Fire fighter will complete the five different events for time. Fire fighter will start by crawling through the low crawl simulator, then drag the dummy for one lap, drive the keiser sled from end to end, carry the hose pack to the top of the tower, and raise the rolled hose up the side of the tower. The time stops when the fire fighter exits the interior stairwell at the base of the tower after completing all drills.

Round 17: Sprinkler Stop

- Timed drill
- One shift member
- Drill ground

1st: 5 points 2nd: 4 points 3rd: 3 point

Fire fighter starts from a designated starting area. Fire fighter then advances to the charged sprinkler and "plugs" it. Time ends when the fire fighter returns to the designated starting area.

Round 18: Bucket Brigade

- Timed drill
- All shift members
- Drill ground

1st: 20 points 2nd: 17 points 3rd: 15 points

Continued

Move water from one point to another with only the use of buckets. Distance of 20 to 25 yards (18 to 23 meters) is utilized. Barrel must be filled to designated marking point and time is called by the barrel judge.

Shift Wars Score Sheet

Shift Wars Score Sheet	**A**	**B**	**C**
Round 1 Academic Quiz (7/5/3)	______	______	______
Round 2 Hose Rolls (5/4/3)	______	______	______
Round 3 FF Rapid Dress (5/3/2)	______	______	______
Round 4 Knots Shuttle Race (7/5/3)	______	______	______
Round 5 Salvage Cover (5/3/2)	______	______	______
Round 6 Rapid Dress Relay (10/7/4)	______	______	______
Round 7 Kieser Sled Relay (7/5/3)	______	______	______
Round 8 Tool Hoist (5/3/2)	______	______	______
Round 9 High Rise Pack (5/3/2)	______	______	______
Round 10 Dummy Drag (4/3/2)	______	______	______
Round 11 Hose Pull (10/7/5)	______	______	______
Round 12 Ladder Deployment (7/5/3)	______	______	______
Round 13 Burst Hose (10/7/5)	______	______	______
Round 14 Hoseline Advancement (7/5/4)	______	______	______
Round 15 Hose Load (5/4/3)	______	______	______
Round 16 **Bonus Event** (10/7/3)	______	______	______
Round 17 Sprinklers (5/4/3)	______	______	______
Round 18 Bucket Brigade (20/17/15)	______	______	______
TOTALS			

Key Points

- Training is key to any fire fighter's success and should be a staple in their daily routine.
- Any time fire fighters are training, safety must be prepared for, and all hazards mitigated prior to starting the activity.
- Challenge yourself and your team by trying different drills. New drills will engage fire fighters and keep them interested.
- With all drills, be sure to provide a post-training critique to identify areas of improvement, however also discuss the areas where your team succeeded.
- Participate in various training sessions outside of your own department. This will introduce you to new ideas, experiences, and opportunities. While not all training may be relevant for every department, there will always be tips and tricks that can be incorporated.

References

Bombardiere, B. and R. Rau. *Conquering the Entanglement Hazard.* Fire Engineering. May 1996. http://www.fireengineering.com/articles/print/volume-149/issue-5/departments/training-notebook/conquering-the-entanglement-hazard.html. Accessed October 30, 2014.

Chubb, M. and J. E. Caldwell. *Tragedy in a Residential High-Rise, Memphis, Tennessee.* Fire Engineering. March 1995. http://www.fireengineering.com/articles/print/volume-148/issue-3/features/tragedy-in-a-residential-high-rise-memphis-tennessee.html. Accessed October 30, 2014.

National Fire Protection Association (NFPA) 1403, *Standard on Live Fire Training Evolutions.* 2012. http://www.nfpa.org/codes-and-standards/document-information-pages?mode=code&code=1403. Accessed October 22, 2014.

Tennessee Task Force One. http://www.tntf1.org/. Accessed October 30, 2014.

Ward, B. *Quick Training: Standpipe Drill.* Fire Engineering. January 1, 2012. http://www.fireengineering.com/articles/print/volume-165/issue-1/departments/training-notebook/quick-drill-standpipe-training-p1.html. Accessed October 22, 2014.

APPENDIX A

Sample Training Schedules and Lesson Plans

The following is a sample training schedule based upon typical requirements of the Insurance Safety Office and most state OSHA organizations. These requirements are based upon data compiled in 2010 and are subject to change at any time. This schedule is based upon a combination and/or volunteer department meeting once a week and one weekend a month; however, this can be easily modified to accommodate career departments.

- This schedule is based upon meeting one night a week for 3 hours of training and one Saturday a month for 8 hours.
- A recruit academy or specific recruit training still needs to be held for all new members. To meet ISO requirements this should be 240 hours of training. A sample recruit academy schedule follows this one in Appendix B.
- Additionally, several of these trainings should be conducted with your mutual aid partners to obtain familiarization with their operations and to obtain full ISO credit.

Sample Training Schedule

Date:	Class:	Requirement Filled:
Jan 4	Fire Prevention	OSHA-ISO
Jan 11	Blood-Borne Pathogens	OSHA-ISO
Jan 14	Fire Fighter Safety	ISO
Jan 18	HAZMAT Refresher – Decontamination Drill	OSHA- ISO
Jan 25	Fire Department Communications	ISO
Feb 1	Hearing Protection	OSHA-ISO
Feb 8	Officer Training – NIMS	ISO-Grant
Feb 15	CPR – AED Part 1	ISO

Continued

Date:	Class:	Requirement Filled:
Feb 18	Public Fire Education	ISO
Feb 22	CPR – AED Part 2	ISO
Mar 1	Power Operated Hand Tools	OSHA-ISO
Mar 8	Apparatus Familiarization/Operational Training	OSHA-ISO
Mar 15	Fire Alarm Systems	ISO
Mar 18	Structure Fire Drill – Night Drill	ISO
Mar 22	Fire Sprinklers	ISO
Mar 29	Cause and Origin Familiarization Structure	ISO
Apr 5	Accident Prevention	OSHA-ISO
Apr 12	Fire Behavior	ISO
Apr 15	Building Construction	ISO
Apr 19	Confined Space Operations – Drill	OSHA - ISO
Apr 26	Personal Protective Equipment	OSHA - ISO
May 3	Search and Rescue	ISO
May 10	Ventilation	ISO
May 17	Rehab	ISO
May 20	Wildland Safety Class – Refresher	ISO - Local
May 24	Wildland Fire Suppression – Drill	ISO - Local
May 31	Wildland Fire Cause and Origin Refresher	ISO - Local
June 7	Ladders	OSHA-ISO
June 14	Forcible Entry – Tools	ISO
June 17	Apparatus Familiarization/Hydraulics	OSHA-ISO
June 21	Live Fire Training – Multi Company Drill	ISO
June 28	Forcible Entry – Techniques	ISO
July 5	Fire Streams/Hydraulics	ISO
July 12	Ropes and Knots	ISO
July 19	Natural Disaster Preparedness	ISO
July 22	Natural Disaster – Multi Company Drill	ISO
July 26	Fire Control	ISO
Aug 2	Water Supply	ISO
Aug 9	Extrication – Classroom	ISO
Aug 16	Vehicle Fires – Classroom	ISO
Aug 19	Extrication/Vehicle Fire – Multi Company Drill	ISO
Aug 23	Salvage and Overhaul	ISO
Aug 30	LODD – Safety Lessons Learned/Remembered	ISO

Continued

Date:	Class:	Requirement Filled:
Sept 6	Apparatus Familiarization/Operational Training	OSHA-ISO
Sept 13	Calling the Mayday/Beginning RIT Ops	ISO
Sept 16	Structure Fire Drill-Multi Company Night Drill	ISO
Sept 20	RIT Team Operations	ISO
Sept 27	Fire Hose	ISO
Oct 4	Fire Hose Practicals	ISO
Oct 11	Officer Training – Pre-Incident Planning	OSHA-ISO
Oct 18	Fire Protection – PPE	OSHA-ISO
Oct 21	Basement Fires – Drill	ISO
Oct 25	Cold Weather Operations	ISO
Nov 1	Fire Extinguishers	OSHA-ISO
Nov 8	Officer Training – Size up	ISO
Nov 15	Toxic Substances in the Workplace	OSHA-ISO
Nov 18	Winter Driving Operations – Drill	ISO
Nov 29	Foam Fire Streams	ISO
Dec 6	SCBA/Respiratory Protection	OSHA-ISO
Dec 13	Apparatus Familiarization/Emergency Driving	OSHA-ISO
Dec 16	SCBA/Search and Rescue – Drill	OSHA-ISO
Dec 20	Hand Tools	ISO

Sample Recruit Schedule

Date:	Class:
Mar 7	Orientation/By Laws and Equipment Checkout/Fit Testing
Mar 9	SOPs
Mar 11	**Nutrition/Physical Assessment**
Mar 14	Communications
Mar 16	Fire Prevention – Home Safety
Mar 21	Public Fire Education
Mar 23	Fire Alarms
Mar 25	**PPE – SCBA Begin Practicals**
Mar 28	Fire Sprinklers
Mar 30	Fire Behavior
Apr 4	Building Construction
Apr 6	Search and Rescue
Apr 8	**PPE-SCBA Confidence Course and Evaluations**
Apr 11	Search and Rescue Continued
Apr 13	Ventilation
Apr 18	Forcible Entry Tools
Apr 20	Fire Hose
Apr 22	**Fire Hose Practicals**
Apr 25	Forcible Entry Procedures
Apr 27	Wildland Fires
May 2	Vehicle Fires – Classroom
May 4	Portable Fire Extinguishers
May 6	**Fire Streams/Hydraulics**
May 9	Ropes and Knots
May 11	Ropes and Knots Practical
May 16	Fire Control – Classroom
May 18	Water Supply
May 23	CPR/AED
May 25	CPR/AED Continued
May 27	**Vehicle Extrication**
May 28	**Vehicle Extrication**
May 30	Salvage and Overhaul
June 1	HAZMAT Awareness
June 6	HAZMAT Operations
June 8	HAZMAT Operations Continued
June 10	**HAZMAT Operations Practical**
June 13	Ground Ladders
June 15	LODD Presentation
June 20	Calling the Mayday
June 22	RIT Training
June 24	**Live Fire Evolutions**
June 27	Test Review
June 29	Fire Fighter One Written Exam
July 1	**Live Fire Evolutions**
July 6	Practical Skills Evaluation
July 11	Basic First Aid
July 13	Basic First Aid
July 15	**Live Fire Evolutions – Night Drill Propane Fires**
July 18	Fire Prevention Presentations
July 29	**Live Fire Evolutions Including Vehicle Fires**

Notes on the above schedule:

- This is an average of 2 nights a week and every other Saturday.
- This is a big time requirement, and all recruits should be advised of it, prior to enrolling on the fire department. It is, however, a course they may only need to take once in their careers, as a complete course (refresher training is necessary throughout a firefighters career).
- This does fulfill the ISO requirement of 240 hours of new recruit training. Additionally it covers the NFPA requirements of training prior to conducting live fire evolutions. If personnel complete the course, prior to operating on the fireground it will cover that NFPA requirement as well.
- It also leaves the second half of the year open for other specialized classes, such as EMT, Basic Emergency Care Course, Apparatus Driver Operator, or Officer Training, that your agency may want to offer.
- Each weeknight class is scheduled for 4 hours, and a recommended start time is 6:00 PM.
- Each weekend class is scheduled for 8 hours and a recommended start time is 8:00 AM.
- Items that are bold indicate a weekend session.

Sample Lesson Plan Summary

Lesson Plan Summary

Course Title: ____________________

Lesson Title: ____________________

Instructor(s): ____________________ Time: ____________________

Type of Lesson: ____________________

Instructional Methods:

____

____

____

References:

____

____

____

Training Aids - Equipment, Supplies, Media, etc.

Preparation:

____

____

____

Assignment:

____

____

____

Objectives:

1.

2.

Sample Lesson Plan

Fire Prevention Division Fire Safety Curriculum

Lesson Plan 2.1

Objectives:

The student will understand what constitutes a tool.

- The student will understand that some tools are only safe for a grown-up to use.
- The student will understand that matches and lighters are "Adult only Tools."
- The student will tell a grown-up about accessible matches and lighters.
- The student will understand what constitutes fireworks.
- The student will be able to identify reasons for not using fireworks.
- The student will be able to identify the safe way to enjoy fireworks.

Time: 30 Minutes

Materials Needed: [PREPARATION]

Picture of Matches and Lighter
Fireworks Pictures
Sample Fireworks
Handouts - Match is a tool - Fireworks
Parent Letter
Chalkboard

Activity: Approx. Time:[PRESENTATION]

1. Begin discussion by defining the word tool.

   Ask the class to name examples of simple tools. Write the examples on board. Mention that some tools are only for use by grown-ups. Now ask for examples of tools that are for grown-ups only. 5 Minutes

2. Hold up pictures of matches and lighters and ask what they are. Tell the students they are also tools - tools that are for grown-ups only.

   On the chalkboard create two columns GOOD and BAD.

   Ask the students to name some of the things, both good and bad, that these "Tools" are used for. Write the students' responses in the appropriate column. 5 Minutes

3. Ask the students if they have any younger brothers or sisters. Ask what is the best way to prevent them from playing with adult tools like matches or lighters. Lead their answers to: Keeping these things out of reach and out of sight.

   Now ask what the student should do if they find matches or a lighter.

   Stress that they should NOT pick them up. They should tell a grown-up.

   Ask who are some of the grown-ups they might tell. Write list on the board. 5 Minutes

4. Tell students you are going to show them another picture. Hold up pictures of fireworks. Ask if any of the students have used these kinds of fireworks in the past. Show the sparkler and other fireworks items. Ask students if these things are also fireworks.

   Explain that all of these things are fireworks and can be very dangerous. 5 Minutes

5. Ask if any of the students have been hurt by fireworks or knows of someone who has. After several students give examples, discuss two recent case histories.

   1. A fireworks plant explosion in Ohio killed 2 workers.
   2. Two young boys tossed a firecracker into a tent to scare a friend. The tent caught fire and the friend died

   Ask students where fireworks usually hurt people. Knowing this, ask students "what is the ONLY safe way is to enjoy fireworks. WATCH THEM AT A DISPLAY WHERE PROFESSIONALS SET THEM OFF." 8 Minutes

6. Distribute hand-out material. Ask students to take them and discuss them with their parents. 2 Minutes **[APPLICATION]**

Evaluation: [EVALUATION]

Classroom Discussion
Response to Question

Sample Training Safety Plan

This plan shall be followed in its entirety. Deviation from this plan has the potential to cause injury to the course participants. As a supervisor you have the responsibility to take every precaution reasonable in the circumstances for the protection of a worker (student). This plan shall work in conjunction with the training / teaching plan and lesson plan.

■ Instructors:

Instructors are selected using the following criteria:

Trainer Facilitator or equivalent
Minimum five years in the fire service
Completed similar training and competent to deliver training on this subject

■ Training Safety Officer (TSO):

The Training Safety Officer is qualified due to a firm understanding of the subject and the knowledge, training and experience to organize this work and its performance. The TSO will be assigned to overall safety of the area, all procedures, and any prop use. The TSO shall be known by all participants in the training and will review this plan and update if necessary. The TSO will conduct a Safety Briefing prior to the commencement of the training, providing any warnings and potential hazards.

Classroom / Lecture Safety Concerns:
Training Site Safety Concerns:
Emergency Action Plan: The emergency action plan will consist of:

Use of the Incident Management System
First aid equipment with trained personnel on site
Emergency stop phrase "FREEZE" to immediately stop work
Phone or radio available to call an ambulance
Emergency contact numbers for students' family
Department emergency contact numbers
No work/training continues until situation is stabilized and thoroughly investigated

Practical evolution task specific safety plan: Risk Level **HIGH** **LOW**

Skill:
Safety Concern:
Preventative Measure:

Practical evolution task specific safety plan: Risk Level **HIGH** **LOW**

Skill:
Safety Concern:
Preventative Measure:

I have reviewed and approve this Training Safety Plan:

Training Safety Officer ______________________ ______________________
Name Signature

Lead Instructor ______________________ ______________________
Name Signature

Training Date: ____

Training Location: ____

Sample Acting Officer Course Outline and Schedule

Acting Officer 2012-1					
Date	**Shift**	**Subject**	**Instructor**	**Equipment**	**Confirmed**
April 25/26	B/C	**Courage to be Safe**	Ward	PPT	Yes
May 2/3	C/A	**Strategy & Tactics** Standard Response Approach, The Basics, Medical, MVC, Car Fires, Alarms & Fireground Basics	Peters	PPT	Yes
May 9/10	A/B	**Strategy & Tactics** Single Family & Multi-Family Dwellings	Wilson	PPT / Simulator	Yes
May 16/17	B/C	**Strategy & Tactics** Specialty Teams	Jones	PPT	Yes
		Atypical Calls	Peters	PPT	Yes
May 23/24	C/A	**MDT & Incident Reporting**	Peebles / Kelley	MDTs / PPT	Yes
May 30/31	A/B	**Drug Alcohol Awareness & Harrasment**	GJAC	Paperwork / PPT	Yes
		High Rise Operations	Mooney	PPT / Simulator	Yes
June 6/7	B/C	**Timesheets, OJI, Accidents & FMLA**	HQ	Paperwork / PPT	Yes
		Daily Operations	Peters	PPT	Yes
June 13/14	C/A	**Building Construction**	Ward	Props / PPT	Yes
		Arson Dectection & Reporting	Investigations	PPT	Yes
June 20/21	A/B	**Incident Command**	Chadwick	Worksheets / PPT	Yes
June 27/28	B/C	**Incident Command**	Chadwick	Worksheets / PPT	Yes
July 9-13		Practical Week	Staff / Adjuncts	Props / Simulator	Yes
July 16-20		Practical Week	Staff / Adjuncts	Props / Simulator	Yes
July 23-27		Practical Week	Staff / Adjuncts	Props / Simulator	Yes

Sample Acting Captain's Course Outline

The Acting Captain program designed for the newly promoted Lieutenant. The class is 7 days and will include the following topics:

Day 1 & 2	Day 3	Day 4	Day 5,6, & 7
Values, vision, & loyalty	Risk management	Corrective actions	NIMS
Total quality management	FMLA	Evaluations	ICS SOPs
Modern management	Timesheets	Call in policy	Strategy/Tactics
Development	Exposure paperwork	Daily manpower	Simulations
Responsibilities	Auto loss	24-7	
Safety	Harassment/drug policy	Core competencies	
Teamwork & cooperation	Loss and liability	Monthly reports	
Developing personal skills:		Weekly reports	
Motivation/Delegation		Transferring personnel	
Decision making			
Communication			
Training			
Resource management			
Time management			
Discipline			
Coaching/Counseling			
Leadership types of power			
Leadership continuum			
Duties and responsibilities			
General knowledge			
Interpersonal skills			

Sample Acting Battalion Chiefs Course Outline

The Acting Battalion Chief program is designed for the newly promoted Captain. This is a 9-day class and will include the following topics:

Day 1 & 2	Day 3 & 4	Day 5
IS 300: Expanding Incidents	Incident Safety Officer	Media Relations
Day 6	**Day 7**	**Day 8**
Defensive Driving	Strategy and Tactics	Incident Simulations
Day 9		
Daily Operations/SOPs		

APPENDIX

B Sample Reports and Records

ABC FIRE DEPARTMENT

Course #__________

Training Division Drill Report

Station #__________ Shift #__________ Date __________ 20____

Time ________ to ________ Total time drilled ________ Hrs. ________ Min.

Description of Drill	Time (Hrs. and Min.)

Continued

Names of Fire Fighters Drilled

Objectives

Comments____

____

Equipment Used

Feet of Booster Hose	Feet of 1 ¾" Hose	Feet of 2" or Larger Hose	Ladders Used	Number of Salvage Covers	Gallons of Water	Gallons of Foam

INDIVIDUAL ANNUAL TRAINING RECORD

Name:	Year:	Department:	Station:
Rank:	**Training Officer:**	**Supervisor:**	**Social Security Number:**

Place the date/day you took the training and the number of hours in the boxes provided.

Required Subjects:	Jan.	Feb.	Mar.	April	May	June	July	Aug.	Sep.	Oct.	Nov.	Dec.	Total Hours for Year
Hearing Protection													
Hazardous Materials													
Accident Prevention													
Fire Protection													
Portable Fire Extinguishers													
Bloodborne Pathogens													
Confined Space													
Toxic Substances In Workplace													
NIMS/ICS													
Wildland Safety													
Officer Training* 2													
Company Drills* 8													
Multicompany Drills* 4													
Night Drills* 2													
Apparatus Operation* 4 if applicable													
Emergency Plans 3 yr													
Respiratory Protection 3 yr													
Power Operated Hand Tools 3 yr													
Ladders 3 yr													
Hearing Protection													
Hazardous Materials													

The previous form is an excellent overall, quick reference to ensure that fire fighters have met the appropriate OSHA and some of the ISO requirements each year. This does not specifically address the general ISO requirement of 20 hours of training a month, but could be used to track requirements in these areas.

ABC FIRE DEPARTMENT
INDIVIDUAL FIRE DEPARTMENT TRAINING RECORD

Name			Rank/Position			Date Employed	Social Security Number				Station			
								EXAM			TYPE			
Date	Subject	Location	Academic	Practical	Hours	Instructor	Evaluation Y/N	Written	Oral	Practical	Local	In House	State	National

This record provides the individual fire fighter a means to keep track of their training by subject and hours. It also provides an excellent cross-reference for the training officer to use for the master training record.

FIRE DEPARTMENT TRAINING RECORD

Subject Taught:	**Date/Hours:**
Objectives Met (From Instructor Reference Manual):	**Tests on Subject (please indicate written, practical, or oral):**
Location of Training:	**Mileage to Training Location:**

Required Signatures
(The instructor who taught the course and the students in the class who took the training should verify all training. If this training form is computerized, signatures are still required and a printed copy of this form should be available for review by the training division.)

By signing below, I attest that I have given/received the training on the above subject.

____

Student's Name (legibly printed)

____

Student's Original Signature

____

Student's Social Security Number

____

Instructor's Original Signature or Initials

PROGRESS RECORD

Name:	**Rank/Grade/Position (Pencil):**	**Social Security Number:**	**Date Entered Fire Service:**	**Current Assignment/Location (Pencil):**
Year of Formal Education (Pencil):	**Physical Condition or Restrictions (Pencil):**	**Date Achieved Current Fire Certification (Pencil):**	**Enter Training Program for (Pencil):**	**Estimated Completion Date (Pencil):**

FIRE SERVICE TRAINING RECORD

Title	Date Entered	Date Completed	Results	Training Officers Signature
Fire Fighter I				
Fire Fighter II				
Apparatus Driver Operator (Pumper)				
Apparatus Driver Operator (Aerial)				
Airport Fire Fighter				
Fire Officer I				
Fire Officer II				
Fire Officer III				
Fire Inspector I				

Continued

Fire Inspector II				
Fire Inspector III				
Fire Service Instructor I				
Fire Service Instructor II				
Fire Service Instructor III				
Hazardous Materials Awareness				
Hazardous Materials Operations				
Hazardous Materials Technician				
Hazardous Materials Incident Command				
Incident Safety Officer				
Origin and Cause Technician				
Public Fire Educator I				
Public Fire Educator II				
Public Fire Educator III				

Course Lists (college courses, technical training, seminars, or specialty courses)

Course Title or Subject	Date Completed	Name of School or Location	Grade or Completion	Notes

RECORD OF INDIVIDUAL TRAINING PROGRESS

Name of Trainee	Social Security Number	Signature of Trainee			Initials	Signature of Certified Instructor			Initials
Subject		Fire Fighter I				Fire Fighter II			
Course Outline/Objectives	**NFPA 1001 Section**	**Date**	**Hours**	**Instructor Initials**	**Trainee Initials**	**Date**	**Hours**	**Instructor Initials**	**Trainee Initials**
Orientation 1) OUTLINE OF TRAINING PROGRAM	3-1.3								
a) Materials to be covered: Needs, objectives, purpose, and scope of course, course content, schedule, instructors, expectations and materials									
b) Class scheduling and processes, definitions of cognitive and psychomotor behavioral objectives, training credit									
c) Program curricula, training aids, development and evaluation									
2) DUTIES AND RESPONSIBILITIES FOR FIRE FIGHTERS	3-1.3								
a) Fire department rules, regulations, policies and procedures									
b) Student responsibilities as established by the authority having jurisdiction									
c) Performance observation, appraisal and re-training as required by local department									
3) TESTING PROCEDURES	3-1.3								

The previous form is a great resource for individual classes that are part of an overall program (like Fire Fighter I). The instructor should not initial each item until the student has been instructed, and has shown proficiency in it. This form ensures the student is held accountable for the class, and the instructor is held accountable for the proficiency of the student. Each overall section of a large program would have a sheet like this. This form also shows the individual fire department is training to the standard.

Special Course Record

Name: ______________________________ Social Security Number: ____________________________

Special Courses

Date	School and Location	Subject	Hours
Example	Mid-Winter Fire School	Positive Pressure Ventilation	4

This form is to provide a different way to track the special classes an individual may have taken.

Sample Practical Exercise Matrix

Practical Exercise Matrix 2012-1									
	Scenario 1			1st Floor Fire			3/22/12		
Rotation	1	2	3	4	5	6		Group 1	
E-1								Group 2	
E-2									
TK-3								Group 3	
Rehab									
Air Bottles								Group 4	
Rehab									
								Group 5	
								Group 6	
Rotation	1	2	3	4	5	6			
Recorder									
Inc.Comm.									
E-1-									
E-2-									
T-3-									
Ext. Safety									
RIT/Backup									
Safety 1									
Safety 2									
Ignition 1									
E-1 Driver									
E-2 Driver									
Tx									
Rehab									

Reference

Wyoming Department of Fire Prevention and Electrical Safety, *Train the Fire Trainer Guide*, July 2008.

APPENDIX C

Website Resource List

Building Construction
American Forest and Paper Products
Green Maltese
Structural Building Components Association
WoodAware: Awareness Level Information for the Fire Service

Fire Service History and Culture
Everyone Goes Home
Fire Department of New York (Origin of Maltese Cross)
Fire Department—UShistory.org

Fire Service Law
University of Cincinnati, Engineering and Applied Science, Department of Aerospace Engineering & Engineering Mechanics: Fire & EMS and Safety Law Newsletter
U.S. Department of Labor: Occupational Safety & Health Administration

Hazardous Materials
CSX: Training Materials for Emergency Responders
U.S. Department of Transportation: Pipeline and Hazardous Materials Administration

Health
CrossFit
International Association of Fire Fighters: Health, Safety & Medicine
National Volunteer Fire Council: Heart Healthy Firefighter
Runner's World

Incident Case Studies
CDC: Workplace Safety & Health Topics: Fire Fighter Fatality Investigation and Prevention Program
Firefighter Close Calls
Firefighter Near-Miss Reporting System
Homeland Security: Lessons Learned Information Sharing
National Fallen Firefighters Foundation

Instructor Materials
Chabot College Fire Technology Program
FireServices SLT
International Society of Fire Service Instructors
Oregon Fire Instructors Association

National Incident Management System
FEMA: Emergency Management Institute
FEMA: Independent Study Program (ISP), National Incident Management System
South Carolina Emergency Management Division

Networking
Fire Engineering Community
International Society of Fire Service Instructors
NFA TRADE Virtual Trading Post

Research
FEMA: Learning Resource Center
Fire Engineering Archives
National Institute of Standards and Technology
Underwriters Laboratory Fire Service Safety Institute

Simulations/Scenario-Based Training
Blue Card Command
Fire Engineering Firefighting Simulations
Fully Involved Fire: Professional Training for Fire Professionals
Georgia Smoke Diver Association

Sprinklers and Fire Prevention
United States Sprinkler Fitter Association

Standards
Factory Mutual Global
ISO Mitigation Online
National Fire Protection Association
U.S. Fire Administration

Strategy and Tactics
Kill the Flashover
Modern Fire Behavior
Vent Enter Search

Tools and Equipment
Animated Knots by Grog:
Blue Ridge Mountain Rescue Group (Knots)
U.S. Department of Labor, OSHA: Portable Fire Extinguishers

Vehicle Extrication
FireGraphics, Hybrid Vehicle ERGs
Holmatro Rescue Equipment Training Resources
National Highway Traffic and Safety Administration
National Alternative Fuels Training and Consortium: AFV First Responder Safety Training

APPENDIX D

Excerpt from *ISFSI Fire Department Required Training Self-Assessment Guide*

The following is an excerpt from the *ISFSI Fire Department Required Training Self-Assessment Guide*.[1] This document in its entirety can be found at ISFSI.org.

Fire Department Required Training Self-Assessment Guide

ISFSI

ISO, OSHA, NFPA...How does it all fit together?

[1]International Society of Fire Service Instructors (ISFSI). *ISFSI Fire Department Required Training Self-Assessment Guide.* 2011. http://isfsi.org/. Accessed November 20, 2014.

Every fire department training officer that has been tasked with satisfying requirements of ISO, OSHA and the NFPA has faced the reality of how confusing that can be. This document is just another instructor's attempt at creating a tool that will the training officer to understand the requirements of each of these organizations, how they fit together and what is needed by the agency to comply. YOU should take the time to review each supporting document to determine your agency requirements. Don't let this sit on the shelf! Use it as a tool to help improve you training program.

GOOD LUCK!

2011

Introduction

This document is intended to be used as a planning and assessment tool for initial and ongoing training of fire personnel. Every effort has been made to include all applicable standards and regulations; however the document should be routinely evaluated by the user for thoroughness. The latest editions of the National Fire Protection Association, Occupational Safety and Health Administration and Insurance Service Organization have been referenced throughout this document. Updates to this document should coincide with updates by each organization.

It is important to note that specific statements regarding initial and ongoing training have been included in each section; however the specific standard or regulation should be reviewed when determining level of compliance with each section of this document. It is the responsibility of the user to thoroughly review the standard or regulation prior to completing the assessment.

This document has been organized into categories of like topics in order to make the review of the fire department's training program easier. The categories include:

- **General Occupational Safety**
- **Fireground Operations**
- **Technical Rescue**
- **Emergency Vehicle Operations**
- **Incident Management**

It should be noted that many of these topics are interdependent on one another and may be able to be fulfilled alongside of one another. Respiratory Protection training, for instance, can be accomplished as part of Company Level Training.

Each topic provides the user the opportunity to assess initial and ongoing requirements. Simply read the section and answer the questions. The boxes provided will expand to accept your text. At the end of the document, there is a Self-assessment Summary to provide the user with a "snapshot" of the fire department's training strengths and weaknesses.

Online Resources

Visit the following website for more information on standards and regulations contained within:

- Occupational Safety and Health Administration
- National Fire Protection Association
- Insurance Service Organization

Glossary of Terms

The following are definitions and explanations of some terms you will find in various standards, regulations and documents relating to fire service training. These terms are provided in an effort to assist in making sense of all you will come across in creating a training plan.

1. **Annual** - covering the period of a year
2. **Authority Having Jurisdiction**- An organization, office, or individual responsible for enforcing the requirements of a code or standard, or for approving equipment, materials, an installation, or a procedure
3. **Certification** - genuine or authentic; certifications are issued by the authority having jurisdiction
4. **Commensurate** - equal in measure, or extent; corresponding in size, extent, amount, or degree
5. **Company** – a dedicated group of fire fighters working together
6. **Degradation** - decline to a low, destitute, or demoralized state; specifically referring to "skills degradation"
7. **Driver** - the operator of a motor vehicle
8. **Duties** - a task or service allocated to somebody, especially in the course of work
9. **Fire Department, Fire District, Fire Territory or Fire Brigade**- agencies that are legally obligated to provide fire protection to a specific geographic area
10. **Functions** - an activity or role assigned to somebody or something
11. **Inadequacies** - failure to be of good enough quality
12. **Initial** - coming first, or present at the beginning of an event or process
13. **Likely to Respond** - any member who can and does respond to all calls received by the Department
14. **Multicompany** - more than one fire company
15. **Operator** - somebody who operates machinery, an instrument, or other equipment
16. **OSHA State** – a State that has opted to allow the Occupational Safety and Health Administration to remain the as enforcement agency for occupational safety
17. **OSHA Plan State** – a State that has opted to accept enforcement responsibilities for occupational safety by either adopting federal OSHA

regulations or creating regulations that are "at least as effective as" comparable federal standards. OSHA approves and monitors State plans

18. **Proficiency Opportunities** – the chance to practice the skills that have been taught
19. **Reoccurring Proficiency Cycle**- schedule
20. **Requisite** - necessary or appropriate for a specific purpose
21. **Shall** - a mandatory part of the standard
22. **Should** - a part of the standard that is encouraged but not required
23. **Skills Check** - a test of a members practical abilities
24. **Unfamiliar** - not previously known or recognized

General Occupational Safety and Health

Blood-Borne Pathogens

Regulations and standards:

- OSHA 1910.1030 Blood-Borne Pathogens
- NFPA 1581 Standard on Fire Department Infection Control Program

Initial requirement:

- At the time of initial assignment to tasks where occupational exposure may take place. (OSHA)
- The fire department shall conduct initial and annual training and education programs for all members in accordance with state, provincial, or federal regulations. (NFPA)

Frequency:

- At least annually (within one year of their previous training) thereafter. (OSHA/ NFPA)
- An opportunity for interactive questions and answers with the person conducting the training session. (OSHA)

Self-assessment: blood-borne pathogens:

1. Does the FD meet all initial requirements?
2. Does the FD meet all ongoing requirements?

Emergency Action Plans

Regulations and standards:

- OSHA 1910.38 Emergency Action Plans

Initial requirement:

- An employer must designate and train employees to assist in a safe and orderly evacuation of other employees.
- When the plan is developed or the employee is assigned initially to a job. (OSHA)

Frequency:

- When the employee's responsibilities under the plan change. (OSHA)
- When the plan is changed. (OSHA)

Self-assessment: emergency action plans:

1. Does the FD meet all initial requirements?
2. Does the FD meet all ongoing requirements?

Fit Testing

Regulations and standards:

- OSHA 1910.134 Respiratory Protection Standard
- NFPA 1500 Standard on Fire Department Occupational Safety and Health Program
- OSHA 1910.156 Fire Brigade Standard

Initial requirement:

- The employer shall ensure that an employee using a tight-fitting face piece respirator is fit tested prior to initial use of the respirator (OSHA 1910.134)
- Whenever a different respirator face piece (size, style, model or make) is used. (OSHA 1910.134)
- The fit of the respiratory protection equipment of each new member shall be tested before the members are permitted to use respiratory protection equipment in a hazardous atmosphere. (NFPA 1500)
- Only members with a properly fitting face piece shall be permitted by the fire department to function in a hazardous atmosphere with respiratory protection equipment. (NFPA 1500)
- The employer shall ensure that an employee using a tight-fitting face piece respirator is fit tested prior to initial use of the respirator (OSHA 1910.156)
- Whenever a different respirator face piece (size, style, model or make) is used. (OSHA 1910.156)

Frequency:

- At least annually thereafter. (OSHA 1910.134)
- The face piece seal capability of each member qualified to use respiratory protection equipment shall be verified by qualitative or quantitative fit testing on an annual basis (NFPA 1500)
- Whenever new types of respiratory protection equipment or face pieces are issued. (NFPA 1500)
- At least annually thereafter. (OSHA 1910.156)

Self-assessment: fit testing:

1. Does the FD meet all initial requirements?
2. Does the FD meet all ongoing requirements?

Personal Protective Equipment

Regulations and standards:

- OSHA 1910.132 Personal Protective Equipment
- NFPA 1500 Standard on Fire Department Occupational Safety and Health Program, 2007 Edition

Initial requirement:

- The employer shall provide training to each employee who is required by this section to use PPE. (OSHA)
- Each affected employee shall demonstrate an understanding of the training specified in paragraph (f) (1) of this section, and the ability to use PPE properly, before being allowed to perform work requiring the use of PPE. (OSHA)
- The fire department shall provide all members with a training and education program that covers the operation, limitation, maintenance, and retirement criteria for all assigned personal protective equipment (PPE) expected to be utilized by members.

Frequency:

- When the employer has reason to believe that any affected employee who has already been trained does not have the understanding and skill required by paragraph (f) (2) of this section, the employer shall retrain each such employee. Circumstances where retraining is required include, but are not limited to, situations where:
  - Changes in the workplace render previous training obsolete; or
  - Changes in the types of PPE to be used render previous training obsolete; or
  - Inadequaciesin an affected employee's knowledge or use of assigned PPE indicate that the employee has not retained the requisite understanding or skill. (OSHA)

Self-assessment: personal protective equipment:

1. Does the FD meet all initial requirements?
2. Does the FD meet all ongoing requirements?

Respiratory Protection

Regulations and standards:

- NFPA 1404 Standard for Fire Service Respiratory Protection
- OSHA 1910.134 Respiratory Protection Standard
- NFPA 1500 Standard on Fire Department Occupational Safety and
- Health

Initial requirement:

- The fire department shall establish training and education programs that provide new members initial training, proficiency opportunities, and a method of skill and knowledge evaluation for duties assigned to the member prior to engaging in emergency operations (1500). (NFPA 1404)
- The employer shall provide the training prior to requiring the employee to use a respirator in the workplace. (OSHA)
- The fire department shall provide training and education for all department members commensurate with the duties and functions that they are expected to perform. (NFPA 1500)
- The fire department shall establish training and education programs that provide new members initial training, proficiency opportunities, and a method of skill and knowledge evaluation for duties assigned to the member prior to engaging in emergency operations. (NFPA 1500)

Frequency:

- Retraining shall be administered annually (NFPA 1404)
- Inadequacies in the members' knowledge or use of the respirator indicate that the member has not retained the requisite understanding or skill (NFPA 1404)
- Any other situation arises in which re-training appears necessary to ensure safe respirator use (NFPA 1404)
- Retraining shall be administered annually (OSHA)
- Changes in the workplace or the type of respirator render previous training obsolete (OSHA)
- Inadequacies in the employee's knowledge or use of the respirator indicate that the employee has not retained the requisite understanding or skill (OSHA)
- Any other situation arises in which retraining appears necessary to ensure safe respirator use. (OSHA)

- The fire department shall develop a reoccurring proficiency cycle with the goal of preventing skill degradation and potential for injury and death of members. (NFPA 1500)
- The fire department shall provide an annual skills check to verify minimum professional qualifications of its members. (NFPA 1500)
- Members shall be tested and certified at least annually in the safe and proper use of respiratory protection equipment that they are authorized to use. (NFPA 1500)

Self-assessment: respiratory protection:

1. Does the FD meet all initial requirements?
2. Does the FD meet all ongoing requirements?

■ Hazard Communication

Regulations and standards:

- OSHA 1910.1200, Hazard Communication

Initial requirement:

- Employers shall provide employees with effective information and training on hazardous chemicals in their work area at the time of their initial assignment (OSHA)

Frequency:

- Whenever a new physical or health hazard the employees have not previously been trained about is introduced into their work area. (OSHA)

Self-assessment: hazard communications:

1. Does the FD meet all initial requirements?
2. Does the FD meet all ongoing requirements?

■ Hearing Protection

Regulations and standards:

- OSHA 1910.95, Hearing Protection Standard

Initial requirement:

- The employer shall train each employee who is exposed to noise at or above an 8-hour time weighted average of 85 decibels in accordance with the requirements of this section. The employer shall institute a training program and ensure employee participation in the program.

Frequency:

- The training program shall be repeated annually for each employee included in the hearing conservation program. Information provided in the training program shall be updated to be consistent with changes in protective equipment and work processes.

Self-assessment:hearing protection:

1. Does the FD meet all initial requirements?
2. Does the FD meet all ongoing requirements?

■ Portable Fire Extinguishers

Regulations and standards:

- OSHA 1910.157 Portable Fire Extinguishers

Initial requirement:

- The employer shall provide the education required in paragraph (g) (1) of this section upon initial employment… (OSHA)

Frequency:

- The employer shall provide the education required in paragraph (g) (1) of this section upon initial employment and at least annually thereafter. (OSHA)

Self-assessment:portable extinguishers:

1. Does the FD meet all initial requirements?
2. Does the FD meet all ongoing requirements?

Fireground Operations

■ Fire Fighter I/II

Regulations and standards:

- NFPA 1001 Standard for Fire Fighter Professional Qualifications

Initial requirement:

- Prior to being certified at the Fire Fighter I level; the fire fighter candidate shall meet the general knowledge and skills requirements and the job performance requirements of Chapter 5. (NFPA 1001)
- The job performance requirements shall be accomplished in accordance with the requirements of the authority having jurisdiction and NFPA 1500. (NFPA 1001)
- The fire department shall establish training and education programs that provide new members initial training, proficiency opportunities, and a method of skill and knowledge evaluation for duties assigned to the member prior to engaging in emergency operations. (NFPA 1001)

Frequency:

- The fire department shall develop a reoccurring proficiency cycle with the goal of preventing skill degradation and potential for injury and death of members (1500). (NFPA 1001)
- The fire department shall provide an annual skills check to verify minimum professional qualifications of its members. (NFPA 1001)

Self-assessment: Fire Fighter I/II:

1. Does the FD meet all initial requirements?
2. Does the FD meet all ongoing requirements?

■ Structural Firefighting

Regulations and standards:

- OSHA 1910.156 Fire Brigade Standard
- ISO Fire Suppression Rating Schedule
- Indiana Title 655 Board of Firefighting Personnel Standards and Education

Initial requirement:

- The employer shall provide training and education for all fire brigade members commensurate with those duties and functions that fire brigade members are expected to perform. (OSHA)
- Such training and education shall be provided to fire brigade members before they perform fire brigade emergency activities. (OSHA)
- 240 hours entry level training. (ISO)

Frequency:

- The employer shall assure that training and education is conducted frequently enough to assure that each member of the fire brigade is able to perform the member's assigned duties and functions satisfactorily and in a safe manner so as not to endanger fire brigade members or other employees. (OSHA)
- All fire brigade members shall be provided with training at least annually. (OSHA)
- In addition, fire brigade members who are expected to perform interior structural fire fighting shall be provided with an education session or training at least quarterly. (OSHA)
- 20 hours per member per month continuing training. (ISO)

Self-assessment:structural firefighting:

1. Does the FD meet all initial requirements?
2. Does the FD meet all ongoing requirements?

■ Company Level Training

Regulations and standards:

- ISO Fire Suppression Rating Schedule
- NFPA 1410 Standard on Training for Initial Emergency Scene Operations, 2010 Edition

Frequency:

- Company training at fire stations; 20 Hours per member per month. (Night Drills and Half Day Drills may count toward single company drills)
- NFPA 1410 highlights evaluations to be conducted on first responding units to fire scenes. While this standard does not state frequency for evaluations to be performed, NFPA 1500 generally states evaluations should be performed annually.

Self-assessment:company level training:

1. Does the FD meet all initial requirements?
2. Does the FD meet all ongoing requirements?

■ Night Drills

Regulations and standards:

- ISO Fire Suppression Rating Schedule

Requirement and frequency:

- (2) Two drills per year (3) hours each. (ISO) (May count toward a single company or multicompany drill)

Self-assessment:night drills:

1. Does the FD meet all initial requirements?
2. Does the FD meet all ongoing requirements?

■ Half-Day Drills

Regulations and standards:

- ISO Fire Suppression Rating Schedule

Requirement and frequency:

- (8) Eight drills each per year (3) hours each. (ISO) (May count toward a single company or multicompany drills)

Self-assessment:half-day drills:

1. Does the FD meet all initial requirements?
2. Does the FD meet all ongoing requirements?

■ Multicompany Drills

Regulations and standards:

- ISO Fire Suppression Rating Schedule

- NFPA 1410 Standard on Training for Initial Emergency Scene Operations, 2010 Edition

Requirement and frequency:

- Four drills each per year (3) hours each. (ISO) (Night Drills and Half Day Drills may count toward multicompany drill)
- NFPA 1410 highlights evaluations to be conducted on first responding units to fire scenes. While this standard does not state frequency for evaluations to be performed, NFPA 1500 generally states evaluations should be performed annually.

Self-assessment:multicompanydrills:

1. Does the FD meet all initial requirements?
2. Does the FD meet all ongoing requirements?

Technical Rescue

Confined Space Training

Regulations and standards:

- OSHA 1910.146 Permit Required Confined Spaces
- NFPA1670 Standard on Operations and Training for Technical Search and Rescue Incidents, 2009 Edition

Initial requirement:

- The employer shall provide training so that all employees whose work is regulated by this section acquire the understanding, knowledge, and skills necessary for the safe performance of the duties assigned under this section. (OSHA 1910.146[g][1])
- Train affected employees in basic first-aid and cardiopulmonary resuscitation (CPR). The employer shall ensure that at least one member of the rescue team or service holding a current certification in first aid and CPR is available. (OSHA)
- Training shall be provided to each affected employee:
  - Before the employee is first assigned duties under this section;
  - Before there is a change in assigned duties;
  - Whenever there is a change in permit space operations that presents a hazard about which an employee has not previously been trained;
  - Whenever the employer has reason to believe either that there are deviations from the permit space entry procedures required by paragraph (d)(3) of this section or that there are inadequacies in the employee's knowledge or use of these procedures. (OSHA)
- Each member of the rescue service shall be trained to perform the assigned rescue duties corresponding to his or her designated level of competency. (NFPA)

Frequency:

- Ensure that affected employees practice making permit space rescues at least once every 12 months, by means of simulated rescue operations in which they remove dummies, manikins, or actual persons from the actual permit spaces or from representative permit spaces. (OSHA)
- Representative permit spaces shall, with respect to opening size, configuration, and accessibility, simulate the types of permit spaces from which rescue is to be performed. (OSHA)
- Review the permit space program, using the canceled permits retained under paragraph (e)(6) of this section within 1 year after each entry and revise the program as necessary, to ensure that employees participating in entry operations are protected from permit space hazards.

Note: Employers may perform a single annual review covering all entries performed during a 12-month period. If no entry is performed during a 12-month period, no review is necessary.

Self-assessment:confined space

1. Does the FD meet all initial requirements?
2. Does the FD meet all ongoing requirements?

Hazardous Materials Operations Training

Regulations and standards:

- OSHA 1910.120 Hazardous Waste Operations and Emergency Response
- ISO Fire Suppression Rating Schedule
- NFPA 1500 Standard on Fire Department Occupational Safety and Health Program

Initial requirement:

- General site workers engaged in hazardous substance removal or other activities which expose or potentially expose workers to hazardous substances and health hazards shall receive a minimum of 40 hours of instruction off the site (OSHA)

- And a minimum of three days actual field experience under the direct supervision of a trained experienced supervisor. (OSHA)
- None given (ISO)
- The fire department shall provide specific training to members who are likely to respond to special operations incidents in a support role to special operations technicians. (NFPA 1500)

Frequency:

- Employees specified in paragraph (e) (1) of this section, and managers and supervisors specified in paragraph (e) (4) of this section, shall receive eight hours of refresher training annually on the items specified in paragraph (e) (2) and/or (e) (4) of this section, any critique of incidents that have occurred in the past year that can serve as training examples of related work, and other relevant topics. (OSHA)
- One half day (3 hours) per member per year (ISO)
- The fire department shall provide training and education events as required to support minimum qualifications and certifications expected of its members. (NFPA 1500)
- Members shall practice assigned skill sets on a regular basis but not less than annually. (NFPA 1500)

Self-assessment: hazardous materials operations:

1. Does the FD meet all initial requirements?
2. Does the FD meet all ongoing requirements?

The Control of Hazardous Energy (Lock Out/ Tag Out)

Regulations and standards:

- OSHA 1910.147, The Control of Hazardous Energy (Lock Out/ Tag Out)

Initial requirement:

- The employer shall provide training to ensure that the purpose and function of the energy control program are understood by employees and that the knowledge and skills required for the safe application, usage, and removal of the energy controls are acquired by employees. (OSHA)
- Each affected employee shall be instructed in the purpose and use of the energy control procedure. (OSHA)

Frequency:

- Additional retraining shall also be conducted whenever a periodic inspection under paragraph (c)(6) of this section reveals, or whenever the employer has reason to believe that there are deviations from or inadequacies in the employee's knowledge or use of the energy control procedures. (OSHA)
- The retraining shall reestablish employee proficiency and introduce new or revised control methods and procedures, as necessary. (OSHA)
- The employer shall certify that employee training has been accomplished and is being kept up to date. The certification shall contain each employee's name and dates of training.

Self-assessment: lock out/tag out:

1. Does the FD meet all initial requirements?
2. Does the FD meet all ongoing requirements?

Technical Rescue Awareness

Regulations and standards:

- NFPA 1500 Standard on Fire Department Occupational Safety and Health Program
- OSHA 1910.156 Fire Brigade Standard

Initial requirement:

- The fire department shall provide specific training to members who are likely to respond to special operations incidents in a support role to special operations technicians. (NFPA 1500)
- The employer shall provide training and education for all fire brigade members commensurate with those duties and functions that fire brigade members are expected to perform. (OSHA)
- Such training and education shall be provided to fire brigade members before they perform fire brigade emergency activities. (OSHA)

Frequency:

- The fire department shall provide training and education events as required to support minimum qualifications and certifications expected of its members. (NFPA 1500)
- Members shall practice assigned skill sets on a regular basis but not less than annually. (NFPA 1500)
- The employer shall assure that training and education is conducted frequently enough to assure

that each member of the fire brigade is able to perform the member's assigned duties and functions satisfactorily and in a safe manner so as not to endanger fire brigade members or other employees. (OSHA)

- All fire brigade members shall be provided with training at least annually. (OSHA)
- In addition, fire brigade members who are expected to perform interior structural fire fighting shall be provided with an education session or training at least quarterly. (OSHA)

Self-assessment: technical rescue awareness:

1. Does the FD meet all initial requirements?
2. Does the FD meet all ongoing requirements?

Emergency Vehicle Operations

Driver Training

Regulations and standards:

- NFPA 1451 Standard for a Fire Service Vehicle Operations Training
- Program
- NFPA 1500 Standard on Fire Department Occupational Safety and
- Health
- NFPA 1002 Fire Apparatus Driver / Operator Professional Qualifications
- ISO Fire Suppression Rating Schedule

Initial requirement:

- Members shall be provided with driver training and education appropriate for their duties and responsibilities before being permitted to operate fire department vehicles or apparatus. (NFPA 1451)
- The fire department shall establish training and education programs that provide new members initial training, proficiency opportunities, and a method of skill and knowledge evaluation for duties assigned to the member prior to engaging in emergency operations. (NFPA 1451)
- The fire department shall provide training and education for all department members commensurate with the duties and functions that they are expected to perform (1500). (NFPA 1002)
- The fire department shall establish training and education programs that provide new members initial training, proficiency opportunities, and a method of skill and knowledge evaluation for duties assigned to the member prior to engaging in emergency operations (1500). (NFPA 1002)
- 40 hours for new drivers. (ISO)
- The fire department shall establish training and education programs that provide new members initial training, proficiency opportunities, and a method of skill and knowledge evaluation for duties assigned to the member prior to engaging in emergency operations. (NFPA 1500)

Frequency:

- The fire department shall develop a reoccurring proficiency cycle with the goal of preventing skill degradation and potential for injury and death of members (1500). (NFPA 1451)
- The fire department shall provide an annual skills check to verify minimum professional qualifications of its members (1500). (NFPA 1451)
- Driver training shall be provided for all members as often as necessary to meet the applicable requirements of this chapter, but not less than twice each year. (NFPA 1451)
- The fire department shall develop a reoccurring proficiency cycle with the goal of preventing skill degradation and potential for injury and death of members. (NFPA 1500)
- The fire department shall provide an annual skills check to verify minimum professional qualifications of its members. (NFPA 1500)
- The fire department shall develop a reoccurring proficiency cycle with the goal of preventing skill degradation and potential for injury and death of members (1500). (NFPA 1002)
- The fire department shall provide an annual skills check to verify minimum professional qualifications of its members (1500). (NFPA 1002)
- (4) Four half day (3 hours) sessions per year. (ISO)
- Annual driver training shall include hands-on exercises, excluding virtual reality driver training simulator (DTS), using actual fire apparatus. (NFPA 1451)
- Whenever new or unfamiliar vehicles are placed into service, training and education relating to those vehicles shall be provided for all affected members. (NFPA 1451)

Self-assessment:driver training:

1. Does the FD meet all initial requirements?
2. Does the FD meet all ongoing requirements?

■ Pump Operator

Regulations and standards:

- NFPA 1002 Standard on Fire Apparatus Driver/ Operator Professional Qualifications
- ISO Fire Suppression Rating Schedule

Initial requirement:

- Must meet the job performance requirements defined in Sections 5.1 and 5.2 shall be met prior to certification as a driver / operator - pumper. (NFPA 1002)
- 40 hours for new drivers. (ISO)

Frequency:

- The fire department shall develop a reoccurring proficiency cycle with the goal of preventing skill degradation and potential for injury and death of members. (NFPA 1002)
- The fire department shall provide an annual skills check to verify minimum professional qualifications of its members. (NFPA 1002)
- (4) Four half day (3 hours) sessions per year. (ISO)

Self-assessment:pump operator:

1. Does the FD meet all initial requirements?
2. Does the FD meet all ongoing requirements?

Incident Management

■ Incident Command Training

Regulations and standards:

- NFPA 1500 Standard on Fire Department Occupational Safety andHealth Program
- NFPA 1021 Standard for Fire Officer Professional Qualifications
- NFPA 1001 Standard for Fire Fighter Professional Qualifications
- National Incident Management System
- OSHA 1910.120 Hazardous Materials

Initial requirement:

- All members who are likely to be involved in emergency operations shall be trained in the incident management and accountability system used by the fire department. (NFPA 1500)
- For certification at the Fire Officer Level I, the candidate shall meet the requirements of Fire Fighter II as defined in NFPA 1001, Fire Instructor I as defined in NFPA 1041, and the job performance requirements defined in Sections 4.2 through 4.7 of this standard. (NFPA 1021)
- For certification at Level II, the Fire Fighter I shall meet the general knowledge requirements in 6.1.1.1, the general skill requirements in 6.1.1.2, and the job performance requirements defined in Sections 6.2 through 6.5 of this standard. (NFPA 1001)
- Responsibilities of the Fire Fighter II in assuming and transferring command within an incident management system, performing assigned duties in conformance with applicable NFPA and other safety regulations and authority having jurisdiction procedures, and the role of a Fire Fighter II within the organization. (NFPA 1001)
- Homeland Security Presidential Directive 5 (NIMS)
- Officers and leaders who may be expected to be in charge at an incident should be fully knowledgeable of their company's incident command system. They should know where and how to obtain additional assistance and be familiar with the local district's emergency response plan and the state emergency response plan. (OSHA)

Frequency:

- The fire department shall provide training and education events as required to support minimum qualifications and certifications expected of its members. (NFPA 1500)
- Members shall practice assigned skill sets on a regular basis but not less than annually. (NFPA 1500)
- None Given (NFPA 1021)
- The fire department shall develop a reoccurring proficiency cycle with the goal of preventing skill degradation and potential for injury and death of members (1500). (NFPA 1001)
- The fire department shall provide an annual skills check to verify minimum professional qualifications of its members. (NFPA 1001)
- Preparedness activities conducted on a "steady-state" basis. (NIMS)

Self-assessment: incident command:

1. Does the FD meet all initial requirements?
2. Does the FD meet all ongoing requirements?

Officer Training

Regulations and standards:

- NFPA 1021 Standard for Fire Officer Professional Qualifications
- ISO Fire Suppression Rating Schedule

Initial requirements:

- Fire Fighter II as defined in NFPA 1001 (NFPA 1021)
- Fire Instructor I as defined in NFPA 1041 (NFPA 1021)
- And the job performance requirements defined in Sections 4.2 through 4.7 of this standard. (NFPA 1021)
- None given. (ISO)

Frequency:

- None given. (NFPA 1021)

Self-assessment: officer training:

1. Does the FD meet all initial requirements?
2. Does the FD meet all ongoing requirements?

Self-Assessment Summary

Indicate whether or not your agency complies completely with the all standards within this document for each topic.

General Occupational Safety	Initial	Ongoing
Blood-Borne Pathogens		
Emergency Action Plans		
Fit Testing		
Hazard Communications		
Hearing Protection		
Personal Protective Equipment		
Respiratory Protection		
Portable Fire Extinguishers		
Fireground Operations	**Initial**	**Ongoing**
Company Level Training		
Fire Fighter I/II		
Half-Day Drills		
Multicompany Drills		
Night Drills		
Structural Firefighting		
Technical Rescue	**Initial**	**Ongoing**
Confined Space Training		
Hazardous Materials Operations		
Lock Out/Tag Out		
Technical Rescue Awareness		

Continued

Emergency Vehicle Operations	Initial	Ongoing
Driver Training		
Pump Operator		
Incident Management	**Initial**	**Ongoing**
Incident Command		
Officer Training		

Once you complete the Self-Assessment Summary, add pages to this document and create goals for improvement. Provide yourself target dates to achieve your goals. This will help keep you on task. Take it one step at a time, improve little by little and don't try to fix it all at once. Keep in mind that you will always be working to improve your plan. This should be an annual assessment of your training program.

Train them like their lives depend on it!

Stay Safe...

APPENDIX E

Excerpt from OSHA 1910.120 Appendix E: *Training Curriculum Guidelines*

The following is an excerpt from OSHA 1910.120 Appendix E: *Training Curriculum Guidelines*.[1] This section was included to assist in developing training curriculum per OSHA standards for applicable departments and organizations. It also provides insight into OSHA's methodology and understanding of what defines training.

The following nonmandatory general criteria may be used for assistance in developing site-specific training curriculum used to meet the training requirements of 29 CFR 1910.120(e); 29 CFR 1910.120(p)(7), (p)(8)(iii); and 29 CFR 1910.120(q)(6), (q)(7), and (q)(8). These are generic guidelines and they are not presented as a complete training curriculum for any specific employer. Site-specific training programs must be developed on the basis of a needs assessment of the hazardous waste site, RCRA/TSDF, or emergency response operation in accordance with 29 CFR 1910.120.

It is noted that the legal requirements are set forth in the regulatory text of § 1910.120. The guidance set forth here presents a highly effective program that in the areas covered would meet or exceed the regulatory requirements. In addition, other approaches could meet the regulatory requirements.

Suggested General Criteria

Definitions:

"Competent" means possessing the skills, knowledge, experience, and judgment to perform assigned tasks or activities satisfactorily as determined by the employer.

"Demonstration" means the showing by actual use of equipment or procedures.

"Hands-on training" means training in a simulated work environment that permits each student to have experience performing tasks, making decisions, or using equipment appropriate to the job assignment for which the training is being conducted.

"Initial training" means training required prior to beginning work.

"Lecture" means an interactive discourse with a class lead by an instructor.

"Proficient" means meeting a stated level of achievement.

"Site-specific" means individual training directed to the operations of a specific job site.

"Training hours" means the number of hours devoted to lecture, learning activities, small

[1]U.S. Department of Labor, Occupational Safety & Health Administration website. https://www.osha.gov/pls/oshaweb/owadisp.show_document?p_table=STANDARDS&p_id=9770. Accessed November 6, 2014.

group work sessions, demonstration, evaluations, or hands-on experience.

Suggested core criteria:

1. **Training facility**
   The training facility should have available sufficient resources, equipment, and site locations to perform didactic and hands-on training when appropriate. Training facilities should have sufficient organization, support staff, and services to conduct training in each of the courses offered.
2. **Training Director**
   Each training program should be under the direction of a training director who is responsible for the program. The Training Director should have a minimum of 2 years of employee education experience.
3. **Instructors**
   Instructors should be deem competent on the basis of previous documented experience in their area of instruction, successful completion of a "train-the-trainer" program specific to the topics they will teach, and an evaluation of instructional competence by the Training Director.

   Instructors should be required to maintain professional competency by participating in continuing education or professional development programs or by completing successfully an annual refresher course and having an annual review by the Training Director.

   The annual review by the Training Director should include observation of an instructor's delivery, a review of those observations with the trainer, and an analysis of any instructor or class evaluations completed by the students during the previous year.
4. **Course materials**
   The Training Director should approve all course materials to be used by the training provider. Course materials should be reviewed and updated at least annually. Materials and equipment should be in good working order and maintained properly.

   All written and audio-visual materials in training curricula should be peer reviewed by technically competent outside reviewers or by a standing advisory committee.

   Reviews should possess expertise in the following disciplines were applicable: occupational health, industrial hygiene and safety, chemical/environmental engineering, employee education, or emergency response. One or more of the peer reviewers should be an employee experienced in the work activities to which the training is directed.
5. **Students**
   The program for accepting students should include:
   a. Assurance that the student is or will be involved in work where chemical exposures are likely and that the student possesses the skills necessary to perform the work.
   b. A policy on the necessary medical clearance.
6. **Ratios**
   Student-instructor ratios should not exceed 30 students per instructor. Hands-on activity requiring the use of personal protective equipment should have the following student- instructor ratios. For Level C or Level D personal protective equipment the ratio should be 10 students per instructor. For Level A or Level B personal protective equipment the ratio should be 5 students per instructor.
7. **Proficiency assessment**
   Proficiency should be evaluated and documented by the use of a written assessment and a skill demonstration selected and developed by the Training Director and training staff. The assessment and demonstration should evaluate the knowledge and individual skills developed in the course of training. The level of minimum achievement necessary for proficiency shall be specified in writing by the Training Director.

   If a written test is used, there should be a minimum of 50 questions. If a written test is used in combination with a skills demonstration, a minimum of 25 questions should be used. If a skills demonstration is used, the tasks chosen and the means to rate successful completion should be fully documented by the Training Director.

   The content of the written test or of the skill demonstration shall be relevant to the objectives of the course. The written test and skill demonstration should be updated as necessary to reflect changes in the curriculum and any update should be approved by the Training Director.

   The proficiency assessment methods, regardless of the approach or combination of approaches used, should be justified, documented and approved by the Training Director.

The proficiency of those taking the additional courses for supervisors should be evaluated and documented by using proficiency assessment methods acceptable to the Training Director. These proficiency assessment methods must reflect the additional responsibilities borne by supervisory personnel in hazardous waste operations or emergency response.

8. **Course certificate**
   Written documentation should be provided to each student who satisfactorily completes the training course. The documentation should include:
   a. Student's name.
   b. Course title.
   c. Course date.
   d. Statement that the student has successfully completed the course.
   e. Name and address of the training provider.
   f. An individual identification number for the certificate.
   g. List of the levels of personal protective equipment used by the student to complete the course.

   This documentation may include a certificate and an appropriate wallet-sized laminated card with a photograph of the student and the above information. When such course certificate cards are used, the individual identification number for the training certificate should be shown on the card.
9. **Record keeping**
   Training providers should maintain records listing the dates courses were presented, the names of the individual course attenders, the names of those students successfully completing each course, and the number of training certificates issued to each successful student. These records should be maintained for a minimum of five years after the date an individual participated in a training program offered by the training provider. These records should be available and provided upon the student's request or as mandated by law.
10. **Program quality control**
    The Training Director should conduct or direct an annual written audit of the training program. Program modifications to address deficiencies, if any, should be documented, approved, and implemented by the training provider. The audit and the program modification documents should be maintained at the training facility.

Suggested Program Quality Control Criteria

Factors listed here are suggested criteria for determining the quality and appropriateness of employee health and safety training for hazardous waste operations and emergency response.

A. **Training plan**
   Adequacy and appropriateness of the training program's curriculum development, instructor training, distribution of course materials, and direct student training should be considered, including:
   1. The duration of training, course content, and course schedules/agendas.
   2. The different training requirements of the various target populations, as specified in the appropriate generic training curriculum.
   3. The process for the development of curriculum, which includes appropriate technical input, outside review, evaluation, program pretesting.
   4. The adequate and appropriate inclusion of hands-on, demonstration, and instruction methods.
   5. Adequate monitoring of student safety, progress, and performance during the training.

B. **Program management, Training Director, staff, and consultants**
   Adequacy and appropriateness of staff performance and delivering an effective training program should be considered, including:
   1. Demonstration of the training director's leadership in assuring quality of health and safety training.
   2. Demonstration of the competency of the staff to meet the demands of delivering high quality hazardous waste employee health and safety training.
   3. Organization charts establishing clear lines of authority.
   4. Clearly defined staff duties including the relationship of the training staff to the overall program.
   5. Evidence that the training organizational structure suits the needs of the training program.
   6. Appropriateness and adequacy of the training methods used by the instructors.
   7. Sufficiency of the time committed by the training director and staff to the training program.

8. Adequacy of the ratio of training staff to students.
9. Availability and commitment of the training program of adequate human and equipment resources in the areas of:
   a. Health effects
   b. Safety
   c. Personal protective equipment (PPE)
   d. Operational procedures
   e. Employee protection practices/procedures
10. Appropriateness of management controls.
11. Adequacy of the organization and appropriate resources assigned to assure appropriate training.
12. In the case of multiple-site training programs, adequacy of satellite centers management.

C. Training facilities and resources

Adequacy and appropriateness of the facilities and resources for supporting the training program should be considered, including:

1. Space and equipment to conduct the training.
2. Facilities for representative hands-on training.
3. In the case of multiple-site programs, equipment and facilities at the satellite centers.
4. Adequacy and appropriateness of the quality control and evaluations program to account for instructor performance.
5. Adequacy and appropriateness of the quality control and evaluation program to ensure appropriate course evaluation, feedback, updating, and corrective action.
6. Adequacy and appropriateness of disciplines and expertise being used within the quality control and evaluation program.
7. Adequacy and appropriateness of the role of student evaluations to provide feedback for training program improvement.

D. Quality control and evaluation

Adequacy and appropriateness of quality control and evaluation plans for training programs should be considered, including:

1. A balanced advisory committee and/or competent outside reviewers to give overall policy guidance.
2. Clear and adequate definition of the composition and active programmatic role of the advisory committee or outside reviewers.
3. Adequacy of the minutes or reports of the advisory committee or outside reviewers' meetings or written communication.
4. Adequacy and appropriateness of the quality control and evaluations program to account for instructor performance.
5. Adequacy and appropriateness of the quality control and evaluation program to ensure appropriate course evaluation, feedback, updating, and corrective action.
6. Adequacy and appropriateness of disciplines and expertise being used within the quality control and evaluation program.
7. Adequacy and appropriateness of the role of student evaluations to provide feedback for training program improvement.

E. Students

Adequacy and appropriateness of the program for accepting students should be considered, including:

1. Assurance that the student already possesses the necessary skills for their job, including necessary documentation.
2. Appropriateness of methods the program uses to ensure that recruits are capable of satisfactorily completing training.
3. Review and compliance with any medical clearance policy.

F. Institutional environment and administrative support

The adequacy and appropriateness of the institutional environment and administrative support system for the training program should be considered, including:

1. Adequacy of the institutional commitment to the employee training program.
2. Adequacy and appropriateness of the administrative structure and administrative support.

G. Summary of evaluation questions

Key questions for evaluating the quality and appropriateness of an overall training program should include the following:

1. Are the program objectives clearly stated?
2. Is the program accomplishing its objectives?
3. Are appropriate facilities and staff available?
4. Is there an appropriate mix of classroom, demonstration, and hands-on training?
5. Is the program providing quality employee health and safety training that fully meets the intent of regulatory requirements?

6. What are the program's main strengths?
7. What are the program's main weaknesses?
8. What is recommended to improve the program?
9. Are instructors instructing according to their training outlines?
10. Is the evaluation tool current and appropriate for the program content?
11. Is the course material current and relevant to the target group?

Suggested Training Curriculum Guidelines

The following training curriculum guidelines are for those operations specifically identified in 29 CFR 1910.120 as requiring training. Issues such as qualifications of instructors, training certification, and similar criteria appropriate to all categories of operations addressed in 1910.120 have been covered in the preceding section and are not re-addressed in each of the generic guidelines. Basic core requirements for training programs that are addressed include:

1. General hazardous waste operations.
2. RCRA operations—treatment, storage, and disposal facilities.
3. Emergency response.

A. General hazardous waste operations and site-specific training

1. Off-site training. Training course content for hazardous waste operations, required by 29 CFR 1910.120(e), should include the following topics or procedures:

a. Regulatory knowledge:
 (1) An review of 29 CFR 1910.120 and the core elements of an occupational safety and health program.
 (2) The content of a medical surveillance program as outlined in 29 CFR 1910.120(f).
 (3) The content of an effective site safety and health plan consistent with the requirements of 29 CFR 1910.120(b)(4)(ii).
 (4) Emergency response plan and procedures as outlined in 29 CFR 1910.38 and 29 CFR 1910.120(l).
 (5) Adequate illumination.
 (6) Sanitation recommendation and equipment.
 (7) Review and explanation of OSHA's hazard-communication standard (29 CFR 1910.1200) and lock-out-tag-out standard (29 CFR 1910.147).
 (8) Review of other applicable standards including but not limited to those in the construction standards (29 CFR Part 1926).
 (9) Rights and responsibilities of employers and employees under applicable OSHA and EPA laws.

b. Technical knowledge:
 (1) Type of potential exposures to chemical, biological, and radiological hazards; types of human responses to these hazards and recognition of those responses; principles of toxicology and information about acute and chronic hazards; health and safety considerations of new technology.
 (2) Fundamentals of chemical hazards including but not limited to vapor pressure, boiling points, flash points, pH, other physical and chemical properties.
 (3) Fire and explosion hazards of chemicals.
 (4) General safety hazards such as but not limited to electrical hazards, powered equipment hazards, motor vehicle hazards, walking-working surface hazards, excavation hazards, and hazards associated with working in hot and cold temperature extremes.
 (5) Review and knowledge of confined space entry procedures in 29 CFR 1910.146.
 (6) Work practices to minimize employee risk from site hazards.
 (7) Safe use of engineering controls, equipment, and any new relevant safety technology or safety procedures.
 (8) Review and demonstration of competency with air sampling and monitoring equipment that may be used in a site monitoring program.
 (9) Container sampling procedures and safeguarding; general drum and container handling procedures including

special requirement for laboratory waste packs, shock-sensitive wastes, and radioactive wastes.
  - (10) The elements of a spill control program.
  - (11) Proper use and limitations of material handling equipment.
  - (12) Procedures for safe and healthful preparation of containers for shipping and transport.
  - (13) Methods of communication including those used while wearing respiratory protection.
- c. Technical skills:
  - (1) Selection, use maintenance, and limitations of personal protective equipment including the components and procedures for carrying out a respirator program to comply with 29 CFR 1910.134.
  - (2) Instruction in decontamination programs including personnel, equipment, and hardware; hands-on training including level A, B, and C ensembles and appropriate decontamination lines; field activities including the donning and doffing of protective equipment to a level commensurate with the employee's anticipated job function and responsibility and to the degree required by potential hazards.
  - (3) Sources for additional hazard information; exercises using relevant manuals and hazard coding systems.
- d. Additional suggested items:
  - (1) A laminated, dated card or certificate with photo, denoting limitations and level of protection for which the employee is trained should be issued to those students successfully completing a course.
  - (2) Attendance should be required at all training modules, with successful completion of exercises and a final written or oral examination with at least 50 questions.
  - (3) A minimum of one-third of the program should be devoted to hands-on exercises.
  - (4) A curriculum should be established for the 8-hour refresher training required by 29 CFR 1910.120(e)(8), with delivery of such courses directed toward those areas of previous training that need improvement or reemphasis.
  - (5) A curriculum should be established for the required 8-hour training for supervisors. Demonstrated competency in the skills and knowledge provided in a 40-hour course should be a prerequisite for supervisor training.

2. **Refresher training.** The 8-hour annual refresher training required in 29 CFR 1910.120(e)(8) should be conducted by qualified training providers. Refresher training should include at a minimum the following topics and procedures:
   - a. Review of and retraining on relevant topics covered in the 40-hour program, as appropriate, using reports by the students on their work experiences.
   - b. Update on developments with respect to material covered in the 40-hour course.
   - c. Review of changes to pertinent provisions of EPA or OSHA standards or laws.
   - d. Introduction of additional subject areas as appropriate.
   - e. Hands-on review of new or altered PPE or decontamination equipment or procedures. Review of new developments in personal protective equipment.
   - f. Review of newly developed air and contaminant monitoring equipment.

3. **On-site training.**
   - a. The employer should provide employees engaged in hazardous waste site activities with information and training prior to initial assignment into their work area, as follows:
     - (1) The requirements of the hazard communication program including the location and availability of the written program, required lists of hazardous chemicals, and safety data sheets.
     - (2) Activities and locations in their work area where hazardous substance may be present.
     - (3) Methods and observations that may be used to detect the present or release of a hazardous chemical in the work area (such as monitoring conducted

by the employer, continuous monitoring devices, visual appearances, or other evidence [sight, sound, or smell] of hazardous chemicals being released), and applicable alarms from monitoring devices that record chemical releases.

(4) The physical and health hazards of substances known or potentially present in the work area.

(5) The measures employees can take to help protect themselves from work-site hazards, including specific procedures the employer has implemented.

(6) An explanation of the labeling system and safety data sheets and how employees can obtain and use appropriate hazard information.

(7) The elements of the confined space program including special PPE, permits, monitoring requirements, communication procedures, emergency response, and applicable lock-out procedures.

b. The employer should provide hazardous waste employees information and training and should provide a review and access to the site safety and plan as follows:

(1) Names of personnel and alternate responsible for site safety and health.

(2) Safety and health hazards present on the site.

(3) Selection, use, maintenance, and limitations of personal protective equipment specific to the site.

(4) Work practices by which the employee can minimize risks from hazards.

(5) Safe use of engineering controls and equipment available on site.

(6) Safe decontamination procedures established to minimize employee contact with hazardous substances, including:

(A) Employee decontamination.

(B) Clothing decontamination.

(C) Equipment decontamination.

(7) Elements of the site emergency response plan, including:

(A) Pre-emergency planning.

(B) Personnel roles and lines of authority and communication.

(C) Emergency recognition and prevention.

(D) Safe distances and places of refuge.

(E) Site security and control.

(F) Evacuation routes and procedures.

(G) Decontamination procedures not covered by the site safety and health plan.

(H) Emergency medical treatment and first aid.

(I) Emergency equipment and procedures for handling emergency incidents.

c. The employer should provide hazardous waste employees information and training on personal protective equipment used at the site, such as the following:

(1) PPE to be used based upon known or anticipated site hazards.

(2) PPE limitations of materials and construction; limitations during temperature extremes, heat stress, and other appropriate medical considerations; use and limitations of respirator equipment as well as documentation procedures as outlined in 29 CFR 1910.134.

(3) PPE inspection procedures prior to, during, and after use.

(4) PPE donning and doffing procedures.

(5) PPE decontamination and disposal procedures.

(6) PPE maintenance and storage.

(7) Task duration as related to PPE limitations.

d. The employer should instruct the employee about the site medical surveillance program relative to the particular site, including:

(1) Specific medical surveillance programs that have been adapted for the site.

(2) Specific signs and symptoms related to exposure to hazardous materials on the site.

(3) The frequency and extent of periodic medical examinations that will be used on the site.

(4) Maintenance and availability of records.

(5) Personnel to be contacted and procedures to be followed when signs and symptoms of exposures are recognized.

e. The employees will review and discuss the site safety plan as part of the training program. The location of the site safety plan and all written programs should be discussed with employees including a discussion of the mechanisms for access, review, and references described.

B. RCRA operations training for treatment, storage and disposal facilities

1. As a minimum, the training course required in 29 CFR 1910.120 (p) should include the following topics:

(a) Review of the applicable paragraphs of 29 CFR 1910.120 and the elements of the employer's occupational safety and health plan.

(b) Review of relevant hazards such as, but not limited to, chemical, biological, and radiological exposures; fire and explosion hazards; thermal extremes; and physical hazards.

(c) General safety hazards including those associated with electrical hazards, powered equipment hazards, lock-out-tag-out procedures, motor vehicle hazards and walking-working surface hazards.

(d) Confined-space hazards and procedures.

(e) Work practices to minimize employee risk from workplace hazards.

(f) Emergency response plan and procedures including first aid meeting the requirements of paragraph (p)(8).

(g) A review of procedures to minimize exposure to hazardous waste and various type of waste streams, including the materials handling program and spill containment program.

(h) A review of hazard communication programs meeting the requirements of 29 CFR 1910.1200.

(i) A review of medical surveillance programs meeting the requirements of 29 CFR 1910.120(p)(3) including the recognition of signs and symptoms of overexposure to hazardous substance including known synergistic interactions.

(j) A review of decontamination programs and procedures meeting the requirements of 29 CFR 1910.120(p)(4).

(k) A review of an employer's requirements to implement a training program and its elements.

(l) A review of the criteria and programs for proper selection and use of personal protective equipment, including respirators.

(m) A review of the applicable appendices to 29 CFR 1910.120.

(n) Principles of toxicology and biological monitoring as they pertain to occupational health.

(o) Rights and responsibilities of employees and employers under applicable OSHA and EPA laws.

(p) Hands-on exercises and demonstrations of competency with equipment to illustrate the basic equipment principles that may be used during the performance of work duties, including the donning and doffing of PPE.

(q) Sources of reference, efficient use of relevant manuals, and knowledge of hazard coding systems to include information contained in hazardous waste manifests.

(r) At least 8 hours of hands-on training.

(s) Training in the job skills required for an employee's job function and responsibility before they are permitted to participate in or supervise field activities.

2. The individual employer should provide hazardous waste employees with information and training prior to an employee's initial assignment into a work area. The training and information should cover the following topics:

(a) The Emergency response plan and procedures including first aid.

(b) A review of the employer's hazardous waste handling procedures including the materials handling program and elements of the spill containment program, location of spill response kits or equipment, and the names of those trained to respond to releases.

(c) The hazardous communication program meeting the requirements of 29 CFR 1910.1200.

(d) A review of the employer's medical surveillance program including the recognition of signs and symptoms of exposure to relevant hazardous substance including known synergistic interactions.

(e) A review of the employer's decontamination program and procedures.

(f) An review of the employer's training program and the parties responsible for that program.

(g) A review of the employer's personal protective equipment program including the proper selection and use of PPE based upon specific site hazards.

(h) All relevant site-specific procedures addressing potential safety and health hazards. This may include, as appropriate, biological and radiological exposures, fire and explosion hazards, thermal hazards, and physical hazards such as electrical hazards, powered equipment hazards, lock-out-tag-out hazards, motor vehicle hazards, and walking-working surface hazards.

(i) Safe use engineering controls and equipment on site.

(j) Names of personnel and alternates responsible for safety and health.

C. Emergency response training

Federal OSHA standards in 29 CFR 1910.120(q) are directed toward private sector emergency responders. Therefore, the guidelines provided in this portion of the appendix are directed toward that employee population. However, they also impact indirectly through State OSHA or USEPA regulations some public sector emergency responders. Therefore, the guidelines provided in this portion of the appendix may be applied to both employee populations.

States with OSHA state plans must cover their employees with regulations at least as effective as the Federal OSHA standards. Public employees in states without approved state OSHA programs covering hazardous waste operations and emergency response are covered by the U.S. EPA under 40 CFR 311, a regulation virtually identical to § 1910.120.

Since this is a non-mandatory appendix and therefore not an enforceable standard, OSHA recommends that those employers, employees or volunteers in public sector emergency response organizations outside Federal OSHA jurisdiction consider the following criteria in developing their own training programs. A unified approach to training at the community level between emergency response organizations covered by Federal OSHA and those not covered directly by Federal OSHA can help ensure an effective community response to the release or potential release of hazardous substances in the community.

a. General considerations:
Emergency response organizations are required to consider the topics listed in § 1910.120(q)(6). Emergency response organizations may use some or all of the following topics to supplement those mandatory topics when developing their response training programs. Many of the topics would require an interaction between the response provider and the individuals responsible for the site where the response would be expected.

(1) Hazard recognition, including:

(A) Nature of hazardous substances present.

(B) Practical applications of hazard recognition, including presentations on biology, chemistry, and physics.

(2) Principles of toxicology, biological monitoring, and risk assessment.

(3) Safe work practices and general site safety.

(4) Engineering controls and hazardous waste operations.

(5) Site safety plans and standard operating procedures.

(6) Decontamination procedures and practices.

(7) Emergency procedures, first aid, and self-rescue.
(8) Safe use of field equipment.
(9) Storage, handling, use, and transportation of hazardous substances.
(10) Use, care, and limitations of personal protective equipment.
(11) Safe sampling techniques.
(12) Rights and responsibilities of employees under OSHA and other related laws concerning right-to-know, safety and health, compensations and liability.
(13) Medical monitoring requirements.
(14) Community relations.

b. Suggested criteria for specific courses:
  (1) First responder awareness level.
    (A) Review of and demonstration of competency in performing the applicable skills of 29 CFR 1910.120(q).
    (B) Hands-on experience with the U.S. Department of Transportation's Emergency Response Guidebook (ERG) and familiarization with OSHA standard 29 CFR 1910.1201.
    (C) Review of the principles and practices for analyzing an incident to determine both the hazardous substances present and the basic hazard and response information for each hazardous substance present.
    (D) Review of procedures for implementing actions consistent with the local emergency response plan, the organization's standard operating procedures, and the current edition of DOT's ERG including emergency notification procedures and follow-up communications.
    (E) Review of the expected hazards including fire and explosions hazards, confined space hazards, electrical hazards, powered equipment hazards, motor vehicle hazards, and walking-working surface hazards.
    (F) Awareness and knowledge of the competencies for the First Responder at the Awareness Level covered in the National Fire Protection Association's Standard No. 472, Professional Competence of Responders to Hazardous Materials Incidents.
  (2) First responder operations level.
    (A) Review of and demonstration of competency in performing the applicable skills of 29 CFR 1910.120(q).
    (B) Hands-on experience with the U.S. Department of Transportation's Emergency Response Guidebook (ERG), manufacturer safety data sheets, CHEMTREC/CANUTEC, shipper or manufacturer contacts, and other relevant sources of information addressing hazardous substance releases. Familiarization with OSHA standard 29 CFR 1910.1201.
    (C) Review of the principles and practices for analyzing an incident to determine the hazardous substances present, the likely behavior of the hazardous substance and its container, the types of hazardous substance transportation containers and vehicles, the types and selection of the appropriate defensive strategy for containing the release.
    (D) Review of procedures for implementing continuing response actions consistent with the local emergency response plan, the organization's standard operating procedures, and the current edition of DOT's ERG including extended emergency notification procedures and follow-up communications.
    (E) Review of the principles and practice for proper selection and use of personal protective equipment.

(F) Review of the principles and practice of personnel and equipment decontamination.

(G) Review of the expected hazards including fire and explosions hazards, confined space hazards, electrical hazards, powered equipment hazards, motor vehicle hazards, and walking-working surface hazards.

(H) Awareness and knowledge of the competencies for the First Responder at the Operations Level covered in the National Fire Protection Association's Standard No. 472, Professional Competence of Responders to Hazardous Materials Incidents.

(3) Hazardous materials technician.

(A) Review of and demonstration of competency in performing the applicable skills of 29 CFR 1910.120(q).

(B) Hands-on experience with written and electronic information relative to response decision making including but not limited to the U.S. Department of Transportation's *Emergency Response Guidebook* (ERG), manufacturer safety data sheets, CHEMTREC/ CANUTEC, shipper or manufacturer contacts, computer data bases and response models, and other relevant sources of information addressing hazardous substance releases. Familiarization with OSHA standard 29 CFR 1910.1201.

(C) Review of the principles and practices for analyzing an incident to determine the hazardous substances present, their physical and chemical properties, the likely behavior of the hazardous substance and its container, the types of hazardous substance transportation containers and vehicles involved in the release, the appropriate strategy for approaching release sites and containing the release.

(D) Review of procedures for implementing continuing response actions consistent with the local emergency response plan, the organization's standard operating procedures, and the current edition of DOT's ERG including extended emergency notification procedures and follow-up communications.

(E) Review of the principles and practice for proper selection and use of personal protective equipment.

(F) Review of the principles and practices of establishing exposure zones, proper decontamination and medical surveillance stations and procedures.

(G) Review of the expected hazards including fire and explosions hazards, confined space hazards, electrical hazards, powered equipment hazards, motor vehicle hazards, and walking-working surface hazards.

(H) Awareness and knowledge of the competencies for the Hazardous Materials Technician covered in the National Fire Protection Association's Standard No. 472, Professional Competence of Responders to Hazardous Materials Incidents.

(4) Hazardous materials specialist.

(A) Review of and demonstration of competency in performing the applicable skills of 29 CFR 1910.120(q).

(B) Hands-on experience with retrieval and use of written and electronic information relative to response decision making including but not limited to the U.S. Department of Transportation's *Emergency Response Guidebook* (ERG), manufacturer safety data sheets, CHEMTREC/ CANUTEC, shipper or manufacturer contacts, computer data bases and response models, and other relevant sources of information addressing hazardous substance releases. Familiarization with OSHA standard 29 CFR 1910.1201.

(C) Review of the principles and practices for analyzing an incident to determine the hazardous substances present, their physical and chemical properties, and the likely behavior of the hazardous substance and its container, vessel, or vehicle.

(D) Review of the principles and practices for identification of the types of hazardous substance transportation containers, vessels and vehicles involved in the release; selecting and using the various types of equipment available for plugging or patching transportation containers, vessels or vehicles; organizing and directing the use of multiple teams of hazardous material technicians and selecting the appropriate strategy for approaching release sites and containing or stopping the release.

(E) Review of procedures for implementing continuing response actions consistent with the local emergency response plan, the organization's standard operating procedures, including knowledge of the available public and private response resources, establishment of an incident command post, direction of hazardous material technician teams, and extended emergency notification procedures and follow-up communications.

(F) Review of the principles and practice for proper selection and use of personal protective equipment.

(G) Review of the principles and practices of establishing exposure zones and proper decontamination, monitoring and medical surveillance stations and procedures.

(H) Review of the expected hazards including fire and explosions hazards, confined space hazards, electrical hazards, powered equipment hazards, motor vehicle hazards, and walking-working surface hazards.

(I) Awareness and knowledge of the competencies for the Off-site Specialist Employee covered in the National Fire Protection Association's Standard No. 472, Professional Competence of Responders to Hazardous Materials Incidents.

(5) Incident commander.

The incident commander is the individual who, at any one time, is responsible for and in control of the response effort. This individual is the person responsible for the direction and coordination of the response effort. An incident commander's position should be occupied by the most senior, appropriately trained individual present at the response site. Yet, as necessary and appropriate by the level of response provided, the position may be occupied by many individuals during a particular response as the need for greater authority, responsibility, or training increases. It is possible for the first responder at the awareness level to assume the duties of incident commander until a more senior and appropriately trained individual arrives at the response site.

Therefore, any emergency responder expected to perform as an incident commander should be trained to fulfill the obligations of the position at the level of response they will be providing including the following:

(A) Ability to analyze a hazardous substance incident to determine the magnitude of the response problem.

(B) Ability to plan and implement an appropriate response plan within the capabilities of available personnel and equipment.

(C) Ability to implement a response to favorably change the outcome of the incident in a manner consistent with the local emergency response plan and the organization's standard operating procedures.

(D) Ability to evaluate the progress of the emergency response to ensure that the response objectives are being met safely, effectively, and efficiently.

(E) Ability to adjust the response plan to the conditions of the response and to notify higher levels of response when required by the changes to the response plan.

[54 FR 9317, Mar. 6, 1898, as amended at 55 FR 14073, Apr. 13, 1990; 56 FR 15832, Apr. 18, 1991; 59 FR 43268, Aug. 22, 1994; 61 FR 9227, March 7, 1996; 78 FR 9313, Feb. 8, 2013]

Full document is available at the U.S. Department of Labor, Occupational Safety & Health Administration website. https://www.osha.gov/pls/oshaweb/owadisp.show_document?p_table=STANDARDS&p_id=9770. Accessed November 6, 2014.

INDEX

Figures and tables are indicated by "f" and "t" following page numbers.

B

C

D

E

F

Q

R

S

T

U

V

W

Y

Z